Rodney D. Geisert • Thomas E. Spencer
Editors

Placentation in Mammals

Tribute to E.C. Amoroso's Lifetime Contributions to Viviparity

Springer

Editors
Rodney D. Geisert
Division of Animal Sciences
University of Missouri
Columbia, MO, USA

Thomas E. Spencer
Division of Animal Sciences
University of Missouri
Columbia, MO, USA

ISSN 0301-5556 ISSN 2192-7065 (electronic)
Advances in Anatomy, Embryology and Cell Biology
ISBN 978-3-030-77359-5 ISBN 978-3-030-77360-1 (eBook)
https://doi.org/10.1007/978-3-030-77360-1

This Springer imprint is published by the registered company Springer Nature Switzerland AG.
The registered company address is: Gewerbestrasse 11, 6330 Cham, Switzerland

In Memoriam

William Richard (Twink) Allen, CBE, FRCVS, ScD
29 August 1940–6 June 2021

Over the past 50 years, Twink Allen, as he was universally known, was arguably the most influential individual worldwide in the field of equine reproduction, a status achieved through a combination of original, first-class scientific investigations and practical application of the results of his own research efforts and those of others. Those practical advances have so greatly improved the efficiency of horse breeding that reproductive problems are no longer at the top of the list of challenges in the equine industry. Twink's contributions to this volume on placentation in the horse and elephant bookend the final chapter of a remarkable career. It is fitting that this book reflects also on the life and work of E. C. Amoroso, who was a friend and mentor to Twink in his early years in Cambridge.

An Antipodean by birth, Twink was raised in New Zealand and completed his veterinary training at Sydney University in Australia. The dark cloud of a serious automobile accident during his first year of veterinary practice in New Zealand had a silver lining. A full year in hospital led Twink to explore postgraduate study and resulted in a scholarship to undertake a PhD in England under the mentorship of Roger Short, FRS. As part of his PhD thesis research at the University of Cambridge, Twink developed an assay for equine Chorionic Gonadotrophin (eCG) that became the international standard for many years (Allen 1969b). He used this assay to ascertain the levels of eCG in the blood of horse mares and donkey jennies carrying intra- and interspecies conceptuses (horse, donkey, mule, and hinny). This work was published as a single author paper in Nature *(Allen 1969a)—quite extraordinary for a young PhD candidate, even at Cambridge! The data were amazing—they showed a pattern of hormone levels in the four types of pregnancy that can best be explained by the theory of genomic imprinting, a phenomenon that was not elucidated for more than two decades after this publication. This is an example of how Twink Allen's work was often ahead of its time, too often for it to be by chance alone. Rather, it reflects Twink's insight, drive, and even his impatience with the deliberate pace of scientific investigation. Twink always worked with a sense of urgency that propelled his career and those of his students and colleagues.*

Following completion of his PhD, Twink remained in England with his rapidly growing family. He was soon named the inaugural Director of the British Thoroughbred Breeders' Association Equine Fertility Unit, located initially on the grounds of the British government's Animal Research Station on the outskirts of Cambridge. In this outstanding scientific environment, Twink attracted a bevy of veterinary surgeons and scientists to his research group, and they took full advantage of the expertise in reproductive biology around them.

In the first years of his appointment, Twink made the seminal discovery that the equine endometrial cups, which are the sole source of eCG, are of fetal, and not maternal, origin, as had been thought for the previous half-century (Allen et al. 1973). The endometrial cup trophoblast cells originate from the differentiation of the invasive trophoblast cells of the chorionic girdle. Twink championed the idea that the maternal lymphocyte response to the endometrial cups represents a form of cellular immunity directed against paternally inherited antigens expressed by the trophoblast cells of the cups.

Twink pioneered techniques of surgical and nonsurgical embryo transfer in equids (Allen and Rowson 1975) and used them to investigate maternal anti-fetal immune responses in interspecies (i.e., mule and hinny) and extraspecies (e.g., donkey-in-horse) pregnancies, studies which rank among the most imaginative experiments in pregnancy immunology undertaken in any species. This research extended the bounds of maternal immunological tolerance of the fetus and at the same time enabled the use of embryo transfer to propagate endangered species (Allen and Short 1997).

Twink's expertise in embryo transfer allowed him to achieve several "firsts," including the first equine identical twins produced by embryo splitting and the first successful birth of a horse following international transport of a frozen-thawed embryo. Twink also popularized the commercial use of equine embryo transfer, and it is now in widespread use globally in the horse industry, increasingly in combination with somatic cell nuclear transfer to produce identical clones of elite performance horses.

In the 1970s Twink developed simple, robust protocols for the use of prostaglandins to manipulate the estrus cycle of mares (Allen and Rossdale 1973). This increased the overall fertility rate of mares at commercial stud farms by allowing more matings in fertile estrus cycles in a single breeding season. Prostaglandins are now in daily use by equine reproductive specialists across the globe.

In the early 1980s Twink adapted early-design ultrasound instruments from human medicine for use in horses (Simpson et al. 1982). His promotion and improvement of this important technology revolutionized equine breeding management by allowing visualization of the reproductive tract, and in particular the developing follicles of the ovary and very early horse embryos (day 10–14). Thus, determination of ovulation, pregnancy diagnosis, and detection of twin pregnancies (a very bad outcome in horse breeding) have become routine and relatively simple.

In the 1990s, Twink designed an outstanding new research facility in Newmarket that supported his Equine Fertility Unit and allowed expansion into new areas, including fetal programming and stem cell biology. In conjunction with this move, Twink was appointed as the Jim Joel Professor of Equine Reproduction in the Cambridge University Veterinary School.

Twink also made enormous contributions as an organizer of scientific meetings and editor of proceedings from those conferences and workshops. Two examples in particular stand out: the International Symposia on Equine Reproduction, which have been held every 4 years following the inaugural meeting in Cambridge in the early 1970s, and the Equine Embryo Transfer Workshops convened by the Havemeyer Foundation since the mid-1980s. Twink was the key individual responsible for the success of these two series, which bring together veterinary scientists and clinicians, and which foster rapid adoption of scientific advances into practice.

Twink was a spell-binding lecturer and communicator of science who brought alive the excitement of biology and its application to problems in the real world. This role brought its share of tensions and controversies, and Twink never shied away from them. He was often caught between the natural conservatism of Thoroughbred racehorse breeders and his own quest for advancing reproductive efficiency through the application of assisted reproductive technologies in the horse.

Twink's interests spanned the entire range of reproductive biology in the horse, and this short memorial touches on only a few of the early highlights. Twink's fascination with the placenta and his great love for Africa and its wildlife led him to comparative studies of placentation in elephant, giraffe, zebra, wildebeest, impala, and hedgehog. Twink was also the driving force for the establishment of the Camel Reproduction Centre in Dubai. And it was in the desert of the United Arab Emirates where Twink spent his final years, directing research at the Equine Reproduction Centre of the Sharjah Equine Hospital.

Prof. W. R. Allen's research achievements over a 50-year career place him in the highest tier of investigators in veterinary science. Twink was a pioneer of heroic, insightful experiments. His generosity in sharing ideas and expertise earned him countless friends across the globe in a life fully lived.

Douglas F. Antczak and Sandra Wilsher

Allen WR (1969a) Factors influencing pregnant mare serum gonadotrophin production. Nature 223:64–65

Allen WR (1969b) The immunological measurement of pregnant mare serum gonadotrophin. J Endocrinol 43:593–598

Allen WR, Rossdale PD (1973) A preliminary study upon the use of prostaglandins for inducing oestrus in non-cycling thoroughbred mares. Equine Vet J 5:137–140

Allen WR, Rowson LE (1975) Surgical and non-surgical egg transfer in horses. J Reprod Fertil Suppl 525–530

Allen WR, Short RV (1997) Interspecific and extraspecific pregnancies in equids: anything goes. J Hered 88:384–392

Allen WR, Hamilton DW, Moor RM (1973) The origin of equine endometrial cups. II. Invasion of the endometrium by trophoblast. Anat Rec 177:485–501
Simpson DJ, Greenwood RE, Ricketts SW, Rossdale PD, Sanderson M, Allen WR (1982) Use of ultrasound echography for early diagnosis of single and twin pregnancy in the mare. J Reprod Fertil Suppl 32:431–439

Contents

Mammalian Placentation: A Tribute to E.C. Amoroso's Contributions to Placenta Development

Rodney D. Geisert and Thomas E. Spencer

Abstract Establishment of viviparity in mammals evolved through not only the long-term retainment of the fetus within the maternal uterus but differentiation and expansion of cell layers to form functional membranes to exchange O_2/CO_2 and nutrients between the placenta and maternal circulations. Development of a fetal placental vascular circulation to interact with the maternal uterus is critical to the survival of all species. However, the fascination with the mammalian placenta is the robust variation in types, form, attachment, invasiveness, structure, cell differentiation, endocrine function, and regulation of the maternal immune system. Despite the obvious role of the placenta to support fetal development, mammals have evolved multiple strategies to give live birth at term. The placenta and the maternal–fetal interface during pregnancy can be quite simple to very complex. Professor E.C. Amoroso contributed greatly to the study of comparative placentation in animals. His paper "Placentation" in Marshall's Physiology of Reproduction published in 1952 remains the standard for comparative placental anatomy today. The present volume on "Mammalian Placentation" brings together current reviews for leading experts to diversity of placentation in a number of mammalian species. Chapters will discuss viviparity, blastocyst formation, and placentation in the cow, pig, horse, mouse, dog, primate, human, elephant, and marsupials.

Keywords Implantation · Placentation · Mammal · Viviparity

The development of viviparity in mammals as well as some invertebrate species required the adaptation of the placenta to serve as a functional conduit for interplay

R. D. Geisert (✉)
Division of Animal Sciences, S121 Animal Science Research Center, University of Missouri, Columbia, MO, USA
e-mail: geisertr@missouri.edu

T. E. Spencer
Division of Animal Sciences, S135 Animal Science Research Center, University of Missouri, Columbia, MO, USA
e-mail: spencerte@missouri.edu

R. D. Geisert, T. E. Spencer (eds.), *Placentation in Mammals*, Advances in Anatomy, Embryology and Cell Biology 234, https://doi.org/10.1007/978-3-030-77360-1_1

between the semi-allograftic fetus with the maternal uterus. Although primarily recognized for his work on pulmonary circulation in the sixteenth century, the Italian anatomist and surgeon, Realdo Columbo utilized the Latin word for cake "*placenta uterina*" (Greek *plakoenta*) which means flat or plate to describe the large round vascular structure he observed attached to the uterus during pregnancy of women. Although the "placenta" protects the fetus from maternal immune rejection and provides oxygen and nutrient flow to support a fetus to term across all the species; structural differentiation of this fetal–maternal interface is very varied among species.

In the absence of an adequate supply of yolk to support embryo development in oviparous and ovoviviparous species, mammalian viviparity evolved through the differentiation of extraembryonic cells which contribute to placental formation. Initiation of placental development occurs before hatching from the zona pellucida when the outermost cells of a morula form tight junctions to establish cellular polarity. Compaction of the morula establishes outer and inner cells which positions the Na^+/K^+ ATPase system to move fluid into the extracellular space of the compact morula to forming a blastocyst (Fig. 1a). Cellular positioning within a morula induces differential transcriptional activity of the Hippo and Ras signaling pathways to dictate formation of the inner cell mass (ICM) and trophoblast of the preimplantation embryo (Senner and Hemberger 2010; Cockburn and Rossant 2010; Artus and Chazaud 2014). Blastocyst development is somewhat conserved across viviparous mammals and is the direct prelude to formation of the placenta. However, depending upon the species, blastocyst hatching from its zona pellucida can either lead to attachment and invasion through the surface epithelium to implant within the endometrium (examples: rodents, primates, humans) or continued growth (Fig. 1b) and expansion (Fig. 1c) throughout the uterine lumen (examples: ruminants, pig, horse) where trophoblast (Greek trephein meaning to feed; and blastos meaning germinato) comes into direct contact with endometrial epithelium and secretions to support development to term. Continued cellular differentiation and expansion of the ICM (epiblast) provides the extraembryonic mesoderm and endoderm (hypoblast) layers, from its posterior region (Perry 1981), that combine with the trophectoderm to contribute to formation of the classical placenta membranes (Fig. 1d). Mammalian viviparity requires not only the development of the placenta but also the retention of the embryo within the uterus for an extended period before birth, maintenance of pregnancy via progesterone production by the maternal corpus luteum and later the placenta, secretions from the uterus including endometrial glands, O_2, and nutrient exchange between the maternal and placental vasculatures, and modification of the maternal immune system at the placental/uterine interface.

Placentation of mammals can be classified by gross morphology (placental shape) and the number of cellular layers (cellular microstructure) that separate the placenta and uterine vasculature (Grosser 1909; Mossman 1937). Although placental development undergoes gross morphological changes from early to late gestation, the mammalian placenta is classified into four basic groups: Diffuse (pig, horse), Cotyledonary (ruminants), Zonary (dog, cat, elephant), and Discoid (mouse, rat, rabbit, primate, and human). However, the amount of placental erosion or cellular

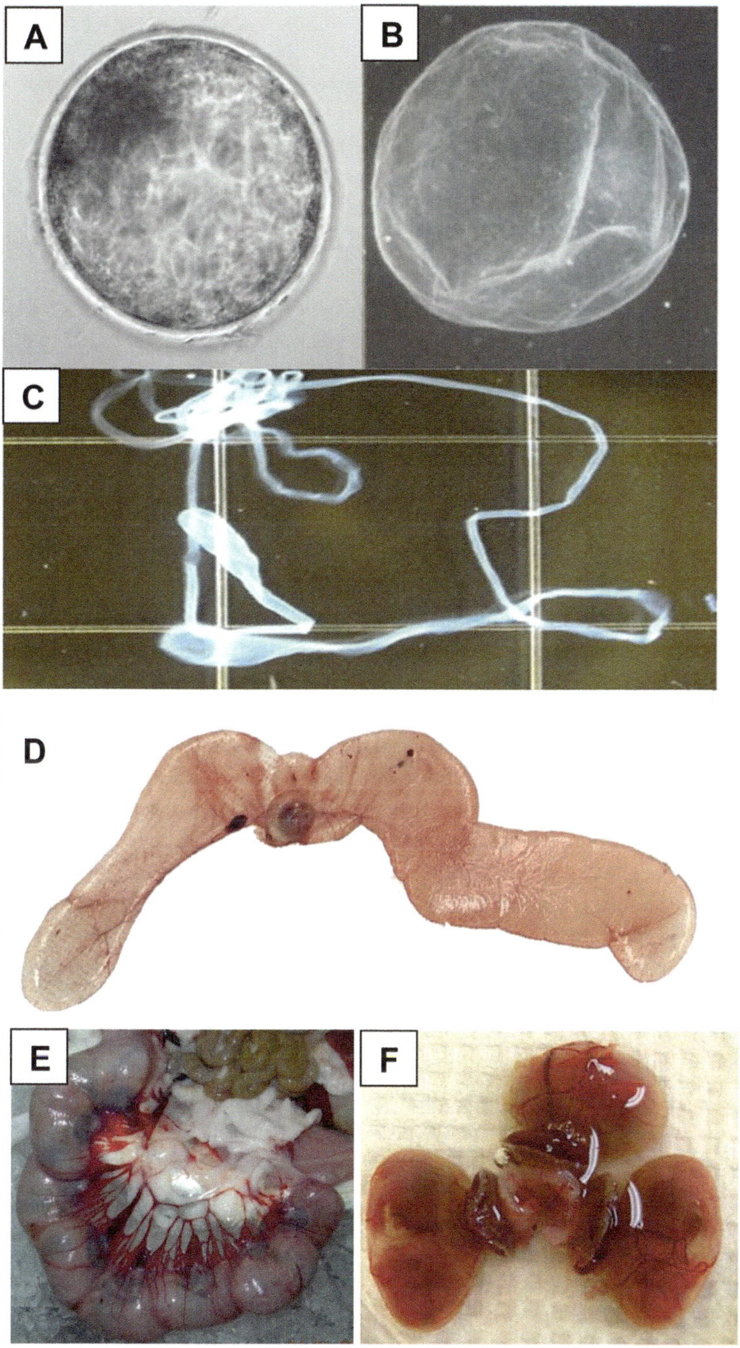

Fig. 1 Transitions in embryo development for formation of the mammalian placenta. (**a**) Day 7 pig blastocyst enclosed in the zona pellucida. (**b**) Day 10 spherical (5 mm) pig conceptus. (**c**) Rapidly elongating trophoblast of a day 12 pig conceptus expanding within the uterine horn. (**d**) Diffuse pig placenta (day 35) containing embryo enclosed in the amnion and the fluid filled chorioallantois. (**e**) Pregnant day 18 rat uterus displaying invasive placentation. (**f**) Day 18 rat discoid placentae

Fig. 2 Emmanuel Ciprian Amoroso. (Photograph obtained from RV Short's tribute to EC Amoroso published in Biographical Memoirs of Fellows of the Royal Society (Short 1985))

layer displacement of the maternal endometrium provides a more accurate description of placentation as provided by the impressive illustrations made by EC Amoroso (1952). The pig (Diffuse) has an *epitheliochorial* type of placentation where the expanded chorion only attaches to the surface epithelium of the endometrium but does not erode or invade into the uterus. Although ruminants (Cotyledonary) initially have epitheliochorial type of attachment to the uterine surface epithelium, binucleate cells migrate from the chorion and fuse with the endometrial surface epithelium to form a syncytium which is classified as *syndesmochorial*. Species such as the dog and cat (Zonary) have an endotheliochorial type of placenta that invades into the endometrium eroding all maternal layers but the endothelium of the maternal capillaries is not breached. The *hemochorial* placentation in the mouse, rat (Fig. 1e, f), primate and human represents the most invasive type of placentation where the placenta blood vessels are directly bathed by maternal blood.

E.C. Amoroso, known as Amo by his friends and colleagues, (Fig. 2) contributed so much to our early understanding and knowledge of placentation across a great variety of species. He was born in the Port of Spain, Republic of Trinidad, where his eye sight was affected by exposure to typhoid fever at a young age which makes his work in placenta histology and illustrations even more remarkable. Amoroso spoke seven languages and was renowned for his elegant, captivating, and acclaimed lectures around the world. Although his work was not solely focused on placental development, he studied and published manuscripts investigating over 28 species. Professor Amoroso's classic, seminal chapter "Placentation" published in the 3rd edition of Marshall's Physiology of Reproduction in 1952 followed the work of many German anatomists (O. Grosser, O. Krolling and H. Strahl) and others (R. Assheton, H.W. Mossman, and A. Robertson) who pioneered understanding comparative placentation in mammals. His 184-page chapter (Amoroso 1952) is a masterpiece of intellectual information on comparative placentation. The chapter contains 88 figures, ~124 hand-drawn illustrations, and histological sections with over 434 references. Professor Amoroso's chapter covered adaptations in viviparity and ovoviviparity of fish, amphibians, and reptiles but mainly focused on comparative mammalian placenta including marsupials. The following quote taken from Professor Amoroso's chapter provides his keen interest in the placenta.

> The mammalian ovum, in all except the Monotremata, is small and does not contain a sufficient supply of nutriment for the developing embryo. It is retained for a longer or shorter period in the uterus, where, by special modifications of the uterine mucosa and a part of the ovum, a placenta is formed, and a transmission of nutriment from the mother to the embryo is made possible. Accordingly, the normal mammalian placenta may be defined "as an apposition, or fusion, of the foetal membranes to the uterine mucosa for physiological exchange" (Mossman 1937). The changes in the maternal and embryonic tissues vary greatly in the several orders, and even in the groups of the same order, but in all they are sufficiently complicated to render their explanation a matter of great difficulty. It is doubtful in any anatomical structure has given rise to keener or more prolonged controversies than the placenta (Amoroso 1952).

Professor Amoroso contributed greatly to our understanding of placentation in the pig, horse, cat, dog, rabbit, mouse, rat, guinea pig, primates, and human and contributed to early work in elephant placentation. Prior to his passing in 1982, Professor Amoroso made the following quote to his colleagues:

> Consider the past for what is relevant to the present; consider the present what is relevant to the formation of the future; consider the future for what may most enlarge man's freedom and fulfilment (Weaver 1982).

Several review papers concerning the mammalian placenta have been published since Professor Amoroso's extensive chapter (Perry 1981; Roberts et al. 2016). The following chapters on viviparity, blastocyst formation, and placentation in the cow, pig, horse, mouse, dog, primate, human, elephant, and marsupials attempt to build on the past, understand the present, and continue to progress towards bettering reproduction of animals and mankind honor the enormous contributions made by Professor Amoroso.

References

Amoroso EC (1952) Placentation. In: Parkes AS (ed) Marshall's physiology of reproduction, vol 2, 3rd edn. Longmans, London, pp 127–311

Artus J, Chazaud CA (2014) Close look at the mammalian blastocyst: epiblast and primitive endoderm formation. Cell Mol Life Sci 71:3327–3338

Cockburn K, Rossant J (2010) Making the blastocyst: lessons from the mouse. J Clin Invest 120:995–1003. https://doi.org/10.1172/JCI41229

Grosser O (1909) Vergleichende Anatomie und Entwicklungsgeschichte der Eihaute und der Placenta. W. Braumuller, Vienna and Leipzig

Mossman HW (1937) Comparative morphogenesis of the fetal membranes and accessory uterine structures. Contr Embryol Carnegie Instn 158:129–246

Perry JS (1981) The mammalian fetal membranes. J Reprod Fertil 62:321–335
Roberts RM, Green JA, Schulz LC (2016) The evolution of the placenta. Reproduction 152:R179–R189. https://doi.org/10.1530/REP-16-0325
Senner CE, Hemberger M (2010) Regulation of early trophoblast differentiation—lessons from the mouse. Placenta 31:944–950. https://doi.org/10.1016/j.placenta.2010.07
Short RV (1985) Emmanuel Ciprian Amoroso. Biogr Mem Fellows R Soc 31:2–30. https://doi.org/10.1098/rsbm.1985.0001
Weaver BMQ (1982) Professor E. C. Amoroso, C.B.E. T.C., F.R.S. Proc Assoc Vet Anaest 10:3–6

The Evolution of Viviparity in Vertebrates

Wesley C. Warren and Frank Grutzner

Abstract In the vertebrate tree of life, viviparity or live birth has independently evolved many times, resulting in a rich diversity of reproductive strategies. Viviparity is believed to be a mode of reproduction that evolved from the ancestral condition of oviparity or egg laying, where most of the fetal development occurs outside the body. Today, there is not a simple model of parity transition to explain this species-specific divergence in modes of reproduction. Most evidence points to a gradual series of evolutionary adaptations that account for this phenomenon of reproduction, elegantly displayed by various viviparous squamates that exhibit placentae formed by the appositions of maternal and embryonic tissues, which share significant homology with the tissues that form the placenta in therian mammals. In an era where the genomes of many vertebrate species are becoming available, studies are now exploring the molecular basis of this transition from oviparity to viviparity, and in some rare instances its possible reversibility, such as the Australian three-toed skink (*Saiphos equalis*). In contrast to the parity diversity in squamates, mammals are viviparous with the notable exception of the egg-laying monotremes. Advancing computational tools coupled with increasing genome availability across species that utilize different reproductive strategies promise to reveal the molecular underpinnings of the ancestral transition of oviparity to viviparity. As a result, the dramatic changes in reproductive physiology and anatomy that accompany these parity changes can be reinterpreted. This chapter will briefly explore the vertebrate modes of reproduction using a phylogenetic framework and where possible highlight the role of potential candidate genes that may help explain the polygenic origins of live birth.

W. C. Warren (✉)
Division of Animal Sciences, University of Missouri, 440G/446 Life Sciences Center, Columbia, MO, USA
e-mail: warrenwc@missouri.edu

F. Grutzner
Comparative Genome Biology, Department of Genetics and Evolution, School of Biological Sciences, The University of Adelaide, Adelaide, SA, Australia
e-mail: Frank.grutzner@adelaide.edu.au

R. D. Geisert, T. E. Spencer (eds.), *Placentation in Mammals*, Advances in Anatomy, Embryology and Cell Biology 234, https://doi.org/10.1007/978-3-030-77360-1_2

Keywords Viviparity · Oviparity · Parity evolution

1 Introduction

Charles Darwin's theory of evolution describes a process of natural selection by which allows species to adapt to their environment enhancing reproductive success and survival. In vertebrates, reproductive strategies can be uniform in some lineages and remarkably diverse and flexible in others. Fortunately evolution has left a tantalizing series of clues as to how ancestral physiological, morphological, biochemical, and molecular networks were coopted that led to the species-specific forms of reproductive parity. Firstly, before investigating these hints of nature, simply reproduction can be sexual or asexual, the later a rarity in vertebrates. Decades of literature debate asexual origins, meiotic drivers, and disadvantages to that of sexual reproduction (Fu et al. 2019; Zanders and Unckless 2019) and both modes of reproduction can be associated with viviparity or oviparity. However, the focus of this brief chapter will be the nourishing of a growing fetus by either a viviparous or oviparous system that has left a fascinating legacy of how evolution reshaped the genetic blueprints of life "genomes" toward a preferential mode of parity, oviparity versus viviparity, in particular when altered gene signatures suggest an association with placental development and function.

Viviparity, in which eggs are fertilized and embryos develop inside the maternal reproductive tract or body cavity, as opposed to oviparity, in which females deposit fertilized eggs that develop outside of the parent, can differ in even closely related species. The most frequent examples in nature are squamates. A vast literature strongly suggests that viviparous phenotypes, such as sophisticated and diverse placental structures, have evolved convergently from oviparity many times in nature (Blackburn 1999, 2015). Although, evidence in squamates suggests more malleable modes of parity (Blackburn 2015), viviparity is predominant and nearly exclusive in mammals where placental biology has been most studied (Fig. 1). Still, there is no definitive answer as to why viviparity is almost exclusive in certain vertebrate clades, mammals, while it is overwhelmingly oviparity in others, for example, birds, amphibians, and fishes.

A number of theories have been offered to explain the evolution of this phylogenetic distribution of parity. A species range of mobility and territory span, e.g., limited range mammals and the far spanning habitat of birds, is one such contributing factor that has possibly led to separate clades of parity. However, at least some species in nearly all bird families stay in the same place all year, thus avoiding the rigors of migration. The great horned owl is one such example. Moreover, before human habitat disruption occurred, many mammal species migrated and even today some still do in the continent of Africa. A species mobility does allow the placement of the offspring in the best sites for growth, such as the migration of wild ungulates to rich sources of vegetation phenology. But in viviparity, the developing embryos are protected internally and physiologically maintained with mobility and nutritional costs to the mother. Migratory movements of terrestrial mammals are a phenomenon

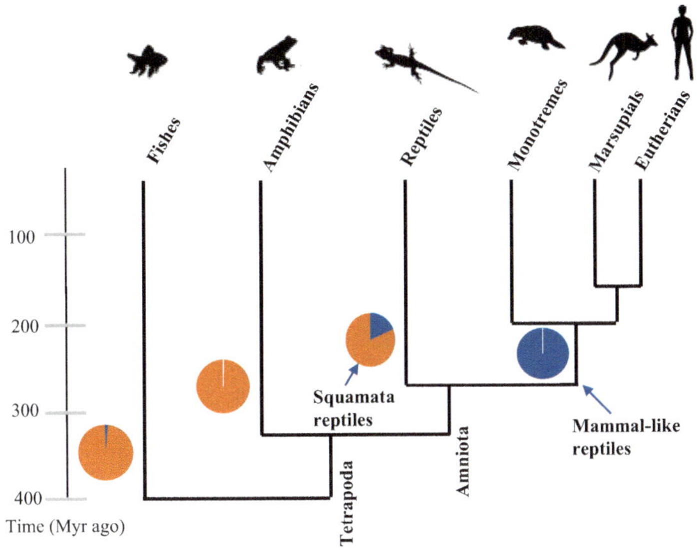

Fig. 1 Evolution of the species-specific reproductive strategies of viviparity and oviparity. A timescale of splits in species clade diversification is estimated. Counts of species practicing viviparity (blue) or oviparity (orange) are plotted by their major lineage branches

critical to their persistence, wherein some ungulate species prior to domestication, seasonally migrated to enhance lifetime reproductive fitness. Some hypotheses suggest this movement dependence on fluctuations in food supply propels as yet an unknown rapid, non-genomic adaptive mechanism that better fits offspring environment (nutritional), thus giving them a reproductive advantage (Sharpe 2018).

A number of preferential and detrimental attributes are associated with oviparity as well. Multiple offspring offset the higher mortality accompanying the external exposure of eggs to predators and their exposure to changing environment (weather, temperature). Higher offspring numbers also serves in increasing germline recombination events, one key process of a species genetic adaptation to the changing environment. Consequently, some genomic regions of reproductive mode relevance are likely subject to intense natural selection or alternatively have evolved sequence structures, such as CG dinucleotide expansions, that allow more fluid gene regulatory modifications. One can summarize these observed "reproductive strategies," as having advantages as well as disadvantages that affect their species-specific evolution (Table 1).

Genome-wide studies have sought sequence structures, mostly genes and their pathways, that display shared features, across species that evolved viviparity or oviparity. As organisms naturally experiment with these new sequence variants, new species emerge and continue to evolve. In the 1850s, when Darwin described this engine of natural selection, the underlying molecular mechanisms were unknown. But over the past century, advances in genetics and molecular biology have outlined a modern, neo-Darwinian theory of how evolution works: DNA

Table 1 Evolutionary advantages and disadvantages of reproductive modes among vertebrates

Oviparity	Viviparity
• Multiple offspring to offset higher observed environmental and predator mortality causes	• Developing embryos protected internally but with mobility and nutritional costs to the mother
• Shared care of offspring	• Full maternal commitment to developing offspring even with lower birth rates
• Parents encumbered with protecting eggs from predators and other environmental sources of mortality	• Higher likelihood of offspring placement in optimal sites for growth
• Multiple offspring offer a larger gene pool for adaptive change	• Live birth is associated with exposure to predators and birth complications

sequences randomly mutate, and organisms with the specific sequences best adapted to the environment will multiply and prevail or disappear. In an era of many vertebrate genomes being available, each with unique reproductive strategies, newly developed computational methods, such as the use of pangenome variant graphs (Hickey et al. 2020), may uncover the complex structural changes that occurred along the oviparity to viviparity trajectory. Not unexpectedly gene expression changes play a role in likely differentiating parity modes (Foster et al. 2020; Gao et al. 2019). Some have been closely associated with viviparity prominently genomic imprinting in mammals (Kaneko-Ishino and Ishino 2019). When within species, occurrence of both oviparous and viviparous reproductive strategies naturally exist, such as the lizard species of *Saiphos equalis* (Smith and Shine 1997), and the resolution of genetic change is even more pronounced. Even seasonal shifts along a oviparity-viviparity continuum in some lizard populations suggest a novel advantage to viviparity under these unique conditions (Shine et al. 2018).

Viviparity has its independent origins across a broad span of the vertebrate phylogenetic tree, including bony and cartilaginous fishes, amphibians, mammals, and squamate reptiles (Fig. 1). Hundreds of studies have documented the anatomical and physiological features of viviparity and oviparity and debated their origins. Is oviparity ancestral, what are the molecular mechanisms responsible for the evolution of viviparity, and are these molecular mechanisms the same for separate origins of viviparity? Thus far, very few studies have asked the question what are the fundamental molecular networks that species have used to evolve to a viviparity state, and possibly reverse itself within certain clades (Gao et al. 2019). New comparative studies at species, genus, and family levels are needed if we are to find the essential genetic components related to the extraordinary reproductive strategy of viviparity. Such transitions of oviparous to viviparous modes of reproduction require many adaptations including reduction in eggshell construction, delayed oviposition, altered maternal immune system responses, not to mention the placenta development of intricate blood supply for fetal nutrition and gas exchange. These questions are best considered using examples of specific transitions among closely-related squamates, fishes, amphibians, and mammals. Each of these life-bearing strategies have been studied from the standpoints of morphology, physiology, endocrinology, ecology, and evolution. In this brief exploration of viviparous origins prior to the terrestrial life of vertebrates, we appropriately look back over 400 million years ago and start with the fishes.

2 Fishes

Among the estimated 32,672 species of fishes (Froese and Pauly 2018), viviparity is a rarity, with approximately 510 classified as viviparous (Parenti and Grier 2004). In addition, some species display an intermediate ovoviviparous mode of reproduction where nutrient supply is deposited before fertilization (lecithotropy) (Macfarlane and Bowers 1995) which allows contrasts to matrotrophic species which have extensive maternal nourishment. In fishes, viviparity is clearly a specialized phenotype, again likely an oviparity derived mode of reproduction. In some viviparous fishes, novel embryo provisioning processes have been observed when compared to the nutrient partitioning in eutherian mammals by blood plasma passing to the fetus via the placenta and umbilical cord (Bell and Ehrhardt 2002). The most studied is the process of vitellogenesis, a provisioning of vitellogenin (*vtg*) proteins during embryonic development. These yolk protein nutrients are lost in eutherian mammals. Some viviparous fish species have maintained the *vtg* proteins that provision the yolk protein in oviparous species, and a large amount of yolk nutrient is contained in their eggs (Sawaguchi et al. 2005; Vega-Lopez et al. 2007). The bony fishes of the teleost family Goodeidae are viviparous, but unlike most all other viviparous species have not lost the *vtg* proteins. In one member of Goodeidae, *Xenotoca eiseni*, the *vtg* proteins are suggested to be a contributing matrotrophic factor supplied to the intraovarian embryo, a nourishing mechanism lost in a majority of viviparous vertebrates. For all other viviparous fishes, the *vtg* genes have been lost through a coevolution mechanism with casein genes (Brawand et al. 2008).

The Poecilid fishes, guppies and mollies, offer an altogether different dichotomy, of placental evolution a clade of entirely live-bearing species. As a result, a theory of genomic conflict has been deliberated, that is the viviparity-driven conflict hypothesis where the variation in placentation among Poecilia is proposed to effect rates of male sexual selection and drive greater speciation (Furness et al. 2019). Pollux et al. (2014), empirically tested if the evolution of the placenta was driving sexual selection in live-bearing fish and concluded that there are no male sexual traits that effect the evolution of the placenta (Pollux et al. 2014); instead a more complex interaction between female reproductive mode, male sexual selection, and the rate of speciation is a better working model (Furness et al. 2019). With high-quality genome assemblies now available for members of the viviparous Poecilid genus, advancing computational methods can delineate gene expansion and contraction more accurately as well as lineage-specific changes among putative regulatory sequences to differentiate this evolved state of viviparity.

More recently, genomic and transcriptomic analysis revealed remarkable analogies in the evolution of live-bearing rockfish (*Sebastes schlegelii*) (He et al. 2019). In these fish, sperm storage precedes a fertilization lengthy development of the embryo in the ovary and the ovarian wall has a presumed role similar to the mammalian uterus. Remarkably during implantation and development in the ovary, an upregulation of a number of genes involved in placentation, cell adhesion, invasion, and blood vessel function as well as oxygen and nutrient transfer occurs. This revealed that zona pellucida proteins function to retain sperm while metalloproteinases, such as high choriolysin enzyme (*hce1-like*) that is also described as hatchling enzyme, may play a role in releasing the embryo from the chorion at pre-hatching. In this first genome-wide study of an unusual viviparous teleost, sperm storage and development of the fry in the ovary suggests a selective advantage over egg-laying due to unpredictable and seasonal changes in nutrient supply. Only continued research of other rare viviparous fishes will further improve our comprehension of live-birth rationale.

3 Amphibians

A viviparous mode of reproduction has evolved independently in all three orders of amphibians—salamanders, frogs, and caecilians. Of the approximately 8000 amphibian species, a vast majority of them are egg layers. The few instances of viviparity are mostly confined to several species of caecilians, although a few frogs and the fire salamander enjoy this distinction. The study of the first "live-bearing" Asian amphibian, *Gegeneophis seshachari*, emphasizes the need to probe deeper into their phylogenetic reproductive strategies since it is the only known caecilian genus to have both oviparous and viviparous species (Gower et al. 2008). Anurans feature an extraordinary variety of reproductive modes and parental care for offspring, for example, gastric brooding (Haddad and Prado 2005). As in other vertebrates, the modes of reproduction are developing eggs retained in the oviducts, oviparity, and viviparity that include oviductal retention of developing embryos with the young born as tadpoles (Iskandar et al. 2014), metamorphosed juveniles, or froglets (Wells 2008). Then again out of the ordinary reproductive approaches in amphibians are observed in intraoviductal cannibalism and retention of externally fertilized ova in the skin of the mother's back. Other stark parity twists are also observed in amphibians. The diversity of amphibian oviduct structure and function is a furtherance of this reproduction epitome (Wake and Dickie 1998). Oviduct morphology is regionally differentiated to secrete varying products that encompass the egg that across all three orders offer species-specific avenues of investigation into ancestral transitions of oviparity to varying forms of viviparity. Our understanding of the developmental and functional bases for fetal adaptations in amphibians is undeveloped (Wake 2015) as very few studies have attempted to identify genes and pathways involved in parity transition in amphibia. Jantra et al. (2007) analyzed immune gene expression patterns in amphibia with different parity modes and found differential expression possibly linked with viviparity as one example.

4 Reptiles

Squamata reptiles include snakes and lizards, representing over 10,700 extant species (www.reptile-database.org) with vastly differing modes of reproduction. The exclusive use of oviparity in some reptile branches exclude any insight about molecular origins of viviparity. In birds as descendants of the dinosaur, oviparity is the only reproductive strategy but with remarkable post egg-laying oddities, such as brood parasitism behavior in some lineages (Feeney and Riehl 2019). Interestingly, a live-bearing species of Archosauromorpha reptile that is today represented by birds and crocodilians suggests an estimated transition to live birth occurred some 50 million years earlier in this clade (Liu et al. 2017). Still, within squamate reptiles, oviparity has transitioned to viviparity over 100 times (Blackburn 2015) that is astounding against a total of ~140 origins across all vertebrates (Blackburn 2000). This makes squamates a prime lineage to investigate the evolution of viviparity from oviparity that is generally thought to originate with lengthened periods of egg retention. This derived nature of viviparity is considered irreversible but theoretical models of its evolution, particularly among squamates, are not always steadfast (Pyron and Burbrink 2015). The reasoning for so-called "switchbacks" among squamates contribute to this debate (Blackburn 2015). Simulations of phylogenetic origins for viviparity are complicated to broadly interpret due to at least ten species that practice reproductive bimodality and imply parity mode is a malleable trait (Pyron and Burbrink 2014). However, there are unifying trait features such as eggshells that significantly share species-wide structure and their oviduct glands that synthesize egg fibers, thus providing strong predictive evidence that oviparity is ancestral. Even so, current squamate genomic studies are just now endeavoring to grasp the beginnings of viviparity.

Some closely related lizards within the genus *Phrynocephalus* are oviparous and viviparous. With an estimated divergence time of 13 million years for this monophylogentic clade of viviparous, lizard species compared to their oviparous relatives' gene expression differences in the maternal oviduct throughout development of the eggs and embryos were explored (Gao et al. 2019). As a result, numerous stage-specific gene regulatory differences were found in both species. A stasis in viviparous gene expression through birth was suggested to encourage prolonged uterine gene expression as opposed to rapid shifts seen in oviparous species. Only 3 of 148 oviparous differentially expressed genes that were shared in the viviparous lizard showing these genes are downregulated in viviparous lizards. Importantly, changes in the underlying gene sequences, a possible result of natural selection, were not found thus suggesting gene expression is the major factor associated with parity differences in these lizard species. Very few candidate genes have been proposed to be associated with placenta development and stability, some include *Hβ58* (Paulesu et al. 2001), various proteases (Cornetti 2018), estrogen receptor (Gao et al. 2019), and a variety of its effected genes (Eyster 2016) and the cadherins role in attachment maintenance (Wu et al. 2011). Of these candidates, estrogen's role in modulating viviparity is the most widely studied (Huet-Hudson et al. 1990; Fujimoto et al. 2005). A recent transcriptome analysis between viviparous and oviparous populations of the lizard *Saiphos equalis* provided a unique opportunity to study the transition between parity modes. The long egg retention populations showed vastly different transcriptome profiles compared to oviparous lizards supporting the idea that long egg retention is an intermediate towards viviparity (Foster et al. 2020). Candidate squamate genes associated with viviparous origins in these few studies offer a unique avenue in understanding how evolution reworked gene networks toward one parity type or the other among amniotes (Fig. 2). As in mammals, the viviparous squamate develops placental membranes that interact with adjacent uterine tissues responsible for adhesion to the uterus, respiratory gas exchange, water transport, and nutrient transport justifying additional comparative investigations (Fig. 2).

5 Mammals

In mammals, the only lineage that evolved or retained oviparity is the monotremes (platypus and echidna). Other mammals feature life birth at different stages, and if eggs and their prolonged retention are an ancient precursor to the mammalian placenta, the marsupials represent an ideal intermediate group with a more simplified intrauterine development for the study of evolutionary paths to eutherian placental structures (Frankenberg and Renfree 2018). Confusion can arise though in the characterization of mammalian placental biology. Eutherian mammals sometimes mistakenly classified as placental mammals are to the exclusion of marsupials and monotremes while in fact all three major groups of mammals feature placentae. Guernsey et al. (2017) describe the marsupial placenta that similarly mediates early embryonic development but despite a shared mechanism of nourishing the infant. Even in the egg-laying monotreme, the embryo is supported by a simple placenta during early development. Both the monotreme hatchling and marsupial newborn rely on a far more complex repertoire of milk proteins for further development. In many viviparous species, the mother provides nutrients to the embryo during gestation, a pattern known as "matrotrophy." Yet, when observing the life history of mammalian viviparity, at the base of mammalian phylogeny one enigma remains. Only five species of the 5416 extant mammals are not viviparous, the oviparous monotremes of echidna (four extant species) and platypus. The discovery of egg-laying in platypus and echidna was a sensation among European Zoologists and was remarkably reported on the same day in 1884 by Haake in echidna and Caldwell in platypus. The various attributes of the monotremes and how their oviparous mode of reproduction, as well as many other unexpected phenotypes, conflict with existing theories of species evolution remains under study. The

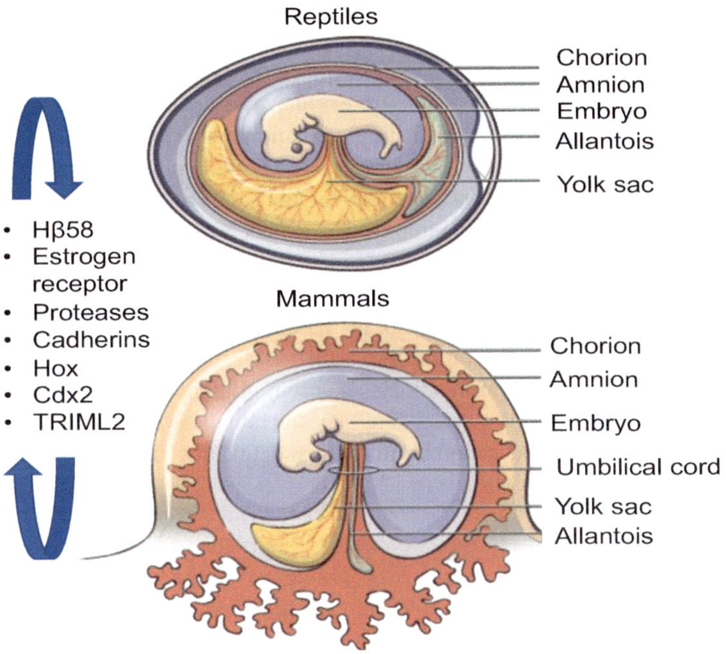

Fig. 2 Extraembryonic membranes of viviparous reptiles and mammals. Both develop membranes that form a placenta during pregnancy to serve the nutritional needs of the developing embryo and fetus. Many ancestral genes, a subset listed here, have been putatively repurposed to facilitate viviparous transitions across vertebrate phylogenies. (Figure modified with permission from shmoop.com)

platypus genome was a harbinger of evidence about what the ancient mammal-like reptile genomes may have lost in the burst of eutherian species. The ancestral descent of known physiological systems that were counterintuitive, such as the ability to lay eggs (Warren et al. 2008), was revealed in a collection of studies. The remnants of genes that are associated with venom (Whittington et al. 2009), lactation, egg-laying, testes descent, sensory gene evolution, gastric function (Ordonez et al. 2008), immune gene family expansions (Whittington et al. 2008), jumping genes (retrotransposed elements), and many others were characterized (Warren et al. 2008). Despite the variety of mammalian placental types and differences among viviparous squamates, all are derived from the trophoectodermal layer of cells in the developing embryo and unique gene adaptations offer partial explanations (Fig. 2). In a search for genes that drive the trophoectoderm differentiation, a precursor to the functioning placenta, the oviparous platypus has offered some transcription factors that putatively control trophoectoderm destiny, specifically POU domain transcription factors (Niwa et al. 2008; Frankenberg et al. 2010). More comparative studies will be needed to firmly associate this unique monotreme regulatory sequence as a key precursor in directing cellular differentiation toward placental structures that establish uterine nourishment of embryos. Other derived eutherian genic examples of placental support also exist. For instance, the proper cis-regulatory gene regulation of prolactin in endometrial cells by *HoxA-11* transcription factors is necessary for proper placentation. These maternal stemmed cis-regulatory elements that control prolactin endometrial expression are exclusive to eutherian mammals (Lynch et al. 2008). Some studies suggest that inflammation has been repurposed as a way for the mother and fetus to communicate. Only eutherians display a dramatic downregulation of inflammation after implantation, thus averting premature parturition (Chavan et al. 2017). Gene duplication events after the divergence of eutherians for the Tripartite-motif family-like genes, in particularly the unique eutherian existence of *TRIML2*, suggests a specific role in the control of placental induced inflammation (Zhang et al. 2020). Also, the parent of origin-dependent monoallelic gene expression or genomic imprinting, which is seen as an important driver of placentation and viviparity on mammals, is absent in monotremes (Keverne 2014). The candidate gene examples noted here and others validate the vibrant use of the marsupial and monotreme genomes to further define the origins of mammalian placental complexity.

6 Conclusion

The evolutionary transition from oviparity to viviparity is now better understood and can occur in intermediate steps via internal incubation of the eggs or more or less extensive nourishment through placental tissues. Vertebrates offer a variety of evolutionary solutions to balancing fecundity, number of offspring, and environmental uncertainty with the challenges of internal development of the fetus (Amoroso 1952). While many gene signaling pathways involved in adhesion, invasion, immunosuppression, and nutrient and gas exchange have been associated with viviparity, different genes among them have been recruited to support parity. Together, a compendium of data supports the evolution of viviparity allowed for the gradual loss of yolk-dependent nourishment of the developing fetus. Illustrations of oviparous coupled genes and their partnering molecular networks that collectively have evolved toward a role in sustaining the developing fetus via the placenta offer exciting expansions in parity knowledge. On the other hand, the immense complexity observed when comparing oviparous and viviparous gene systems intimates uncertainty in this broad statement. Continuing studies that offer compelling counter arguments of parity beginnings will hopefully systemize the gene networks potentially utilized to achieve the appropriate level of reproductive success. Certainly, with many high-quality genomes now available for various oviparous and viviparous species, scientists are empowered to test many gene centric theories on the molecular plan deployed to confer one mode of reproduction over another. Perhaps E.C. Amoroso summarized it simplistically and best when asked to define the placental features of mammals that distinguish them from lower vertebrates when he noted one key word—"protection" and then followed "This protection conferred by viviparity reaches its climax in mammals."

References

Amoroso EC (1952) Placentation. In: Parkes AS (ed) Marshall's physiology of reproduction, vol 2, 3rd edn. Longmans, London, pp 127–311

Bell AW, Ehrhardt RA (2002) Regulation of placental nutrient transport and implications for fetal growth. Nutr Res Rev 15(2):211–230

Blackburn DG (1999) Viviparity and oviparity: evolution and reproductive strategies. In: Encyclopedia of reproduction. Academic Press, New York, pp 994–1003

Blackburn DG (2000) Classification of the reproductive patterns of amniotes. Herpetol Monogr 14:371–377

Blackburn DG (2015) Evolution of viviparity in squamate reptiles: reversibility reconsidered. J Exp Zool B Mol Dev Evol 324(6):473–486

Brawand D, Wahli W, Kaessmann H (2008) Loss of egg yolk genes in mammals and the origin of lactation and placentation. PLoS Biol 6(3):e63

Chavan AR, Griffith OW, Wagner GP (2017) The inflammation paradox in the evolution of mammalian pregnancy: turning a foe into a friend. Curr Opin Genet Dev 47:24–32

Cornetti LEA (2018) Candidate genes involved in the evolution of viviparity: a RAD sequencing experiment in the lizard Zootoca vivipara (Squamata: Lacertidae). Zool J Linnean Soc 183:196–207

Eyster KM (2016) The estrogen receptors: an overview from different perspectives. Methods Mol Biol 1366:1–10

Feeney WE, Riehl C (2019) Monogamy without parental care? Social and genetic mating systems of avian brood parasites. Philos Trans R Soc Lond Ser B Biol Sci 374(1769):20180201

Foster CSP, Thompson MB, Van Dyke JU, Brandley MC, Whittington CM (2020) Emergence of an evolutionary innovation: gene expression differences associated with the transition between oviparity and viviparity. Mol Ecol 29(7):1315–1327

Frankenberg S, Renfree MB (2018) Conceptus coats of marsupials and monotremes. Curr Top Dev Biol 130:357–377

Frankenberg S, Pask A, Renfree MB (2010) The evolution of class V POU domain transcription factors in vertebrates and their characterisation in a marsupial. Dev Biol 337(1):162–170

Froese R, Pauly D (2018) Fish base. World Wide Web Electronic Publication

Fu C, Coelho MA, David-Palma M, Priest SJ, Heitman J (2019) Genetic and genomic evolution of sexual reproduction: echoes from LECA to the fungal kingdom. Curr Opin Genet Dev 58–59:70–75

Fujimoto J, Nakagawa Y, Toyoki H, Sakaguchi H, Sato E, Tamaya T (2005) Estrogen-related receptor expression in placenta throughout gestation. J Steroid Biochem Mol Biol 94 (1–3):67–69

Furness AI, Pollux BJA, Meredith RW, Springer MS, Reznick DN (2019) How conflict shapes evolution in poeciliid fishes. Nat Commun 10(1):3335

Gao W, Sun YB, Zhou WW, Xiong ZJ, Chen L, Li H, Fu TT, Xu K, Xu W, Ma L, Chen YJ, Xiang XY, Zhou L, Zeng T, Zhang S, Jin JQ, Chen HM, Zhang G, Hillis DM, Ji X, Zhang YP, Che J (2019) Genomic and transcriptomic investigations of the evolutionary transition from oviparity to viviparity. Proc Natl Acad Sci U S A 116(9):3646–3655

Gower DJ, Giri V, Dharne MS, Shouche YS (2008) Frequency of independent origins of viviparity among caecilians (Gymnophiona): evidence from the first 'live-bearing' Asian amphibian. J Evol Biol 21(5):1220–1226

Guernsey MW, Chuong EB, Cornelis G, Renfree MB, Baker JC (2017) Molecular conservation of marsupial and eutherian placentation and lactation. Elife 6

Haddad CFB, Prado CPA (2005) Reproductive modes in frogs and their unexpected diversity in the Atlantic forest of Brazil. Bioscience 55:207–217

He Y, Chang Y, Bao L, Yu M, Li R, Niu J, Fan G, Song W, Seim I, Qin Y, Li X, Liu J, Kong X, Peng M, Sun M, Wang M, Qu J, Wang X, Liu X, Wu X, Zhao X, Wang X, Zhang Y, Guo J, Liu Y, Liu K, Wang Y, Zhang H, Liu L, Wang M, Yu H, Wang X, Cheng J, Wang Z, Xu X,

Wang J, Yang H, Lee SM, Liu X, Zhang Q, Qi J (2019) A chromosome-level genome of black rockfish, Sebastes schlegelii, provides insights into the evolution of live birth. Mol Ecol Resour 19(5):1309–1321

Hickey G, Heller D, Monlong J, Sibbesen JA, Siren J, Eizenga J, Dawson ET, Garrison E, Novak AM, Paten B (2020) Genotyping structural variants in pangenome graphs using the vg toolkit. Genome Biol 21(1):35

Huet-Hudson YM, Chakraborty C, De SK, Suzuki Y, Andrews GK, Dey SK (1990) Estrogen regulates the synthesis of epidermal growth factor in mouse uterine epithelial cells. Mol Endocrinol 4(3):510–523

Iskandar DT, Evans BJ, McGuire JA (2014) A novel reproductive mode in frogs: a new species of fanged frog with internal fertilization and birth of tadpoles. PLoS One 9(12):e115884

Jantra S, Bigliardi E, Brizzi R, Ietta F, Bechi N, Paulesu L (2007) Interleukin 1 in oviductal tissues of viviparous, oviparous, and ovuliparous species of amphibians. Biol Reprod 76(6):1009–1015

Kaneko-Ishino T, Ishino F (2019) Evolution of viviparity in mammals: what genomic imprinting tells us about mammalian placental evolution. Reprod Fertil Dev 31(7):1219–1227

Keverne EB (2014) Mammalian viviparity: a complex niche in the evolution of genomic imprinting. Heredity (Edinb) 113(2):138–144

Liu J, Organ CL, Benton MJ, Brandley MC, Aitchison JC (2017) Live birth in an archosauromorph reptile. Nat Commun 8:14445

Lynch VJ, Tanzer A, Wang Y, Leung FC, Gellersen B, Emera D, Wagner GP (2008) Adaptive changes in the transcription factor HoxA-11 are essential for the evolution of pregnancy in mammals. Proc Natl Acad Sci U S A 105(39):14928–14933

Macfarlane R, Bowers M (1995) Matrotrophic viviparity in the yellowtail rockfish Sebastes flavidus. J Exp Biol 198:1197–1206

Niwa H, Sekita Y, Tsend-Ayush E, Grutzner F (2008) Platypus Pou5f1 reveals the first steps in the evolution of trophectoderm differentiation and pluripotency in mammals. Evol Dev 10 (6):671–682

Ordonez GR, Hillier LW, Warren WC, Grutzner F, Lopez-Otin C, Puente XS (2008) Loss of genes implicated in gastric function during platypus evolution. Genome Biol 9(5):R81

Parenti LR, Grier HJ (2004) Evolution and phylogeny of gonad morphology in bony fishes. Integr Comp Biol 44(5):333–348

Paulesu L, Cateni C, Romagnoli R, Chellini F, Angelini F, Guarino FM, Rider V, Imakawa K, Bigliardi E (2001) Evidence of H beta 58, a gene involved in mammalian placental development, in the three-toed skink, Chalcides chalcides (Squamata: Scincidae), a viviparous placentotrophic reptile. Placenta 22(8–9):735–741

Pollux BJ, Meredith RW, Springer MS, Garland T, Reznick DN (2014) The evolution of the placenta drives a shift in sexual selection in livebearing fish. Nature 513(7517):233–236

Pyron RA, Burbrink FT (2014) Early origin of viviparity and multiple reversions to oviparity in squamate reptiles. Ecol Lett 17(1):13–21

Pyron RA, Burbrink FT (2015) Contrasting models of parity-mode evolution in squamate reptiles. J Exp Zool B Mol Dev Evol 324(6):467–472

Sawaguchi S, Koya Y, Yoshizaki N, Ohkubo N, Andoh T, Hiramatsu N, Sullivan CV, Hara A, Matsubara T (2005) Multiple vitellogenins (Vgs) in mosquitofish (Gambusia affinis): identification and characterization of three functional Vg genes and their circulating and yolk protein products. Biol Reprod 72(4):1045–1060

Sharpe RM (2018) Programmed for sex: nutrition-reproduction relationships from an intergenerational perspective. Reproduction 155(3):S1–S16

Shine R, Wapstra E, Olsson M (2018) Seasonal shifts along the oviparity-viviparity continuum in a cold-climate lizard population. J Evol Biol 31(1):4–13

Smith SA, Shine R (1997) Intraspecific variation in reproductive mode within the scincid lizard Saiphos equalis. Aust J Zool 45:435–445

Vega-Lopez A, Ortiz-Ordonez E, Uria-Galicia E, Mendoza-Santana EL, Hernandez-Cornejo R, Atondo-Mexia R, Garcia-Gasca A, Garcia-Latorre E, Dominguez-Lopez ML (2007) The role of

vitellogenin during gestation of Girardinichthys viviparus and Ameca splendens; two goodeid fish with matrotrophic viviparity. Comp Biochem Physiol A Mol Integr Physiol 147(3):731–742
Wake MH (2015) Fetal adaptations for viviparity in amphibians. J Morphol 276(8):941–960
Wake MH, Dickie R (1998) Oviduct structure and function and reproductive modes in amphibians. J Exp Zool 282(4–5):477–506
Warren WC, Hillier LW, Marshall Graves JA, Birney E, Ponting CP, Grutzner F, Belov K, Miller W, Clarke L, Chinwalla AT, Yang SP, Heger A, Locke DP, Miethke P, Waters PD, Veyrunes F, Fulton L, Fulton B, Graves T, Wallis J, Puente XS, Lopez-Otin C, Ordonez GR, Eichler EE, Chen L, Cheng Z, Deakin JE, Alsop A, Thompson K, Kirby P, Papenfuss AT, Wakefield MJ, Olender T, Lancet D, Huttley GA, Smit AF, Pask A, Temple-Smith P, Batzer MA, Walker JA, Konkel MK, Harris RS, Whittington CM, Wong ES, Gemmell NJ, Buschiazzo E, Vargas Jentzsch IM, Merkel A, Schmitz J, Zemann A, Churakov G, Kriegs JO, Brosius J, Murchison EP, Sachidanandam R, Smith C, Hannon GJ, Tsend-Ayush E, McMillan D, Attenborough R, Rens W, Ferguson-Smith M, Lefevre CM, Sharp JA, Nicholas KR, Ray DA, Kube M, Reinhardt R, Pringle TH, Taylor J, Jones RC, Nixon B, Dacheux JL, Niwa H, Sekita Y, Huang X, Stark A, Kheradpour P, Kellis M, Flicek P, Chen Y, Webber C, Hardison R, Nelson J, Hallsworth-Pepin K, Delehaunty K, Markovic C, Minx P, Feng Y, Kremitzki C, Mitreva M, Glasscock J, Wylie T, Wohldmann P, Thiru P, Nhan MN, Pohl CS, Smith SM, Hou S, Nefedov M, de Jong PJ, Renfree MB, Mardis ER, Wilson RK (2008) Genome analysis of the platypus reveals unique signatures of evolution. Nature 453 (7192):175–183
Wells KD (2008) The ecology and behavior of amphibians. The University of Chicago Press, Chicago
Whittington CM, Papenfuss AT, Bansal P, Torres AM, Wong ES, Deakin JE, Graves T, Alsop A, Schatzkamer K, Kremitzki C, Ponting CP, Temple-Smith P, Warren WC, Kuchel PW, Belov K (2008) Defensins and the convergent evolution of platypus and reptile venom genes. Genome Res 18(6):986–994
Whittington CM, Koh JM, Warren WC, Papenfuss AT, Torres AM, Kuchel PW, Belov K (2009) Understanding and utilising mammalian venom via a platypus venom transcriptome. J Proteome 72(2):155–164
Wu Q, Thompson MB, Murphy CR (2011) Changing distribution of cadherins during gestation in the uterine epithelium of lizards. J Exp Zool B Mol Dev Evol 316(6):440–450
Zanders SE, Unckless RL (2019) Fertility costs of meiotic drivers. Curr Biol 29(11):R512–R520
Zhang X, Pavlicev M, Jones HN, Muglia LJ (2020) Eutherian-specific gene TRIML2 attenuates inflammation in the evolution of placentation. Mol Biol Evol 37(2):507–523

Development of Pre-implantation Mammalian Blastocyst

Bhanu P. Telugu and Laramie Pence

Contents

Abstract The preimplantation mammalian embryo is a simplistic, self-contained, and a superior model for investigating the inherent complexities of cell fate decision mechanisms. All mammals begin their humble journey from a single-cell fertilized zygote contained within a proteinaceous coat called the zona pellucida. The zygote embarks on a series of well-orchestrated events, beginning with the activation of embryonic genome, transition from meiotic to mitotic divisions, spatial organization of the cells, timely differentiation into committed trophectoderm (TE) and primitive endoderm (PrE), and ultimately escape from zona pellucida for implantation into the uterus. The entire development of preimplantation embryo can be studied in vitro using a minimalistic and defined culture system. The ease of culture along with the

B. P. Telugu (✉)
Department of Animal and Avian Sciences, University of Maryland, College Park, MD, USA

Division of Animal Sciences, University of Missouri, Columbia, MO, USA
e-mail: btelugu@umd.edu; telugub@missouri.edu

L. Pence
Department of Animal and Avian Sciences, University of Maryland, College Park, MD, USA
e-mail: lpence@umd.edu

R. D. Geisert, T. E. Spencer (eds.), *Placentation in Mammals*, Advances in Anatomy, Embryology and Cell Biology 234, https://doi.org/10.1007/978-3-030-77360-1_3

ability to manipulate gene expression and image the embryos makes them an ideal model system for investigation into the first two of several cell fate decisions made by the embryo that result in a pluripotent epiblast (EPI) and differentiated TE and PrE lineages. This chapter reviews our latest knowledge of preimplantation embryo development, setting the stage for understanding placental development in subsequent chapters in this Book.

Keywords Inner cell mass · Trophectoderm · Blastocyst · Epiblast · Hippo signaling

Abbreviations

Amot	Angiomotin
aPKC	Atypical protein kinase C
dnYap	Dominant-negative Yap
EPI	Epiblast
ICM	Inner cell mass
Lats-KD	Lats kinase dead
LH	Luteinizing hormone
Na^+/K^+ ATPase	Sodium potassium ATPase
PAR	PDZ-domain-containing scaffold protein
PGC	Primordial germ cell
PLC	Phospholipase C
PrE	Primitive endoderm
TE	Trophectoderm

1 Introduction

The developing mammalian embryo is a unique and valuable model system for studying cell fate decisions. One of the fundamental questions in biology is how a relatively simple single-cell zygote achieves the seemingly impossible task of generating a complex multicellular being. The embryonic cells are tasked with having to make orderly cell fate decisions in a spatiotemporal manner, differentiate into specialized cell types, self-organize into higher order tissues and organ systems, and ultimately generate a functional being. In this regard, the developing mammalian embryo, a "self-contained" system serves as a unique model system for in vitro investigation into genetic, epigenetic, and mechanochemical forces that are at the basis for regulating cell fate (White et al. 2018). Early stages of mammalian embryo development, from maturation of oocytes to blastocyst, can be recapitulated in vitro using a simplistic culture system without the need for external maternal cues (Whitten and Biggers 1968). The recent discovery of genome editors now expands the range of mammalian species (beyond mice) available for genetic manipulation

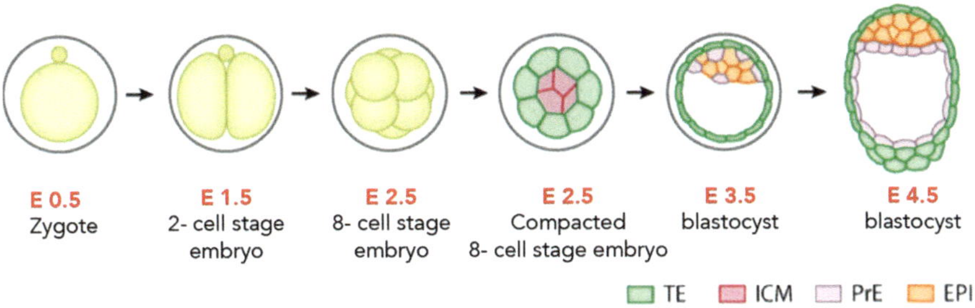

Fig. 1 Mouse embryo preimplantation development. The embryo undergoes successive and timely cleavage divisions resulting in an increased number of totipotent blastomeres with a progressively smaller cell size. In the compact morula, these blastomeres become self-organized into inside and outside cells. By embryonic day (E) 2.5 late eight-cell stage, the boundaries between the outside cells begin to disappear by "compaction." By E3.5, the embryo transforms into a fluid-filled blastocyst with outside trophectoderm (TE) and inner cell mass (ICM). By E4.5, the ICM cells differentiate further into primitive endoderm cells (PrE) lining the blastocoel cavity and pluripotent epiblast (EPI) cells at embryonic pole. TE marks the first visual and deterministic cell fate decision followed by the development of PrE making the second fate decision. In the figure, cell lineages are marked by distinctive colors to visually differentiate them

and for comparative studies. Advances in sequencing technologies, including the ability to sequence single cells, now allow us to capture genetic and epigenetic signature at a higher resolution within the embryos. Finally, breakthroughs in live cell imaging permit tracking of respective morphological fates in vitro (Nagy et al. 2003; Brinster 1963). Cumulatively, the advantages offered by the use of the embryo as a model system and numerous technological advantages are facilitating major breakthroughs in our understanding of first two cell fate decisions: (1) the emergence of trophectoderm (TE) and inner cell mass (ICM) and (2) the differentiation of ICM into pluripotent epiblast (EPI) and committed primitive endoderm (PrE). In this chapter, key landmark events in the development of blastocyst are discussed (Fig. 1). These include awakening of the oocyte by sperm penetration and subsequent transition from meiotic to mitotic divisions (Fig. 2a–f), reductional cleavage divisions resulting in the progressive decline in the size of the embryonic cells and the emergence of heterogeneities in the cleaved blastomeres (Fig. 2g), self-organization of cleaved blastomeres into inside and outside cells (Fig. 2h), compaction of the outer cells and the erasure of distinct cell boundaries (Fig. 2h–j), development of trophectoderm and fluid-filled blastocyst cavity (Fig. 2k), generation of extraembryonic endoderm, and ultimately hatching and implantation of the embryo (Fig. 2l–n). Among the mammalian embryos, the mouse zygote has been studied extensively and will be the basis for discussion below. Several recent articles on mammalian blastocyst development have been published, and the readers are encouraged to peruse the literature for additional information (White et al. 2018; White and Plachta 2020; Zhang and Hiiragi 2018; Leung and Zernicka-Goetz 2015).

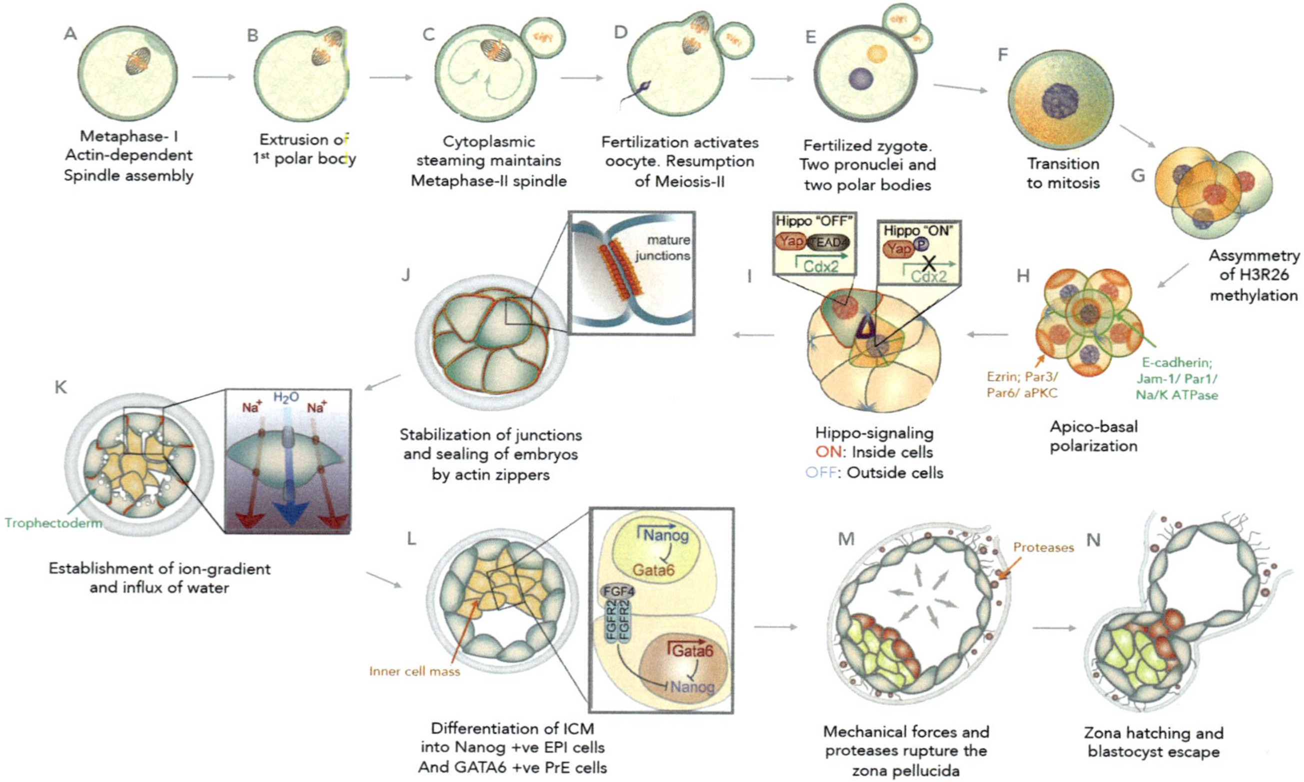

Fig. 2 Key landmark events leading to blastocyst development. (**a**) Oocyte following LH surge resume meiosis I. Actin microfilaments self-organize into meiotic spindles to one side of the embryo resulting from asymmetrical cortical softening. (**b**) Completion of meiosis I results in extrusion of one half of the chromosomes. (**c**) Extruded chromosomes with small amounts of cytoplasm result in the formation of residual cell, first polar body. Cytoplasmic streaming

holds the now metaphase II in place and is arrested at this stage. (**d**) Penetration of sperm following fertilization results in the completion of meiosis II and extrusion of second polar body. (**e**) The now haploid oocyte genome and sperm genomes undergo decondensation, accumulate nuclear membrane, and generate male and female pronuclei. (**f**) The now fertilized zygote begins transition to mitotic cleavage divisions. The maternally deposited messages mediate the transition until zygotic genome activation that takes place at two-cell stage in mouse embryos. (**g**) Asymmetries in epigenetic signatures among blastomeres arise at four-cell stage. Two of the four blastomeres express high levels of H3R26 methylation. (**h**) By late eight-cell stage apicobasal polarization emerges in the outside cells. On the apical surface: microvilli develop on the contact-free cell surface. Actin filaments and actin-binding protein Ezrin localize to the apical surface alongside Par3/Par6/aPkc apical polarization complex. On the basal surface, E-cadherin accumulates at the junctional complex alongside Par1/Jam1 and Na/K ATPase. (**i**) By 16-cell stage post-compaction embryo, Hippo signaling is fixed. In the outside cells, Hippo signaling is inactive, resulting in the upregulation of deterministic TE-specification gene, *Cdx2*. In the inside cells, hippo signaling is activated, resulting in the loss of Cdx2 and upregulation of pluripotency gene, *Pou5f1*. (**j**) The actin rings in the outside cells stabilize and mature junctions in a zipper-like mechanism. (**k**) Activation of sodium pumps resulting in pumping of sodium ions and diffusion of water resulting in progressive accumulation of fluids and fusion of microlumens to form blastocyst cavity. (**l**) Inside pluripotent cells differentiate into Nanog +ve EPI cells (red) and Gata6 +ve PrE cells (yellow) in a characteristic salt-and-pepper fashion. (**m**) The PrE cells via a combination of apoptosis and migration, resulting in the lining of PrE cells towards the blastocyst cavity. The blastocyst via a combination of mechanical forces resulting from expansion of blastocyst and expression of proteases result in the thinning and eventual rupture of zona pellucida. (**n**) Expanded blastocyst escapes from zona pellucida in a process known as zona hatching. (The composite figure has been generated from White et al. (2018) with permission from Elsevier Journal)

2 Setting the Stage: Fertilization Awakens the Oocyte and Jumpstarts Embryonic Development

Embryonic development begins with the fusion of a haploid sperm with the oocyte. The timeline for sperm and egg development is heterogeneous in both sexes. Both gametes start their journey as primordial germ cells (PGC)—a small cluster of cells in the primitive streak at the posterior end of the gastrulating embryo (Ohinata et al. 2009; Saitou and Yamaji 2012; Irie et al. 2014; Ginsburg et al. 1990). Upregulation of a triad of transcription factors *Prdm1* (Ohinata et al. 2005), *Prdm14* (Yamaji et al. 2008, 2013; Ma et al. 2011; Grabole et al. 2013), and *Tfap2c* (Schafer et al. 2011; Weber et al. 2010) in a subset of precursor mesodermal cells results in the suppression of somatic program, epigenetic erasure, and reactivation of pluripotency, ultimately giving rise to PGC. The PGC then make their orderly journey to their eventual resting place, the genital ridge—a mesenchyme abutting the mesonephric system. Key differences between the sexes emerge here. In males, upregulation of Nanos2 prevents progression of male germ cells to meiosis, which is resumed postnatally (Suzuki and Saga 2008). In females, the germ cells enter into prophase of meiosis I and are arrested at the diplotene stage (dictyate stage) at the time of birth and undergo sequential albeit discontinuous progression to a haploid stage. The oocyte maintains close contact with the nurse cells—the cumulus cells via transzonal cellular projections that traverse the width of zona pellucida and provide nutritional and transcriptional support, and regulate meiosis (Barrett and Albertini 2010). Postnatally, a surge in luteinizing hormone (LH) will result in the loss of cumulus-oocyte contacts, breakdown of the nuclear membrane, and resumption of meiosis (Bury et al. 2017). In what could be considered as a first departure from symmetry, the cortex of the oocyte softens at one-end promoting migration and assembly of meiotic spindle, followed by cortical thickening to hold the spindle in place (Fig. 2a) (Chaigne et al. 2013). The oocytes lack centrosomes; therefore, the first few divisions are dependent on self-organization of actin filaments (Fig. 2a) (Dumont et al. 2007; Yi et al. 2013). Asymmetrical positioning of the spindle ensures generation and expulsion of a smaller sister cell (polar body) encompassing one half of the chromosomes and a smaller volume of cytoplasm following meiosis I (Fig. 2b, c) (Yi et al. 2011; Mogessie and Schuh 2017). This ensures that bulk of cytoplasm with transcriptional and translational machinery was preserved in the oocyte for progression to mitotic divisions. Following the first meiotic division culminating in generation of the first polar body, the meiotic spindle is assembled in a similar fashion with polymerization of actin filaments, cytoplasmic streaming (Fig. 2d) (Yi et al. 2011), and is arrested at metaphase stage of meiosis II again, awaiting activation by sperm. Following entry of sperm into the oocyte, a sperm-borne oocyte-activating factor (PLC-zeta) triggers reactivation and resumption of meiosis and expulsion of second polar body (Fig. 2d, e) (Clift and Schuh 2013). This marks the first keystone event.

3 Cleavage, Imprinting Erasure, and Embryonic Genome Activation

Following fertilization, the maternal and paternal chromosomes undergo decondensation, and assemble nuclear membranes forming male and female pronuclei (Fig. 2e). The next immediate hurdle is the transition of oocyte from meiotic to mitotic divisions when the now zygotic genome is transcriptionally silent. The maternally deposited transcripts guide this transition and predominantly code for protein transportation, localization, and cell cycle genes (Fig. 2f) (Li and Albertini 2013). Paternal and maternal DNA undergo histone exchange, global DNA methylation, gamete-specific imprinting erasure, and activation of retrotransposons, culminating in the attainment of a totipotent genome (Habibi and Stunnenberg 2017; Jachowicz et al. 2017; Burton and Torres-Padilla 2014). Even though the mammalian embryo does not go through a predetermined cell fate specification as seen in *Drosophila*, C. elegans, *Xenopus*, and other lower phyla, it is now clear that subtle differences between blastomeres emerge as early as two-cell stage and major differences are evident by four-cell stage (Fig. 2g) (Tabansky et al. 2013; Fujimori et al. 2003; Piotrowska-Nitsche et al. 2005; Littwin and Denker 2011; Antczak and Van Blerkom 1997). Following DNA replication and first mitotic cleavage division, differences in allocation of maternal transcripts between the two blastomeres, especially in the localization of ribosomal RNAs, are evident at two-cell stage (Piotrowska et al. 2001; Piotrowska and Zernicka-Goetz 2001). Likewise, differences in expression of epigenetic modifiers Prdm14 (Burton et al. 2013) and Carm1 (Torres-Padilla et al. 2007) were reported by the four-cell stage, with two of the four cells expressing high levels of Prdm14 and Carm1. Additionally, differences in H3 methylation H3R17 and H3R26 emerge in blastomeres at the four-cell stage, which tend to bias contribution to ICM (Fig. 2g) (Torres-Padilla et al. 2007; Burton and Torres-Padilla 2014). Carm1 regulates H3R26 methylation and increases expression of Sox2, ultimately biasing the cells to pluripotent ICM fate. It is not clear, what regulates increased expression of Carm1 at the four-cell stage. Prior studies that reported no differences in transcriptional profile among blastomeres until the 16-cell stage in the mouse embryo were clearly constrained by the depth and limitations of sequencing (Dietrich and Hiiragi 2008; Guo et al. 2010; Ralston and Rossant 2008; Wicklow et al. 2014; Alarcon and Marikawa 2005; Motosugi et al. 2005; Hiiragi and Solter 2004). In summary, though the blastomeres are overtly homogenous and seemingly identical morphologically, heterogeneities are built into blastomeres by the four-cell stage, setting the stage for lineage specification (Tabansky et al. 2013; Fujimori et al. 2003). Regardless, the fate of the blastomeres remains flexible and is only finalized at the blastocyst stage to accommodate for stochastic errors that are to be expected in the developing embryo.

4 Compaction: First Visual Departure

Following a series of mitotic cleavage events that lead to a progressively smaller and morphologically indistinguishable blastomeres within the zona pellucida, the first overt morphological change that disrupts the uniformity is the compaction of blastomeres soon after the eight-cell stage (Ziomek and Johnson 1980). Blastomeres on the outside establish apicobasal polarity with the accumulation of microvilli, actin-binding protein Ezrin, and actomyosin complex on the contact-free apical cell surface (Fig. 2h) (Ducibella et al. 1977; Louvet et al. 1996; Vinot et al. 2005). On the basolateral surface, E-cadherin accumulates in the junctional complexes (Vestweber et al. 1987; Ziomek and Johnson 1980). It is yet unclear as to what triggers apical polarity at the cell surface, but preliminary reports suggest that actomyosin complex is accumulated at the apical cortex by phospholipase C-mediated hydrolysis of phosphoinol phosphate 2 and activation of protein kinase C (Zhu et al. 2017). Activated protein kinase activates RhoA, which in turn triggers polarization of the actin network, accumulation of the Par3-Par6-aPKC complex (Vinot et al. 2005), and ultimately the formation of an apical domain. In the cell-cell contact basolateral surface, accumulation of Par1 (Vinot et al. 2005), Jam-1 (Fig. 2h) (Thomas et al. 2004), and Na^+/K^+ ATPase (Watson and Kidder 1988) was seen alongside E-cadherin (Fig. 2k). Accumulation of E-cadherin at the basolateral surface increases the surface area of cell contacts, resulting in a gradual flattening of the outer blastomeres and erasure of distinct cell boundaries, and establishment of a primitive epithelium-like structure in the embryo (Fig. 2j) (Johnson 2009; Ducibella and Anderson 1975). It was long believed that accumulation of E-cadherin at junctional complexes and an increase in intercellular adhesion were responsible for morphological changes during compaction. However, recent studies cast doubt on this basic assumption as to whether E-cadherin transligation will yield enough forces to achieve cellular deformation (Samarage et al. 2015; Maitre et al. 2012). Rather, the discovery of filopodia—long membrane protrusions emanating from the apical cell surface and adhering via E-cadherin to the apical surface of adjacent cells—to be likely responsible for characteristically distinct compaction event (Fierro-Gonzalez et al. 2013). The filopodia are connected to the cytoskeleton and myosin motor protein, myosin-10 (Fierro-Gonzalez et al. 2013). Laser ablation of the filopodia results in rapid rounding of the cells and loss of compaction, bringing adhesion-based compaction model into question (Maitre et al. 2015). These observations corroborate a mechanochemical model for compaction, whereby the pulsatile actomyosin contractility linking the neighboring cells is redistributed by junctional E-cadherin away from cell-cell contact surface. Future investigations are aimed at understanding how the E-cadherin-dependent filopodia emerge, and the pulsatile actomyosin contractility is initiated and regulated.

5 Specification of Inner Cell Mass (ICM) and Trophectoderm: The First Cell Fate Decision

Following compaction and between 8- and 16-cell late-morula stage embryo, the cells are organized into outer polar and inner apolar cells, which give rise to TE and ICM lineages, respectively. However, the phenotypes and lineage identities at the 16-cell stage are not stable. Isolated apolar blastomeres can become polarized if placed on the outside position (Ziomek et al. 1982); outside cells of the late morula can produce ICM derivatives when aggregated with earlier embryos (Rossant and Vijh 1980), and cells of the ICM can produce trophoblast tissue (Handyside 1978). It is not until blastocyst formation that the TE and ICM lineages are irreversibly determined. Formation of the blastocyst marks the first visual cell fate decision, with TE cells on the outside committed to placental development, and cells on the inside (ICM) committed to the development of fetus and components of extraembryonic membranes. These events as discussed below are dependent on stochastic and deterministic events involving Hippo signaling system.

5.1 Differentially Expressed Transcription Factors Lead to Lineage Specification

At the late 16-cell stage, when the TE and ICM lineages are set-aside, differential expression of specific transcription factors becomes apparent in the mouse embryo. Inner apolar cells show strong expression of Sox family member *Sox2* (Avilion et al. 2003), the homeobox gene *Nanog* (Chambers et al. 2003; Mitsui et al. 2003), and octamer-binding transcription factor gene *Pou5f1* (Niwa et al. 2000). Sox2 is the first transcription factor to be specifically upregulated in the inner cells starting at the 16-cell stage (Guo et al. 2010). *Nanog* and *Pou5f1* are found uniformly expressed in all cells, with strong expression evident at the eight-cell stage, and later restricted to the ICM of the blastocyst (Fig. 2l) (Palmieri et al. 1994; Dietrich and Hiiragi 2008).

The caudal-related homeobox gene, *Cdx2*, and the zinc-finger transcription factor *Gata3*, have an expression pattern opposite to that of *Sox2*, *Nanog*, and *Pou5f1*. *Cdx2* and *Gata3* are strongly expressed in the outside polarized cell population (Strumpf et al. 2005; Beck et al. 1995; Ralston et al. 2010). It is speculated that an imbalance between Pou5f1 and Cdx2 initiates a reciprocal inhibition system that results in the restricted pattern (Ralston et al. 2010). To support this, TE-like cells found in *Cdx2* null embryos reexpress Pou5f1 (Strumpf et al. 2005), while *Pou5f1* null embryos express TE markers in the ICM (Nichols et al. 1998). However, a more recent study shows that *Pou5f1* levels remain high during the accumulation of Cdx2 and speculates that *Pou5f1* levels therefore do not affect accumulation and maintenance of Cdx2 (Dietrich and Hiiragi 2008). It is still possible however that the reciprocal inhibition pathway is dependent on Cdx2 reaching a certain threshold before it is able to downregulate *Pou5f1*. Either way, single blastomere RNA sequencing has shown that activation of TE-specific genes, including *Cdx2*, shows up as the first overt transcriptional difference between blastomeres (Posfai et al. 2017). To summarize, the currently accepted model for lineage specification in mouse embryo combines gene expression patterning and blastomere position. In this model, molecular differences between blastomeres are established during the eight-cell stage, and then a combination of stochastic processes restrict blastomeres with specific molecular profiles to the inside or outside populations (Dietrich and Hiiragi 2008; Lanner 2014). Following this stochastic allocation, a deterministic event led by Hippo signaling establishes the lineage commitment (Fig. 3) (Nishioka et al. 2008).

5.2 Hippo Signaling Pathway and Lineage Specification

The Hippo signaling pathway was first discovered in *Drosophila* as a tumor suppressor-signaling pathway, but is also conserved in mice and humans (Harvey et al. 2013; Yu and Guan 2013). The Hippo pathway is regulated by a variety of stimuli including cell-cell adhesion mediated by E-cadherin and by cell polarity mediated by angiomotin (Amot) which is thought to be the stimuli that initiates Hippo signaling activity in the mouse embryo (Fig. 3) (Kim et al. 2011; Hirate et al. 2013). The main components of this pathway are the protein kinase Mst1/2 and its co-activator Sav1 that function to activate kinase Lats1/2. In turn, Lats1/2 and its coactivator Mob1a/b phosphorylate Yap1 and Taz, which are transcriptional coactivators of Tead proteins (Yu and Guan 2013). Amot interacts with the E-cadherin-a α/β-catenin complex and serves as a scaffold protein that associates with Yap and Lats proteins (Fig. 3) (Paramasivam et al. 2011). Phosphorylation of Yap1 results in cytoplasmic sequestration and in turn suppresses target gene expression. Therefore, activation of Hippo signaling suppresses expression of *Tead* target genes, and inactivation of the pathway induces the expression of those target genes.

In the mouse embryo, Hippo signaling is an upstream regulator of *Cdx2* expression and is directly involved in ICM vs. TE lineage specification (Yagi et al. 2007). In the inside cells of the mouse embryo destined to become ICM, Hippo signaling is activated as a result of increased cell-cell contacts (Nishioka et al. 2009). This activation is mediated by Lats1 and Lats2 kinase, which phosphorylates and sequesters Yap1 in the cytoplasm, preventing it from coactivating Tead4, a transcription factor that directly regulates *Cdx2* expression (Nishioka et al. 2009; Hirate et al. 2013). Transient upregulation of *Lats2* in the mouse embryo results in reduced nuclear accumulation of Yap1, downregulation of *Cdx2*, and failure to form a blastocoel because of the absence of TE cells (Fig. 3) (Nishioka et al. 2009). This phenotype is also a characteristic of *Tead4* null embryos, indicating that aberrant overexpression of Hippo component Lats2 suppresses differentiation to the TE lineage via its control of Tead4 activation (Yagi et al. 2007). In the outside cells destined to become TE, Hippo signaling is inactive, which allows Yap1 to translocate into the nucleus, activate Tead4, and upregulate *Cdx2* (Fig. 3) (Nishioka et al.

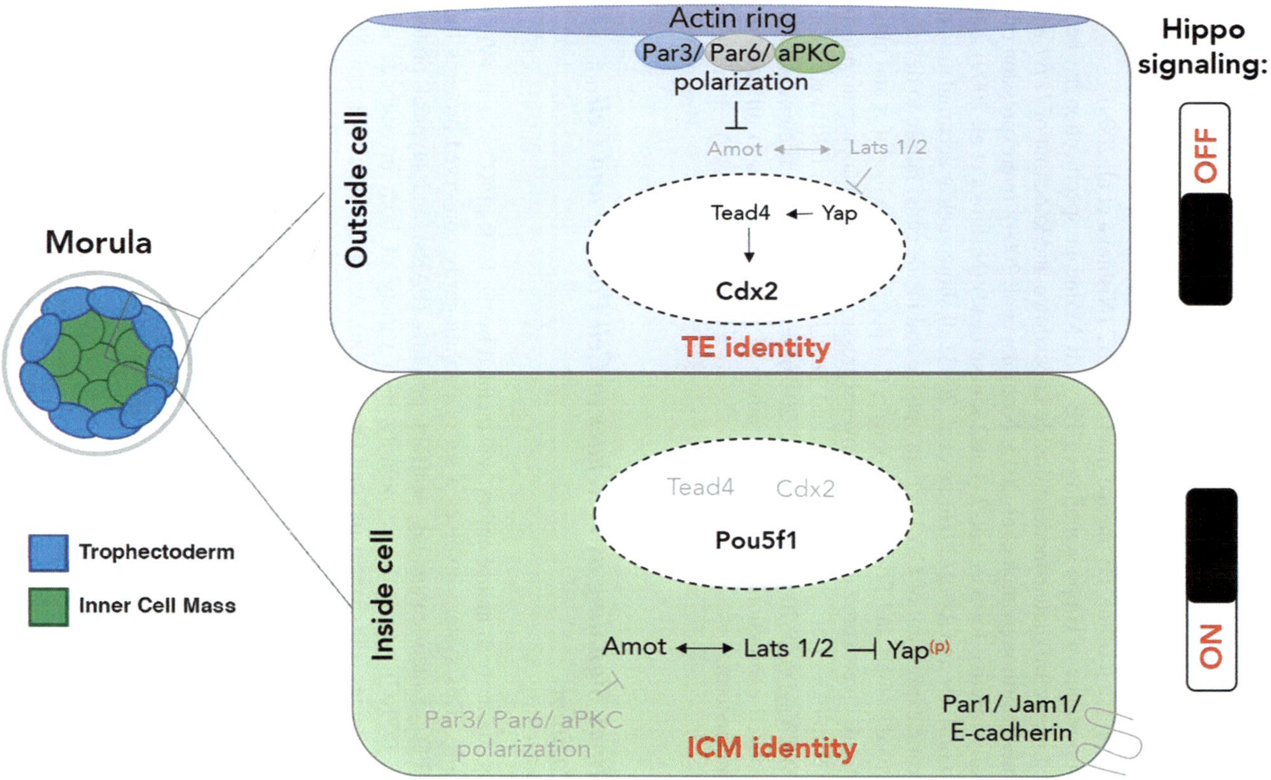

Fig. 3 Hippo signaling in the mouse embryo. Hippo signaling regulates lineage specification in the early mouse embryo. In the outside cells, Hippo signaling is active. Development of polarization complex (Par3/Par6/aPKC) in the outside cells results in tethering of junction-associate protein angiomoitin (Amot) to the apical side and subsequent lack of Lats1/2 activation. Yap protein is available for translocation into the nucleus, which in turn activates Tead4. Tead4 in turn

2008, 2009). Null mutation of either Lats1 or Lats2 does not appear to disrupt normal development during the preimplantation stages (McPherson et al. 2004; St John et al. 1999; Yabuta et al. 2007). However, overexpression of a catalytically inactive form of Lats2 (Lats kinase dead, or Lats-KD) is able to dominantly inhibit both Lats1 and Lat2, resulting in an embryo that expresses Cdx2 in inside cells (Nishioka et al. 2009). Furthermore, a double knockout of both *Lats1* and *Lats2* exhibits the same phenotype, strongly suggesting that Lats1/2 kinase is necessary to activate expression of the TE lineage marker, *Cdx2* (Nishioka et al. 2009).

The importance of Hippo signaling in TE vs. ICM lineage specification has been confirmed in several mouse models: double knockdown and knockout of *Lats1* and *Lats2* genes (Lorthongpanich et al. 2013; Nishioka et al. 2009), overexpression of a catalytically inactive form of Lats2 (Lats2 kinase dead) (Nishioka et al. 2009), and depletion of *Amot* and *Amotl2* (Hirate et al. 2013) all show nuclear accumulation of Yap1, strong expression of *Cdx2* in both outside and inside cells, failure to develop ICM-derived tissues, and TE-like blastomeres that populate both the inner and outer cell positions. Overexpression of Lats2, overexpression of a dominant negative form of Yap (Nishioka et al. 2009) (dnYap), and *Tead4* null embryos (Nishioka et al. 2008, 2009; Yagi et al. 2007; Ralston et al. 2010) show cytoplasmic accumulation of Yap1, reduced expression of *Cdx2*, and failure of embryonic cells to differentiate into trophectoderm.

5.3 *Hippo Signaling as It Relates to Cell Polarity and Cell Position*

Apicobasal polarity in mouse embryos is dictated by the PAR-aPKC system (Vinot et al. 2005). This system involves a set of evolutionarily conserved proteins that include PDZ-domain-containing scaffold proteins (PARs) and atypical protein kinase C (aPKC) which dictate polarity in a variety of both invertebrate and vertebrate species (Fig. 3) (Suzuki and Ohno 2006). As discussed above, polarization begins at the eight-cell stage in the mouse embryo and is marked by apically restricted microvilli (experimentally visualized with phosphorylated-Ezrin) (Louvet et al. 1996) and accumulation of PARD6b and later aPKCζ at the apical surface (Vinot et al. 2005). Establishment of polarity in turn has been shown to directly suppress Hippo signaling activity in embryos (Hirate et al. 2013). Disruption of the PAR-aPKC system via injection of RNAi constructs in mouse zygotes causes Hippo

---

Fig. 3 (continued) activates trophectoderm gene expression, including *Cdx2*. In the inside cells, Hippo signaling is active. Lack of polarization complex allows for Amot to activate Lats kinase. Lats kinases phosphorylate Yap and subsequently sequester Yap protein in the cytoplasm. Lack of Yap translocation in the nucleus results in loss of expression of TE specification genes and upregulation of pluripotent genes including *Pou5f1* and establishment of pluripotent inner cell mass cells

pathway activation as evidenced by disruption of apical localization of Amot, exclusion of Yap1 from the nucleus, and TE development failure (Hirate et al. 2013). However, when the same embryos are dissociated, Hippo signaling is not activated despite the disruption of polarity (Hirate et al. 2013). This suggests that activation of Hippo signaling is not only dependent on the apolar status of individual blastomeres but is also dependent on cell-cell adhesion. Furthermore, in apolar cells where the PAR-aPKC system is inactive, Amot is found associated with basolateral adherens junctions via binding to the E-cadherin complex (Hirate et al. 2013). Activation of Amot at adherens junctions is thought to potentiate the activation/function of Lats kinases and results in Hippo pathway activation (Hirate et al. 2013). Together, these observations indicate that Hippo signaling responds to a combination of inputs: cell-cell adhesion and cell polarization. In the inner cells, Hippo signaling is activated by increased cell-cell contacts whereas in the outer cells, cell-cell contact-dependent Hippo signaling is suppressed by the polarization status of the outside cells mediated by the PAR-aPKC system (Sasaki 2015). To relate this finding to TE-specific expression of *Cdx2*, a study analyzing localization of aPKC in *Cdx2* mutant embryos showed that aPKC localization to the apical surface was not affected by the lack of *Cdx2*. This suggests that cell polarization is independent of Cdx2, and *Cdx2* upregulation is genetically downstream of cell polarization (Ralston and Rossant 2008).

From these studies, progression of events can be summarized as follows: (1) cell polarization begins at the eight-cell stage, (2) from the 16-cell stage onwards final polarity status in individual blastomeres is established, (3) Hippo signaling is suppressed in the polarized outer cells and is activated by cell-cell adhesion in the apolar inner cells (Fig. 3), and (4) this leads to lineage specific gene expression to stabilize the TE and ICM fates in the blastocyst (Fig. 3). Hippo signaling is therefore a key driver for segregation of ICM and TE lineages; however, it is still unknown at what time point Hippo signaling is sufficient to establish cell fate (Posfai et al. 2017).

6 Blastocyst Maturation and Hatching

Following the stabilization of Hippo signaling and the establishment of outer TE cells and ICM cells, the embryo transitions to a fluid-filled blastocyst stage. Blastocyst cavity is formed by active transport of sodium ions across the TE cells which create an osmotic gradient and an influx of liquid to create a fluid-filled blastocyst (Fig. 2k) (Aziz and Alexandre 1991; Benos et al. 1985; Watson and Barcroft 2001). The expanding blastocyst cavity is stabilized by tight junctions between the outer cells resulting in the generation of first embryonic epithelium the TE (Wang et al. 2008). Recent studies have highlighted the emergence of atypical apical actin rings that extend to the junctions between the outer cells and subsequent stabilization of adherens and tight junctions in a zipper-like mechanism in the TE (Fig. 2j) (Zenker et al. 2018). The expanding blastocyst cavity results in marginalization of ICM cells towards one end of the blastocyst (embryonic pole). The pluripotent ICM cells express Pou5f1, Sox2, Nanog, and Gata6 (Fig. 2l) (Ohnishi et al. 2014). Soon after, a reciprocal expression is established between Nanog expressing EPI cells producing FGF4 in an Sox2/Pou5f1-dependent fashion and Gata6 and Ffgr1/2 expressing FGF4-responsive PrE cells—in a characteristic "salt-and-pepper" fashion (Fig. 2l) (Chazaud et al. 2006). Following the establishment of two distinct populations, the PrE cells line up the blastocyst cavity through a combination of migration, apoptosis, and adhesion (Fig. 2m) (Saiz and Plusa 2013). The blastocyst at this stage will need to escape from zona pellucida to initiate implantation with uterine epithelium. The blastocyst escapes from zona in still a poorly understood process but likely includes a combination of mechanical forces from an expanding blastocyst, and proteolytic activity (Fig. 2m) (Seshagiri et al. 2009; Perona and Wassarman 1986). The escape of blastocyst (Fig. 2n) from zona pellucida will herald an implantation phase, and subsequent development is contingent on maternal inputs.

7 Concluding Remarks

More than one-third of pregnancies are lost in the preimplantation phase of embryo development, and close to 50% of human pregnancies fail during the first few weeks of pregnancy (Hyde and Schust 2015). The prevalence is especially worse for recipients in ovum donation programs, and in patients undergoing in vitro fertilization (McDonald et al. 2009; Margalioth et al. 2006; Abdalla et al. 1998). These huge rates in embryonic losses are also true in livestock, where embryonic and preimplantation loses remain major contributors of infertility (Berg et al. 2010). Majority of our current understanding on blastocyst development comes from studies in rodent models. Traditionally, this may have to do with ease of husbandry and the ability to perform sophisticated genetic modifications. However, key differences in the expression of lineage specification genes in non-rodent models remain and necessitate investigation into alternative model species. For example, there is a lack of mutually exclusive and antagonistic expression of *POU5F1* and *CDX2* in TE and ICM cells in embryos from livestock, a hallmark of rodent embryos (Rossant 2011). As discussed in subsequent chapters in this book, key differences also emerge post-hatching in the implantation phase between mouse, human, and other livestock species. In this regard, comparative studies involving livestock and other non-rodent model systems will likely contribute to a greater understanding of lineage specification, and in bridging gaps left by the rodent models. Livestock offer several key advantages, including a well-established culture system, unlimited supply of oocytes (slaughterhouse), and in the post-CRISPR and genome editing era, an ability to perform genetic modification. Ultimately, cross species investigations will unlock conserved mechanisms, highlight key differences, and will have a greater impact on animal and human health.

References

Abdalla HI, Billett A, Kan AK, Baig S, Wren M, Korea L, Studd JW (1998) Obstetric outcome in 232 ovum donation pregnancies. Br J Obstet Gynaecol 105(3):332–337. https://doi.org/10.1111/j.1471-0528.1998.tb10096.x

Alarcon VB, Marikawa Y (2005) Unbiased contribution of the first two blastomeres to mouse blastocyst development. Mol Reprod Dev 72(3):354–361. https://doi.org/10.1002/mrd.20353

Antczak M, Van Blerkom J (1997) Oocyte influences on early development: the regulatory proteins leptin and STAT3 are polarized in mouse and human oocytes and differentially distributed within the cells of the preimplantation stage embryo. Mol Hum Reprod 3(12):1067–1086. https://doi.org/10.1093/molehr/3.12.1067

Avilion AA, Nicolis SK, Pevny LH, Perez L, Vivian N, Lovell-Badge R (2003) Multipotent cell lineages in early mouse development depend on SOX2 function. Genes Dev 17(1):126–140. https://doi.org/10.1101/gad.224503

Aziz M, Alexandre H (1991) The origin of the nascent blastocoele in preimplantation mouse embryos ultrastructural cytochemistry and effect of chloroquine. Roux Arch Dev Biol 200 (2):77–85. https://doi.org/10.1007/BF00637187

Barrett SL, Albertini DF (2010) Cumulus cell contact during oocyte maturation in mice regulates meiotic spindle positioning and enhances developmental competence. J Assist Reprod Genet 27 (1):29–39. https://doi.org/10.1007/s10815-009-9376-9

Beck F, Erler T, Russell A, James R (1995) Expression of Cdx-2 in the mouse embryo and placenta: possible role in patterning of the extra-embryonic membranes. Dev Dyn 204(3):219–227

Benos DJ, Biggers JD, Balaban RS, Mills JW, Overstrom EW (1985) Developmental aspects of sodium-dependent transport processes of preimplantation rabbit embryos. Soc Gen Physiol Ser 39:211–235

Berg DK, van Leeuwen J, Beaumont S, Berg M, Pfeffer PL (2010) Embryo loss in cattle between days 7 and 16 of pregnancy. Theriogenology 73(2):250–260. https://doi.org/10.1016/j.theriogenology.2009.09.005

Brinster RL (1963) A method for in vitro cultivation of mouse ova from two-cell to blastocyst. Exp Cell Res 32:205–208. https://doi.org/10.1016/0014-4827(63)90093-4

Burton A, Torres-Padilla ME (2014) Chromatin dynamics in the regulation of cell fate allocation during early embryogenesis. Nat Rev Mol Cell Biol 15(11):723–734. https://doi.org/10.1038/nrm3885

Burton A, Muller J, Tu S, Padilla-Longoria P, Guccione E, Torres-Padilla ME (2013) Single-cell profiling of epigenetic modifiers identifies PRDM14 as an inducer of cell fate in the mammalian embryo. Cell Rep 5(3):687–701. https://doi.org/10.1016/j.celrep.2013.09.044

Bury L, Coelho PA, Simeone A, Ferries S, Eyers CE, Eyers PA, Zernicka-Goetz M, Glover DM (2017) Plk4 and Aurora A cooperate in the initiation of acentriolar spindle assembly in mammalian oocytes. J Cell Biol 216(11):3571–3590. https://doi.org/10.1083/jcb.201606077

Chaigne A, Campillo C, Gov NS, Voituriez R, Azoury J, Umana-Diaz C, Almonacid M, Queguiner I, Nassoy P, Sykes C, Verlhac MH, Terret ME (2013) A soft cortex is essential for asymmetric spindle positioning in mouse oocytes. Nat Cell Biol 15(8):958–966. https://doi.org/10.1038/ncb2799

Chambers I, Colby D, Robertson M, Nichols J, Lee S, Tweedie S, Smith A (2003) Functional expression cloning of Nanog, a pluripotency sustaining factor in embryonic stem cells. Cell 113 (5):643–655

Chazaud C, Yamanaka Y, Pawson T, Rossant J (2006) Early lineage segregation between epiblast and primitive endoderm in mouse blastocysts through the Grb2-MAPK pathway. Dev Cell 10 (5):615–624. https://doi.org/10.1016/j.devcel.2006.02.020

Clift D, Schuh M (2013) Restarting life: fertilization and the transition from meiosis to mitosis. Nat Rev Mol Cell Biol 14(9):549–562. https://doi.org/10.1038/nrm3643

Dietrich JE, Hiiragi T (2008) Stochastic processes during mouse blastocyst patterning. Cells Tissues Organs 188(1–2):46–51. https://doi.org/10.1159/000118783

Ducibella T, Anderson E (1975) Cell shape and membrane changes in the eight-cell mouse embryo: prerequisites for morphogenesis of the blastocyst. Dev Biol 47(1):45–58. https://doi.org/10.1016/0012-1606(75)90262-6

Ducibella T, Ukena T, Karnovsky M, Anderson E (1977) Changes in cell surface and cortical cytoplasmic organization during early embryogenesis in the preimplantation mouse embryo. J Cell Biol 74(1):153–167. https://doi.org/10.1083/jcb.74.1.153

Dumont J, Million K, Sunderland K, Rassinier P, Lim H, Leader B, Verlhac MH (2007) Formin-2 is required for spindle migration and for the late steps of cytokinesis in mouse oocytes. Dev Biol 301(1):254–265. https://doi.org/10.1016/j.ydbio.2006.08.044

Fierro-Gonzalez JC, White MD, Silva JC, Plachta N (2013) Cadherin-dependent filopodia control preimplantation embryo compaction. Nat Cell Biol 15(12):1424–1433. https://doi.org/10.1038/ncb2875

Fujimori T, Kurotaki Y, Miyazaki J, Nabeshima Y (2003) Analysis of cell lineage in two- and four-cell mouse embryos. Development 130(21):5113–5122. https://doi.org/10.1242/dev.00725

Ginsburg M, Snow MH, McLaren A (1990) Primordial germ cells in the mouse embryo during gastrulation. Development 110(2):521–528

Grabole N, Tischler J, Hackett JA, Kim S, Tang F, Leitch HG, Magnusdottir E, Surani MA (2013) Prdm14 promotes germline fate and naive pluripotency by repressing FGF signalling and DNA methylation. EMBO Rep 14(7):629–637. https://doi.org/10.1038/embor.2013.67

Guo G, Huss M, Tong GQ, Wang C, Li Sun L, Clarke ND, Robson P (2010) Resolution of cell fate decisions revealed by single-cell gene expression analysis from zygote to blastocyst. Dev Cell 18(4):675–685. https://doi.org/10.1016/j.devcel.2010.02.012

Habibi E, Stunnenberg HG (2017) Transcriptional and epigenetic control in mouse pluripotency: lessons from in vivo and in vitro studies. Curr Opin Genet Dev 46:114–122. https://doi.org/10.1016/j.gde.2017.07.005

Handyside AH (1978) Time of commitment of inside cells isolated from preimplantation mouse embryos. J Embryol Exp Morphol 45:37–53

Harvey KF, Zhang X, Thomas DM (2013) The Hippo pathway and human cancer. Nat Rev Cancer 13(4):246–257. https://doi.org/10.1038/nrc3458

Hiiragi T, Solter D (2004) First cleavage plane of the mouse egg is not predetermined but defined by the topology of the two apposing pronuclei. Nature 430(6997):360–364. https://doi.org/10.1038/nature02595

Hirate Y, Hirahara S, Inoue K, Suzuki A, Alarcon VB, Akimoto K, Hirai T, Hara T, Adachi M, Chida K, Ohno S, Marikawa Y, Nakao K, Shimono A, Sasaki H (2013) Polarity-dependent distribution of angiomotin localizes Hippo signaling in preimplantation embryos. Curr Biol 23 (13):1181–1194. https://doi.org/10.1016/j.cub.2013.05.014

Hyde KJ, Schust DJ (2015) Genetic considerations in recurrent pregnancy loss. Cold Spring Harb Perspect Med 5(3):a023119

Irie N, Tang WW, Azim Surani M (2014) Germ cell specification and pluripotency in mammals: a perspective from early embryogenesis. Reprod Med Biol 13(4):203–215. https://doi.org/10.1007/s12522-014-0184-2

Jachowicz JW, Bing X, Pontabry J, Boskovic A, Rando OJ, Torres-Padilla ME (2017) LINE-1 activation after fertilization regulates global chromatin accessibility in the early mouse embryo. Nat Genet 49(10):1502–1510. https://doi.org/10.1038/ng.3945

Johnson MH (2009) From mouse egg to mouse embryo: polarities, axes, and tissues. Annu Rev Cell Dev Biol 25:483–512. https://doi.org/10.1146/annurev.cellbio.042308.113348

Kim NG, Koh E, Chen X, Gumbiner BM (2011) E-cadherin mediates contact inhibition of proliferation through Hippo signaling-pathway components. Proc Natl Acad Sci U S A 108 (29):11930–11935. https://doi.org/10.1073/pnas.1103345108

Lanner F (2014) Lineage specification in the early mouse embryo. Exp Cell Res 321(1):32–39. https://doi.org/10.1016/j.yexcr.2013.12.004

Leung CY, Zernicka-Goetz M (2015) Mapping the journey from totipotency to lineage specification in the mouse embryo. Curr Opin Genet Dev 34:71–76. https://doi.org/10.1016/j.gde.2015.08.002

Li R, Albertini DF (2013) The road to maturation: somatic cell interaction and self-organization of the mammalian oocyte. Nat Rev Mol Cell Biol 14(3):141–152. https://doi.org/10.1038/nrm3531

Littwin T, Denker HW (2011) Segregation during cleavage in the mammalian embryo? A critical comparison of whole-mount/CLSM and section immunohistochemistry casts doubts on segregation of axis-relevant leptin domains in the rabbit. Histochem Cell Biol 135(6):553–570. https://doi.org/10.1007/s00418-011-0816-0

Lorthongpanich C, Messerschmidt DM, Chan SW, Hong W, Knowles BB, Solter D (2013) Temporal reduction of LATS kinases in the early preimplantation embryo prevents ICM lineage differentiation. Genes Dev 27(13):1441–1446. https://doi.org/10.1101/gad.219618.113

Louvet S, Aghion J, Santa-Maria A, Mangeat P, Maro B (1996) Ezrin becomes restricted to outer cells following asymmetrical division in the preimplantation mouse embryo. Dev Biol 177(2):568–579

Ma Z, Swigut T, Valouev A, Rada-Iglesias A, Wysocka J (2011) Sequence-specific regulator Prdm14 safeguards mouse ESCs from entering extraembryonic endoderm fates. Nat Struct Mol Biol 18(2):120–127. https://doi.org/10.1038/nsmb.2000

Maitre JL, Berthoumieux H, Krens SF, Salbreux G, Julicher F, Paluch E, Heisenberg CP (2012) Adhesion functions in cell sorting by mechanically coupling the cortices of adhering cells. Science 338(6104):253–256. https://doi.org/10.1126/science.1225399

Maitre JL, Niwayama R, Turlier H, Nedelec F, Hiiragi T (2015) Pulsatile cell-autonomous contractility drives compaction in the mouse embryo. Nat Cell Biol 17(7):849–855. https://doi.org/10.1038/ncb3185

Margalioth EJ, Ben-Chetrit A, Gal M, Eldar-Geva T (2006) Investigation and treatment of repeated implantation failure following IVF-ET. Hum Reprod 21(12):3036–3043. https://doi.org/10.1093/humrep/del305

McDonald SD, Han Z, Mulla S, Murphy KE, Beyene J, Ohlsson A (2009) Preterm birth and low birth weight among in vitro fertilization singletons: a systematic review and meta-analyses. Eur J Obstet Gynecol Reprod Biol 146(2):138–148. https://doi.org/10.1016/j.ejogrb.2009.05.035

McPherson JP, Tamblyn L, Elia A, Migon E, Shehabeldin A, Matysiak-Zablocki E, Lemmers B, Salmena L, Hakem A, Fish J, Kassam F, Squire J, Bruneau BG, Hande MP, Hakem R (2004) Lats2/Kpm is required for embryonic development, proliferation control and genomic integrity. EMBO J 23(18):3677–3688. https://doi.org/10.1038/sj.emboj.7600371

Mitsui K, Tokuzawa Y, Itoh H, Segawa K, Murakami M, Takahashi K, Maruyama M, Maeda M, Yamanaka S (2003) The homeoprotein Nanog is required for maintenance of pluripotency in mouse epiblast and ES cells. Cell 113(5):631–642

Mogessie B, Schuh M (2017) Actin protects mammalian eggs against chromosome segregation errors. Science 357(6353). https://doi.org/10.1126/science.aal1647

Motosugi N, Bauer T, Polanski Z, Solter D, Hiiragi T (2005) Polarity of the mouse embryo is established at blastocyst and is not prepatterned. Genes Dev 19(9):1081–1092. https://doi.org/10.1101/gad.1304805

Nagy A, Gertsenstein M, Vintersten K, Behringer R (2003) Manipulating the mouse embryo: a laboratory manual, 3rd edn. Cold Spring Harbor Laboratory Press, Cold Spring Harbor, NY

Nichols J, Zevnik B, Anastassiadis K, Niwa H, Klewe-Nebenius D, Chambers I, Scholer H, Smith A (1998) Formation of pluripotent stem cells in the mammalian embryo depends on the POU transcription factor Oct4. Cell 95(3):379–391

Nishioka N, Yamamoto S, Kiyonari H, Sato H, Sawada A, Ota M, Nakao K, Sasaki H (2008) Tead4 is required for specification of trophectoderm in pre-implantation mouse embryos. Mech Dev 125(3–4):270–283. https://doi.org/10.1016/j.mod.2007.11.002

Nishioka N, Inoue K, Adachi K, Kiyonari H, Ota M, Ralston A, Yabuta N, Hirahara S, Stephenson RO, Ogonuki N, Makita R, Kurihara H, Morin-Kensicki EM, Nojima H, Rossant J, Nakao K,

Niwa H, Sasaki H (2009) The Hippo signaling pathway components Lats and Yap pattern Tead4 activity to distinguish mouse trophectoderm from inner cell mass. Dev Cell 16(3):398–410. https://doi.org/10.1016/j.devcel.2009.02.003

Niwa H, Miyazaki J, Smith AG (2000) Quantitative expression of Oct-3/4 defines differentiation, dedifferentiation or self-renewal of ES cells. Nat Genet 24(4):372–376

Ohinata Y, Payer B, O'Carroll D, Ancelin K, Ono Y, Sano M, Barton SC, Obukhanych T, Nussenzweig M, Tarakhovsky A, Saitou M, Surani MA (2005) Blimp1 is a critical determinant of the germ cell lineage in mice. Nature 436(7048):207–213. https://doi.org/10.1038/nature03813

Ohinata Y, Ohta H, Shigeta M, Yamanaka K, Wakayama T, Saitou M (2009) A signaling principle for the specification of the germ cell lineage in mice. Cell 137(3):571–584. https://doi.org/10.1016/j.cell.2009.03.014

Ohnishi Y, Huber W, Tsumura A, Kang M, Xenopoulos P, Kurimoto K, Oles AK, Arauzo-Bravo MJ, Saitou M, Hadjantonakis AK, Hiiragi T (2014) Cell-to-cell expression variability followed by signal reinforcement progressively segregates early mouse lineages. Nat Cell Biol 16 (1):27–37. https://doi.org/10.1038/ncb2881

Palmieri SL, Peter W, Hess H, Scholer HR (1994) Oct-4 transcription factor is differentially expressed in the mouse embryo during establishment of the first two extraembryonic cell lineages involved in implantation. Dev Biol 166(1):259–267

Paramasivam M, Sarkeshik A, Yates JR 3rd, Fernandes MJ, McCollum D (2011) Angiomotin family proteins are novel activators of the LATS2 kinase tumor suppressor. Mol Biol Cell 22 (19):3725–3733. https://doi.org/10.1091/mbc.E11-04-0300

Perona RM, Wassarman PM (1986) Mouse blastocysts hatch in vitro by using a trypsin-like proteinase associated with cells of mural trophectoderm. Dev Biol 114(1):42–52. https://doi.org/10.1016/0012-1606(86)90382-9

Piotrowska K, Zernicka-Goetz M (2001) Role for sperm in spatial patterning of the early mouse embryo. Nature 409(6819):517–521. https://doi.org/10.1038/35054069

Piotrowska K, Wianny F, Pedersen RA, Zernicka-Goetz M (2001) Blastomeres arising from the first cleavage division have distinguishable fates in normal mouse development. Development 128 (19):3739–3748

Piotrowska-Nitsche K, Perea-Gomez A, Haraguchi S, Zernicka-Goetz M (2005) Four-cell stage mouse blastomeres have different developmental properties. Development 132(3):479–490. https://doi.org/10.1242/dev.01602

Posfai E, Petropoulos S, de Barros FRO, Schell JP, Jurisica I, Sandberg R, Lanner F, Rossant J (2017) Position- and Hippo signaling-dependent plasticity during lineage segregation in the early mouse embryo. Elife 6. https://doi.org/10.7554/eLife.22906

Ralston A, Rossant J (2008) Cdx2 acts downstream of cell polarization to cell-autonomously promote trophectoderm fate in the early mouse embryo. Dev Biol 313(2):614–629. https://doi.org/10.1016/j.ydbio.2007.10.054

Ralston A, Cox BJ, Nishioka N, Sasaki H, Chea E, Rugg-Gunn P, Guo G, Robson P, Draper JS, Rossant J (2010) Gata3 regulates trophoblast development downstream of Tead4 and in parallel to Cdx2. Development 137(3):395–403. https://doi.org/10.1242/dev.038828

Rossant J (2011) Developmental biology: a mouse is not a cow. Nature 471(7339):457–458. https://doi.org/10.1038/471457a

Rossant J, Vijh KM (1980) Ability of outside cells from preimplantation mouse embryos to form inner cell mass derivatives. Dev Biol 76(2):475–482

Saitou M, Yamaji M (2012) Primordial germ cells in mice. Cold Spring Harb Perspect Biol 4(11). https://doi.org/10.1101/cshperspect.a008375

Saiz N, Plusa B (2013) Early cell fate decisions in the mouse embryo. Reproduction 145(3):R65–R80. https://doi.org/10.1530/REP-12-0381

Samarage CR, White MD, Alvarez YD, Fierro-Gonzalez JC, Henon Y, Jesudason EC, Bissiere S, Fouras A, Plachta N (2015) Cortical tension allocates the first inner cells of the mammalian embryo. Dev Cell 34(4):435–447. https://doi.org/10.1016/j.devcel.2015.07.004

Sasaki H (2015) Position- and polarity-dependent Hippo signaling regulates cell fates in preimplantation mouse embryos. Semin Cell Dev Biol 47-48:80–87. https://doi.org/10.1016/j.semcdb.2015.05.003

Schafer S, Anschlag J, Nettersheim D, Haas N, Pawig L, Schorle H (2011) The role of BLIMP1 and its putative downstream target TFAP2C in germ cell development and germ cell tumours. Int J Androl 34(4 Pt 2):e152–158; discussion e158–159. https://doi.org/10.1111/j.1365-2605.2011.01167.x

Seshagiri PB, Sen Roy S, Sireesha G, Rao RP (2009) Cellular and molecular regulation of mammalian blastocyst hatching. J Reprod Immunol 83(1–2):79–84. https://doi.org/10.1016/j.jri.2009.06.264

St John MA, Tao W, Fei X, Fukumoto R, Carcangiu ML, Brownstein DG, Parlow AF, McGrath J, Xu T (1999) Mice deficient of Lats1 develop soft-tissue sarcomas, ovarian tumours and pituitary dysfunction. Nat Genet 21(2):182–186. https://doi.org/10.1038/5965

Strumpf D, Mao CA, Yamanaka Y, Ralston A, Chawengsaksophak K, Beck F, Rossant J (2005) Cdx2 is required for correct cell fate specification and differentiation of trophectoderm in the mouse blastocyst. Development 132(9):2093–2102

Suzuki A, Ohno S (2006) The PAR-aPKC system: lessons in polarity. J Cell Sci 119(Pt 6):979–987. https://doi.org/10.1242/jcs.02898

Suzuki A, Saga Y (2008) Nanos2 suppresses meiosis and promotes male germ cell differentiation. Genes Dev 22(4):430–435. https://doi.org/10.1101/gad.1612708

Tabansky I, Lenarcic A, Draft RW, Loulier K, Keskin DB, Rosains J, Rivera-Feliciano J, Lichtman JW, Livet J, Stern JN, Sanes JR, Eggan K (2013) Developmental bias in cleavage-stage mouse blastomeres. Curr Biol 23(1):21–31. https://doi.org/10.1016/j.cub.2012.10.054

Thomas FC, Sheth B, Eckert JJ, Bazzoni G, Dejana E, Fleming TP (2004) Contribution of JAM-1 to epithelial differentiation and tight-junction biogenesis in the mouse preimplantation embryo. J Cell Sci 117(Pt 23):5599–5608. https://doi.org/10.1242/jcs.01424

Torres-Padilla ME, Parfitt DE, Kouzarides T, Zernicka-Goetz M (2007) Histone arginine methylation regulates pluripotency in the early mouse embryo. Nature 445(7124):214–218. https://doi.org/10.1038/nature05458

Vestweber D, Gossler A, Boller K, Kemler R (1987) Expression and distribution of cell adhesion molecule uvomorulin in mouse preimplantation embryos. Dev Biol 124(2):451–456. https://doi.org/10.1016/0012-1606(87)90498-2

Vinot S, Le T, Ohno S, Pawson T, Maro B, Louvet-Vallee S (2005) Asymmetric distribution of PAR proteins in the mouse embryo begins at the 8-cell stage during compaction. Dev Biol 282(2):307–319. https://doi.org/10.1016/j.ydbio.2005.03.001

Wang H, Ding T, Brown N, Yamamoto Y, Prince LS, Reese J, Paria BC (2008) Zonula occludens-1 (ZO-1) is involved in morula to blastocyst transformation in the mouse. Dev Biol 318(1):112–125. https://doi.org/10.1016/j.ydbio.2008.03.008

Watson AJ, Barcroft LC (2001) Regulation of blastocyst formation. Front Biosci 6:D708–D730. https://doi.org/10.2741/watson

Watson AJ, Kidder GM (1988) Immunofluorescence assessment of the timing of appearance and cellular distribution of Na/K-ATPase during mouse embryogenesis. Dev Biol 126(1):80–90. https://doi.org/10.1016/0012-1606(88)90241-2

Weber S, Eckert D, Nettersheim D, Gillis AJ, Schafer S, Kuckenberg P, Ehlermann J, Werling U, Biermann K, Looijenga LH, Schorle H (2010) Critical function of AP-2 gamma/TCFAP2C in mouse embryonic germ cell maintenance. Biol Reprod 82(1):214–223. https://doi.org/10.1095/biolreprod.109.078717

White MD, Plachta N (2020) Specification of the first mammalian cell lineages in vivo and in vitro. Cold Spring Harb Perspect Biol 12(4):a035634. https://doi.org/10.1101/cshperspect.a035634

White MD, Zenker J, Bissiere S, Plachta N (2018) Instructions for assembling the early mammalian embryo. Dev Cell 45(6):667–679. https://doi.org/10.1016/j.devcel.2018.05.013

Whitten WK, Biggers JD (1968) Complete development in vitro of the pre-implantation stages of the mouse in a simple chemically defined medium. J Reprod Fertil 17(2):399–401. https://doi.org/10.1530/jrf.0.0170399

Wicklow E, Blij S, Frum T, Hirate Y, Lang RA, Sasaki H, Ralston A (2014) HIPPO pathway members restrict SOX2 to the inner cell mass where it promotes ICM fates in the mouse blastocyst. PLoS Genet 10(10):e1004618. https://doi.org/10.1371/journal.pgen.1004618

Yabuta N, Okada N, Ito A, Hosomi T, Nishihara S, Sasayama Y, Fujimori A, Okuzaki D, Zhao H, Ikawa M, Okabe M, Nojima H (2007) Lats2 is an essential mitotic regulator required for the coordination of cell division. J Biol Chem 282(26):19259–19271. https://doi.org/10.1074/jbc.M608562200

Yagi R, Kohn MJ, Karavanova I, Kaneko KJ, Vullhorst D, DePamphilis ML, Buonanno A (2007) Transcription factor TEAD4 specifies the trophectoderm lineage at the beginning of mammalian development. Development 134(21):3827–3836. https://doi.org/10.1242/dev.010223

Yamaji M, Seki Y, Kurimoto K, Yabuta Y, Yuasa M, Shigeta M, Yamanaka K, Ohinata Y, Saitou M (2008) Critical function of Prdm14 for the establishment of the germ cell lineage in mice. Nat Genet 40(8):1016–1022. https://doi.org/10.1038/ng.186

Yamaji M, Ueda J, Hayashi K, Ohta H, Yabuta Y, Kurimoto K, Nakato R, Yamada Y, Shirahige K, Saitou M (2013) PRDM14 ensures naive pluripotency through dual regulation of signaling and epigenetic pathways in mouse embryonic stem cells. Cell Stem Cell 12(3):368–382. https://doi.org/10.1016/j.stem.2012.12.012

Yi K, Unruh JR, Deng M, Slaughter BD, Rubinstein B, Li R (2011) Dynamic maintenance of asymmetric meiotic spindle position through Arp2/3-complex-driven cytoplasmic streaming in mouse oocytes. Nat Cell Biol 13(10):1252–1258. https://doi.org/10.1038/ncb2320

Yi K, Rubinstein B, Unruh JR, Guo F, Slaughter BD, Li R (2013) Sequential actin-based pushing forces drive meiosis I chromosome migration and symmetry breaking in oocytes. J Cell Biol 200 (5):567–576. https://doi.org/10.1083/jcb.201211068

Yu FX, Guan KL (2013) The Hippo pathway: regulators and regulations. Genes Dev 27 (4):355–371. https://doi.org/10.1101/gad.210773.112

Zenker J, White MD, Gasnier M, Alvarez YD, Lim HYG, Bissiere S, Biro M, Plachta N (2018) Expanding actin rings zipper the mouse embryo for blastocyst formation. Cell 173(3):776–791.e717. https://doi.org/10.1016/j.cell.2018.02.035

Zhang HT, Hiiragi T (2018) Symmetry breaking in the mammalian embryo. Annu Rev Cell Dev Biol 34:405–426. https://doi.org/10.1146/annurev-cellbio-100617-062616

Zhu M, Leung CY, Shahbazi MN, Zernicka-Goetz M (2017) Actomyosin polarisation through PLC-PKC triggers symmetry breaking of the mouse embryo. Nat Commun 8(1):921. https://doi.org/10.1038/s41467-017-00977-8

Ziomek CA, Johnson MH (1980) Cell surface interaction induces polarization of mouse 8-cell blastomeres at compaction. Cell 21(3):935–942. https://doi.org/10.1016/0092-8674(80)90157 2

Ziomek CA, Johnson MH, Handyside AH (1982) The developmental potential of mouse 16-cell blastomeres. J Exp Zool 221(3):345–355

Placentation in Marsupials

Marilyn B. Renfree and Geoff Shaw

Contents

Abstract It is sometimes implied that marsupials are "aplacental," on the presumption that the only mammals that have a placenta are the eponymous "placental" mammals. This misconception has persisted despite the interest in and descriptions of the marsupial placenta, even in Amoroso's definitive chapter. It was also said that marsupials had no maternal recognition of pregnancy and no placental hormone production. In addition, it was thought that genomic imprinting could not exist in marsupials because pregnancy was so short. We now know that none of these ideas have held true with extensive studies over the last four decades definitively showing that they are indeed mammals with a fully functional placenta, and with their own specializations.

Keywords Marsupial · Tammar wallaby · Placenta · Yolk sac · Trophoblast · Endometrium · Diapause · Inflammation · Transcriptome · Omphalopleure · Genomic imprinting · ALID

M. B. Renfree (✉) · G. Shaw
School of BioSciences, The University of Melbourne, Melbourne, VIC, Australia
e-mail: m.renfree@unimelb.edu.au; g.shaw@unimelb.edu.au

R. D. Geisert, T. E. Spencer (eds.), *Placentation in Mammals*, Advances in Anatomy, Embryology and Cell Biology 234, https://doi.org/10.1007/978-3-030-77360-1_4

1 Introduction

In eutherian mammals, while the allantois forms the definitive placenta, placentation does involve both the yolk sac (the choriovitelline placenta) and allantois (the chorioallantoic placenta), for at least part of pregnancy (Amoroso 1952; Carter 2020; Carter and Enders 2016; Enders and Carter 2006; Freyer and Renfree 2009; Freyer et al. 2002). All marsupials have a choriovitelline placenta but some also have a short-lived chorioallantoic placenta (Freyer et al. 2003; Padykula and Taylor 1976; Renfree 2010; Renfree et al. 2013). Indeed, a vascular chorioallantois responsible for respiratory functions is likely to have been a character of the stem marsupial species (Freyer et al. 2003). These variations highlight that each mammalian group has elaborated the fetal membranes in slightly different ways to gain access to uterine secretions and to manifest "any intimate apposition of fusion of the fetal organs to the maternal (or paternal) tissues for physiological exchange" (Mossman 1937). The marsupial placenta is indeed a fully functional "organ of exchange" but it differs from the familiar human and mouse placenta in many ways.

In eutherian mammals, gestation has been significantly extended beyond the length of the estrous cycle, with birth of relatively large young, so requiring a substantial transfer of nutrients to the developing fetus via a complex placenta. In contrast, all marsupials have comparatively short gestation periods relative to size, ranging from the shortest known duration of any mammalian pregnancy of just 12.5 days in the bandicoots to the longest in the red kangaroo of 35 days (Tyndale-Biscoe and Renfree 1987). Gestation is shorter than the estrous cycle in all but one species studied, the swamp wallaby, *Wallabia bicolor* (Menzies et al. 2020; Tyndale-Biscoe and Renfree 1987). Marsupial young are born at an extremely altricial stage, and complete most of their developmental milestones while firmly attached to a teat, usually in a pouch; they have "traded the umbilical cord for the teat" (Renfree 1983) (Fig. 1). Regardless of the length of pregnancy, all mammals depend on their placenta to obtain nutrients and exchange waste products with the maternal system as well as to give and receive signals between the mother–fetus dyad.

2 The Choriovitelline Placenta

The yolk sac forms the first placental contact in all mammals (King and Enders 1993). It is important for nutrient uptake and is the main source of proteins absorption by the fetus in many species. The rodent visceral yolk sac is an important organ of maternal–embryonic exchange of amino acids, transferrin, vitamin B12, calcium, and other ions (King and Enders 1993; Lloyd et al. 1996) and its protein uptake accounts for around 95% of amino acids incorporated into the embryo during organogenesis. It is the route for passive immunity to the embryo in species such as rabbits; it plays a role in fetal blood formation; and its damage leads to embryonic

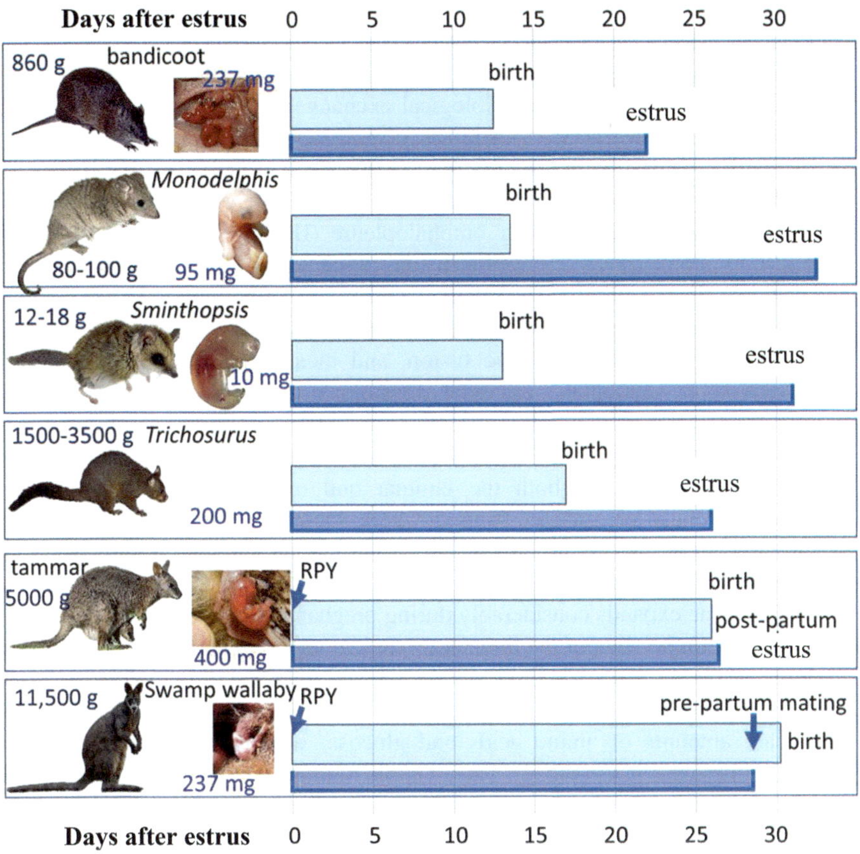

Fig. 1 Gestation and estrous lengths of selected marsupials. In most species, gestation is much shorter than the estrous cycle. In macropodids, gestation normally includes a period of lactational embryonic diapause, and the timings here are given from removal of pouch young (RPY) to birth and estrus. In this group, gestation is extended to almost the duration of the cycle, and in the extraordinary example of the swamp wallaby gestation is longer than the estrous cycle. Swamp wallaby females can mate and ovulate in late gestation so they have a near term fetus in one uterus, and a newly conceived one in the other (Menzies et al. 2020). Marsupial young are tiny at birth. Weights of adult and neonate are given. (Data from Tyndale-Biscoe and Renfree 1987). For most species neonates are illustrated; for bandicoot and tammar, the neonates are in the pouch attached to a teat, with part of a fingernail visible giving a visual scale. The marsupial neonates listed range from 0.005 to 0.1% of maternal weight. For comparison, the corresponding figures for some eutherians are: human 5%; mouse 4.8%; and sheep 8%

malformation. The placental function of the yolk sac early in human pregnancy is often overlooked in view of the highly invasive allantoic placenta that forms later. The human yolk sac is also the initial site of fetal blood formation, is involved in protein biosynthesis (Jauniaux et al. 1994; Jones and Jauniaux 1995), and is an important zone for the nutrient transfer between extraembryonic and embryonic compartments (Burton et al. 2002; Gulbis et al. 1994, 1998). The yolk sac is therefore critical for early pregnancy in eutherian mammals (Beckman et al. 1996; Merad and Wild 1992; Palis and Yoder 2001).

In marsupials, the yolk sac forms the definitive placenta up to birth (Hughes 1974) and is responsible for physiological exchange, biosynthesis, and endocrine function (Freyer et al. 2003; Tyndale-Biscoe and Renfree 1987). The tammar yolk sac consists of two parts: a vascular region, also known as the trilaminar omphalopleure (TOM), that appears to be the primary site of gaseous transfer, and the avascular area, the bilaminar omphalopleure (BOM), that appears to be the primary site of absorptive activity for histotrophic nutrition (Renfree 2010; Renfree and Tyndale-Biscoe 1973b). The relative contribution of both parts differs between species as evident in the relative surface area that is attached to the endometrium, in trophoblast thickness, in yolk sac fusion, and most markedly in the degree of invasiveness. In marsupials, placental physiology has been extensively studied only in the tammar wallaby (Renfree 1973a, c, 2010) and in the short-tailed gray opossum (Freyer et al. 2003, 2007; Zeller and Freyer 2002). There is a similar functional relationship in both the tammar and opossum in that the embryo (s) depends totally on transfer from the yolk sac during the first two-thirds of gestation because there is no attachment to the uterine epithelium until then, and the conceptus floats free in the uterus.

The yolk sac expands considerably during pregnancy (Renfree 1983, 1993). The biochemical components of the three embryonic compartments have been analyzed (Renfree 1973a, c). Yolk sac fluid protein is lower than that of the maternal serum, but much higher than in the amniotic fluid. The marsupial yolk sac fluid also has significant amounts of amino acids and glucose, whereas the amniotic fluid is lacking in these nutrients (Fig. 2). The allantoic fluid is high in urea, not surprisingly as it is the extension of the embryonic bladder (Renfree 1973a). Yolk sac fluid becomes yellow (as a result of breakdown products of bilirubin) while the amniotic fluid is completely clear (Renfree 1973a, 2010). Yolk sac fluid in the tammar contains transferrin (Renfree and Tyndale-Biscoe 1973a) and also metabolic enzymes (Renfree 1973b), as does the exocoelomic fluid in the human conceptus (Jauniaux et al. 1994). In the mouse, transferrin is found in the amniotic fluid but this is of fetal origin, i.e., not transferred from the yolk sac (Renfree and McLaren 1974). Thus in at least these species, there is a gradient from the maternal side to the fetal side of the conceptus, and the presence of metabolites and nutrients is the result of active transfer, not just simple diffusion, which is clearly tightly regulated.

Structural differences between placentas of different marsupial species are reflected in different modes and pathways of embryonic nourishment. The yolk sac placenta in most marsupials is not generally invasive but in the gray short-tailed opossum *Monodelphis domestica* the vascular yolk sac placenta becomes invasive over the last day of the 13.5 days pregnancy (Table 1; Freyer et al. 2007). The nonvascular part of the yolk sac is noninvasive. In the gray short-tailed opossum, there are three main phases of embryonic nourishment. The embryo is nourished by yolk sac fluid that is complemented by uterine secretion during the first phase. Uterine secretion declines during the second phase, as does the protein concentration of secretions, and the embryo relies on the nutrients stored in the

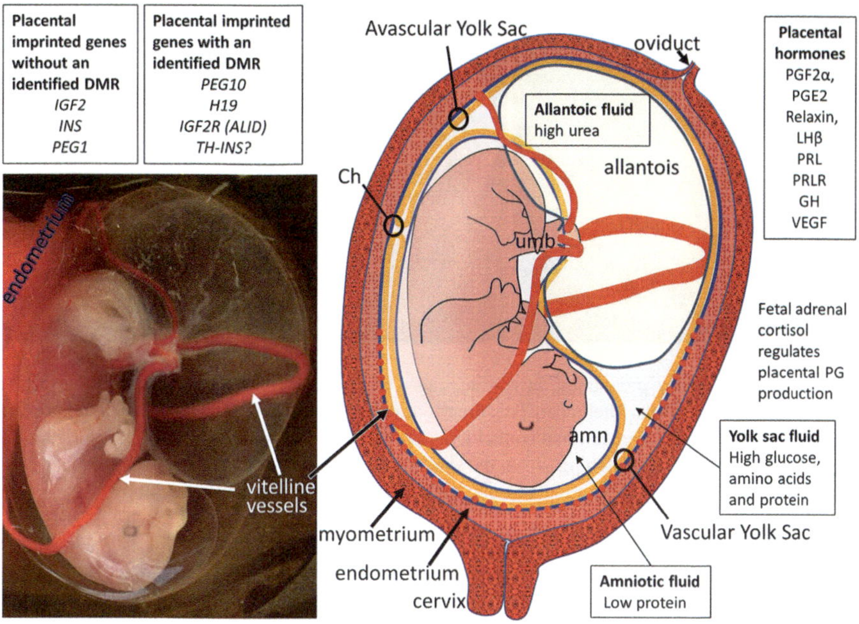

Fig. 2 Placental membranes of a near term tammar wallaby fetus. Left panel: because the placenta does not attach, the fetus and membranes roll free when the uterus is opened. Right panel: a diagrammatic representation of the fetus and membranes in utero. The vascular yolk sac covers about half of the exchange area and is closely apposed to the uterine luminal epithelium with its well-developed subepithelial capillary bed. The placenta is an active endocrine organ and expresses a number of hormones and growth factors and a suite of imprinted genes with and without identified differentially methylated regions (DMR)

Table 1 Summary of the types of placentas and their degree of invasion in each marsupial group

Placenta configuration	Group/species
Type I No allantochorion	*Didelphidae* *Phalangeridae* *Macropodidae* *Acrobatidae*
Type II Temporary allantochorion Allantochorion degenerates	*Dasyurus viverrinus*
Type III Allantochorion without villi Noninvasive, no fusion	*Phascolarctos cinereus* *Vombatus ursinus*
Type IV Allantochorion (placenta) Villi, fusion	*Peramelidae*
Marsupial allantochorion: No mesodermal villi No replacement of yolk sac placenta	

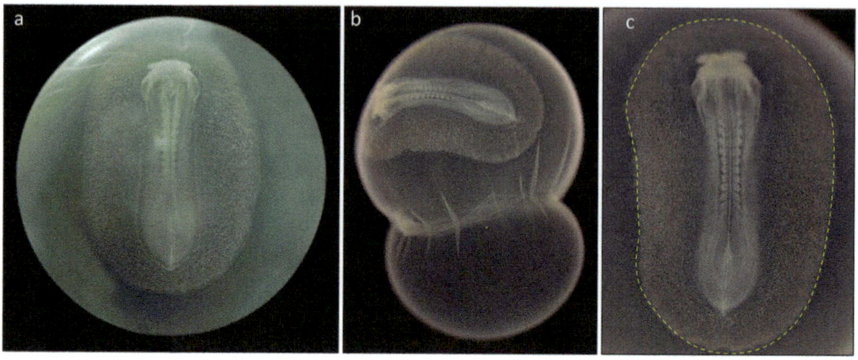

Fig. 3 Through early pregnancy, at somite stages, the conceptus is separated from the uterine epithelium by an acellular shell coat. This breaks down on day 17–18 allowing direct apposition of the yolk sac with the uterine luminal epithelium. (**a**) An 11-somite embryo before shell coat rupture; (**b**) a 13-somite embryo "hatching" from the shell coat after being removed from the uterus; (**c**) the hatched 13 somite embryo with the vascular area indicated by a dashed yellow line

yolk sac fluid until shell coat rupture. There is apparently a concomitant decrease in uterine secretion at this time, and a shift to hemotrophic nutrition (Freyer et al. 2007).

In contrast, the Australian tammar wallaby has an entirely noninvasive yolk sac (Freyer et al. 2002). The shell coat surrounds every marsupial conceptus for about 70–80% of the gestation length and prevents direct contact with the uterine wall until its intrauterine rupture. The shell expands considerably during this time, and its total area increases from ~0.001 m^3 to >0.250 mm^3 by day 18 of the 26-day pregnancy in the tammar (Shaw 1996). Between the formation of the blastocyst and the loss of the shell coat, the vesicle expands due to the uptake of uterine secretions and the embryo differentiates to 12–17 somite stage before "hatching" from the shell coat to make a more intimate apposition to the uterine endometrium at around day 17–18 after diapause (Denker and Tyndale-Biscoe 1986; Menzies et al. 2011; Tyndale-Biscoe and Renfree 1987) (Fig. 3).

In eutherian mammals with invasive placentas, nutrients to the embryo are mainly provided directly from the maternal blood supply. In contrast, in the tammar and opossum, glandular secretion that is complemented by blood components is the main source of nutrients after shell coat rupture up to birth (Freyer et al. 2007; Renfree 1973c). In tammars, the endometrium remains highly secretory to the last days of gestation (Renfree 1973c; Renfree and Tyndale-Biscoe 1973a). The extended secretory phase of the uterus appears to be a derived feature of kangaroo-like marsupials (Freyer et al. 2003).

As in humans and other mammals, multiple conceptuses can occur in normally monovular species, where more than one ovulation can occur or when the conceptus tissue fails to separate as occurs in conjoined twins (Fig. 4). In the monovular tammar, we have recorded many instances of twins as well as of twin or triplet embryos that failed to separate fully (*see* Fig. 4). In those with advanced twin fetuses, the yolk sacs are closely attached to each other in the vascular region. In one twin

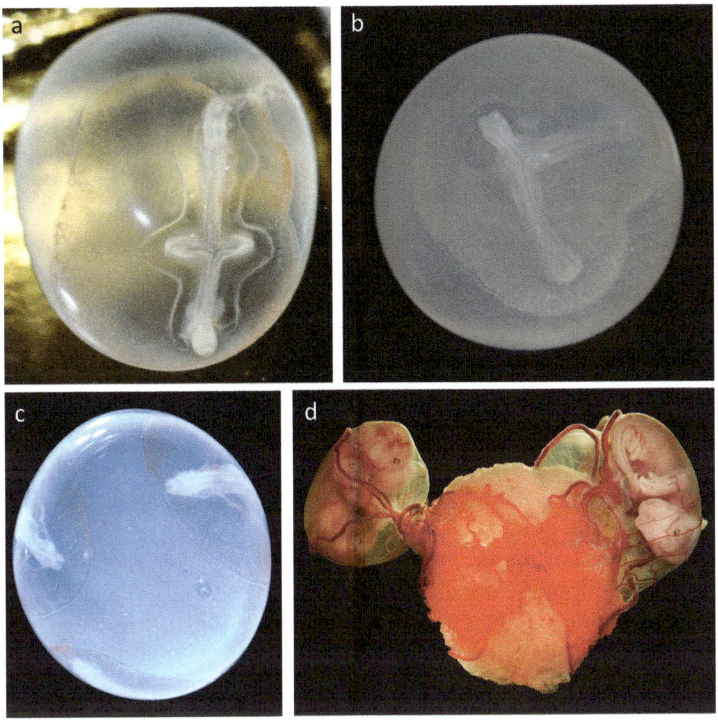

Fig. 4 Twin and triplet tammar embryos. (**a**, **b**) Conjoined twins sharing the same vascular yolk sac area. (**c**) Embryonic vesicle with 3 somite stage embryos with separate vascular yolk sac areas. (**d**) Near term uterus opened to show twin fetuses with separate placentas but fused at the vascular yolk sac: presumably fraternal twins

wombat example, the vascular yolk sacs have fused to form a common yolk sac cavity (Freyer 2003). Yolk sac fusion is an ancestral character in marsupial stem species and occurs in the American opossums and in the Australian dunnarts but in those species the fusion occurs only between the nonvascular regions (Freyer et al. 2003) so it appears that macropodids and vombatids differ from the didelphids and dasyurids in this.

The entire yolk sac of the common wombat (Hughes and Green 1998) and the hairy-nosed wombat is noninvasive (Freyer 2003). In the related koala, the enlarged trophectodermal cells of the nonvascular yolk sac fetal cells were thought to penetrate the uterine epithelium (Hughes 1974) but there is no clear evidence for this in either wombats or koalas (Freyer et al. 2003; Hughes and Green 1998). There is also a well-vascularised allantochorion in these species, but it is noninvasive and never erodes the endometrial epithelium (Table 1).

In the bandicoot family, *Peramelidae*, a functional chorioallantoic placentation develops in addition to the yolk sac placenta (Table 1; Flynn 1923; Hill 1897, 1900). Chorioallantoic fusion is accompanied by a disappearance of the trophoblast layer in late gestation. It is thought that the trophoblast fuses with maternal homokaryons to create maternal–fetal heterokaryons, only at this site (Padykula and Taylor 1976, 1982), as originally suggested by Flynn (1923). Since bandicoots have an ultra-short pregnancy, it is possible that only the initial phases of implantation occur, i.e., cell attachment and cell fusion. The bandicoot placental heterokaryons may represent the survival of an early stage in the evolution of the mammalian chorioallantoic placenta.

3 Endometrial Changes During Pregnancy

There are a number of endometrial genes involved in late pregnancy and parturition that are shared across widely divergent viviparous lineages including mammals, reptiles, and fish. For example, four conserved genes have functions in parturition (*GPER1*), tissue remodelling (*A2M*), cellular adhesion (*DSG2*), and placentation (*Syncytin*), respectively. These genes appear to be ancient conserved genes with an important reproductive function. Syncytins are a class of retroviral genes in eutherian mammals that have a role in placentation. Marsupials also have a *SYNCYTIN* gene of retroviral origin (Cornelis et al. 2015). There is no evidence of syncytium formation in the tammar but the occurrence of a syncytium in the opossum appears to be linked to an exaptation of the fusogenic *syncytin-Opo*1. While there is a conserved retroviral envelope gene in all marsupials (including the opossum and the tammar wallaby), there is only limited expression in the placenta. It would be interesting to examine the bandicoot term placenta in which fetal and maternal nuclei are found within the same cell (Padykula and Taylor 1976; see above).

In the tammar, the endometrium becomes increasingly secretory from the time of reactivation of the diapausing blastocyst in the tammar, reaching a peak of secretory activity in late gestation. Diapause is regulated by these secretions and has been extensively reviewed recently so will not be presented again here (Fenelon et al. 2017; Lefevre et al. 2011; Murphy and Fenelon 2020; Murphy et al. 2000; Renfree and Fenelon 2017; Renfree and Shaw 2000). Interestingly, the molecular factors that control diapause in the mouse, mink, and tammar are remarkably conserved (Cha et al. 2013, 2020; Fenelon and Murphy 2019). Similarly, the changes in the cellular remodeling of the luminal epithelium is conserved during early pregnancy between the dunnart and species with an invasive placenta such as the rat (Dudley et al. 2015, 2017; Laird et al. 2014). In early gestation, protein concentrations in uterine fluid rise dramatically as the embryo reactivates from diapause, including a suite of 128 proteins identified specifically as secretory class, presumably providing histotrophic resources for the embryo (Martin et al. 2016).

While a great deal of attention has been given to endometrial changes during early pregnancy and the reactivation from embryonic diapause in several of the mammalian diapausing species, there has been relatively little attention to those observed in later pregnancy. The functions of the placenta largely depend on the corresponding changes in the uterine endometrium. In the dunnart, the placenta invades on the side of the bilaminar yolk sac (Roberts and Breed 1994). In the tammar and the brushtail possum, during endometrial attachment, there is no invasion but there is still similar remodeling of the uterine epithelium since the secretion of the key basal anchoring molecule, Talin, moves to the base of the uterine epithelium as it occurs in eutherians (Laird et al. 2017a) but there are differences between the possum and the tammar (Laird et al. 2017b). Another protein, desmoglein-2, maintains lateral cell adhesion in the uterine epithelium and is redistributed before invasive implantation in eutherians and is similarly redistributed on the tammar and possum (Laird et al. 2018). By late gestation, the uterine secretome is rich in growth factors and related proteins including IGF2, insulin-like growth factor binding proteins 1, 2, 3, 4, 5, 6, 7, TGFβ, and GDF15 (Martin et al. 2016). At day 24, 2 days before birth, the matrix metalloproteinase MMP2 is also high (Martin et al. 2016), perhaps involved in tissue remodeling or in modulating growth factor action.

4 Maternal Recognition of Pregnancy, Placental Endocrinology, and the Inflammatory Response

Despite the lack of invasion, in the tammar, there is a maternal recognition of pregnancy (Kojima et al. 1993; Renfree 1972, 1973c; Renfree and Blanden 2000). The gravid endometrium becomes more secretory than the non-gravid endometrium, and it is not due to the proximity to the corpus luteum, since blastocyst transfer to the contralateral uterus results in a similar proliferation of the endometrium (Renfree 1972; Renfree and Blanden 2000). It is likely an effect of the trophoblast since non-embryonated vesicles also induce a stimulatory endometrium (Renfree and Tyndale-Biscoe 1973b). This fetal effect is most noticeable in mid-pregnancy in the tammar and is also reflected in the composition of the uterine secretions (Renfree 1973c). The specific uterine proteins and growth factors such as PAF reflect these unilateral responses (Kojima et al. 1993; Renfree 1973c). A similar maternal recognition of pregnancy appears to be widespread in the kangaroo and wallaby family, the *Macropodidae* (Shaw and Rose 1979; Walker and Hughes 1981; Wallace 1981; reviewed in Renfree 2000), and was noted in the historic papers by Owen (1834) and Flynn (1930), although they did not realize that this might be due to the yolk sac placenta. There are also histological differences between the gravid and non-gravid uteri in the dunnart (Cruz and Selwood 1997). In a preliminary study of the tammar placenta, bioassayable chorionic gonadotrophin was detected, (MB Renfree and L. Wide, unpublished results) but this was never investigated further. There are a number of protein hormones including luteinizing hormone β (LH β) genes expressed by the yolk sac placenta, as well as GH, GH-R, IGF-2, PRL (Menzies et al. 2011), IGF2 (Ager et al. 2008b), and relaxin (Parry et al. 1997). Curiously, steroid hormone production is very low in the yolk sac unlike placental production in other (eutherian) mammals (Heap et al. 1980; Renfree and Heap 1977). The controlling factors therefore could be the result of hormone secretion by the trophoblast, although it remains possible that uterine stretching could also play a role (Shaw

1983; Young and Renfree 1979). Using a transcriptomic approach, maternal recognition of pregnancy is now known to occur in the (polyovular) gray opossum (Griffith et al. 2019) in which there are differences between endometrial genes of the gravid uteri compared to that of the estrous cycling endometrium. The endometrial glands of late pregnancy appear to be more secretory than those at the same stage of the estrous cycle. At mid-cycle, the gene expression is not different from that at mid-pregnancy but is markedly different in late pregnancy (Griffith et al. 2019). There are differences in transporter expression between pregnancy and the estrous cycle. Collectively, these changes correlate with morphological differences consistent with an "endometrial" recognition of pregnancy. These changes will be discussed further below.

While there have been several studies of the MHC class I and II genes in the tammar (Cheng et al. 2009; Siddle et al. 2009), the MHC complex at the maternal–fetal interface in marsupials has only been examined in the tammar wallaby (Buentjen et al. 2015). In the tammar, MHC 1 genes have undergone duplication, translocation, and diversification in the genome and are located across seven autosomes (Renfree et al. 2011; Siddle et al. 2010). Classical MHC 1 genes are expressed in the yolk sac placenta as well as in the fetus and the endometrium (Buentjen et al. 2015). The classical sequences in the placenta are most similar to eutherian UC and UA genes, and there are three nonclassical MHC 1 genes (UD, UK, and UM). It is likely that these complexes evolved in the ancestral therian placenta are necessary for modulating the maternal immune response (Buentjen et al. 2015).

More recently, examination of the gene expression during pregnancy in the opossum suggests that the brief inflammatory attachment of the placenta to the endometrium results in parturition shortly after (Chavan et al. 2018; Griffith et al. 2017, 2019). In the opossum, at full term, this response is characterized by an inflammatory response at the maternal–fetal interface consistent with implantation in humans and mice (Griffith et al. 2017). It is also correlated with loss of the shell coat (which is a maternally derived structure). Heparin-binding EGF-like growth factor and Mucin 1 are observed during implantation in eutherians and, after shell coat loss, are also detected in opossum endometrium. The interleukins as inflammatory markers also appear to be especially important in this process (Griffith et al. 2017, 2019). The conserved set of inflammatory markers and immune signaling pathways highlights critical steps taken in the evolution of viviparity in the therian ancestor. The opossum is a basal marsupial species with an ultra-short gestation, but in the macropodids gestation has been extended to occupy the length of the estrous cycle (*see* Fig. 2) and endocrinologically has a very similar regulation of parturition to that of the sheep (Shaw and Renfree 2001; Young et al. 2011). From what we know of pregnancy and parturition in the tammar, it is likely that kangaroos and wallabies have at least some of the features of the eutherian inflammatory responses and that such responses are not restricted to precipitating an early parturition in all marsupials.

4.1 Transcriptomics

Endometrium

It appears that endometrial genes involved in late pregnancy and parturition are shared across widely divergent viviparous lineages including mammals, reptiles, and fish (Whittington et al. 2015; C Whittington, C Murphy, B McAllan, M Thompson, unpublished data). Transcriptomic analysis of early pregnancy in the dunnart *Sminthopsis crassicaudata* identified a large number of differentially expressed genes (Whittington et al. 2018). About 25% of the top expressed 50 genes that are differentially expressed in early pregnancy are those involved in histotrophic nutrition before there is close attachment of the placenta to the endometrium around day 10 of pregnancy. The steroid biosynthesis genes *CYP27A1, HSD3B7, HSDB17B7, AKRIDA* are also in the top 50 differentially upregulated genes. Immune function genes make up 18% of the top 50 downregulated genes, suggesting immunosuppression does occur even before shell coat loss (Whittington et al. 2018). Conversely, a number of immune genes are upregulated (14% in the top 50). Clearly, we have much to still understand about marsupial pregnancy and placentation and the genes and hormones that control it.

Vascular and Avascular Yolk Sac Transcriptome

We have begun next-generation transcriptome sequencing of the endometrium, embryonic, and extraembryonic tissues of the tammar wallaby (SR Frankenberg and MB Renfree, unpublished results). These include yolk sac placentas that were separated into avascular bilaminar omphalopleure (BOM) and the vascular trilaminar omphalopleure (TOM) parts at two different developmental stages: day 18 and day 24 of the 26.5-day pregnancy (Fig. 5). These preliminary sequence data have allowed us to identify some genes that are differentially expressed between the BOM and TOM and so are likely to have important roles specific to those regions. We have begun analysis of our data by comparing the four yolk sac transcriptomes with embryo transcriptomes, allowing identification of genes whose expression is specific to the yolk sac as a whole (Renfree and Frankenberg unpublished data). Although we need more replicates for statistical significance, these preliminary single analyses revealed an enrichment of genes matching gene ontology (GO) terms associated with expected yolk sac functions. Accordingly, GO terms in TOM-specific genes were associated with hematopoiesis and vasculogenesis, while BOM-specific genes were associated with ion transport and water balance (Fig. 4). We also compared the proximal and distal regions of the extraembryonic part of a much earlier day 13 RPY conceptus. These regions later develop into TOM and BOM, respectively. Intriguingly, we found that distal-specific genes at day 13 RPY already show enrichment for ion transport and water balance GO terms,

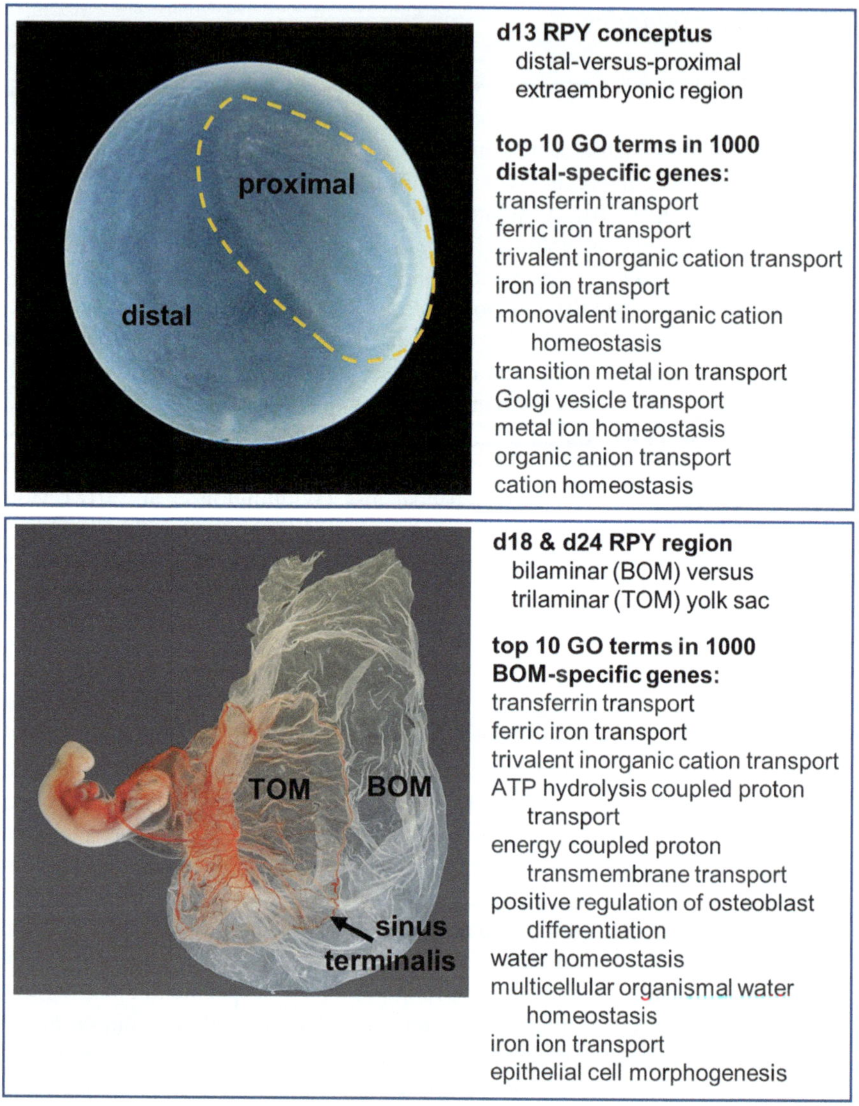

Fig. 5 Differential expression of genes comparing distal vs proximal extraembryonic region (top panel; dashed line indicates where the two areas were separated by microdissection) and bilaminar yolk sac (BOM) vs trilaminar yolk sac (TOM). The top ten gene ontology (GO) terms from 1000 distal specific (top) or BOM specific (bottom) are listed

showing that emergence of BOM-specific gene expression occurs well before overt morphological differentiation of the yolk sac.

Some 115 genes with expression levels of FPKM ≥ 10 are identified from the core term placental transcriptomes of 13 eutherian species that are critical for placental function include vesicular transport, immune regulation, cell fusion and membrane repair, and invasion and may also be associated with the shape of the placenta (Armstrong et al. 2017). Of these, there are 95 genes in the gray short-tailed opossum, *Monodelphis domestica,* showing that the functions of the marsupial placenta have largely been evolutionarily conserved.

All placenta-specific genes are not expressed in all types of placenta and there are relatively few of them (Carter 2018). Another transcriptomic approach reinforces this view (Guernsey et al. 2017). In the placenta of the tammar wallaby, there are differences in gene expression between the trophoblast and endoderm, including a small number of genes of the Igf signaling pathways, for nutrient transfer, placental cytokeratins, *Wnt7b, Stra8, PTN,* and *Gjb3,* and the trophoblast marker *GCM1*, a transcription factor that is important in the formation and development of the placenta. Unexpectedly, there were also markers of eutherian decidua *Cepbp and Vim* despite marsupials lack decidualisation. More surprising, however, was the observation that mouse and wallaby not only shared similar patterns of gene expression in the placenta but also a number of genes expressed in the mammary glands in the tammar that are known to be functionally important in the placenta in eutherians including a high expression of *GCM1* in the mammary gland alveoli and *Gata3* (Guernsey et al. 2017). Marsupials have truly "traded the umbilical cord for the teat" (Renfree 1983).

5 Genomic Imprinting and Evolution of Viviparity

As in eutherian mammals, there is genomic imprinting that appears to be important for placental function since most are imprinted in the placenta. Thus far fewer imprinted genes have been identified in marsupials than in eutherians. The acquisition of noncoding regions (novel CpG islands) appears to have accompanied the evolution of the mammalian genomic imprinting important for regulating fetal and placental growth and viviparity (Suzuki et al. 2011). Of the 18 or so genes that are imprinted in eutherians that have been tested in marsupials, only 8 are imprinted and of these only 3 have a differentially methylated region (DMR): *PEG10* (paternally expressed) and *H19* and *IGF2R,* paternally methylated/maternally expressed. *IGF2, PEG1/MEST,* and *INS* have no DMRs. All of these are expressed in the marsupial placenta.

The first marsupial imprinted gene discovered was IGF2R in the gray opossum (O'Neill et al. 2000) but there was no obvious imprinting control region, the DMR, until an extensive search for allele-specific methylated CpG islands (Lawton et al. 2008). However, the honor of the first marsupial imprinted gene with a DMR went to a retrotransposon-derived imprinted gene (*PEG10*) that is essential for placentation in eutherian mammals and assumed also to be essential for marsupial placentation (Renfree et al. 2011; Suzuki et al. 2007), (Fig. 6). *PEG10* is a retrotransposon-derived imprinted gene controlled by a DMR that was inserted into the genome of the therian ancestor. However, in eutherians, the methylation of the gene extends across to an adjacent gene, *SCGE,* whereas in marsupials, the imprinting is restricted to just the *PEG10* region. Silencing of exogenous DNA (for example, a

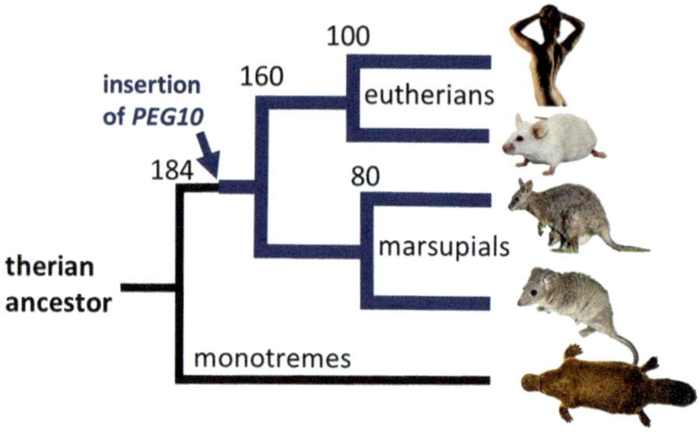

Fig. 6 Imprinting of the placental gene *PEG10* and the time of the insertion of the retrotransposon PEG10 in therian mammals. The short boxes indicate the adjacent gene *SGCE* (that is not imprinted in marsupials) and the long boxes represent the *PEG10* region. The black lollipops indicate DNA methylation of CpG sites, and the gray boxes represent the silent state of transcription. (Adapted from Suzuki et al. (2007))

retrotransposon) can therefore clearly drive the evolution of genomic imprinting (Suzuki et al. 2007, 2018).

IGF2/IGF2 is a critical gene and growth factor in the placenta of both marsupial and eutherian mammals and is one of the best known imprinted genes (Renfree et al. 2009). The *IGF2-H19* imprinted regions are conserved in both groups (Smits et al. 2008; Suzuki et al. 2005) but the long noncoding RNA *Airn* in intron 2 of *IGF2R* that controls the domain is absent in marsupials. Instead, we have found a novel lncRNA, *ALID,* in intron 12 that may perform a similar function (Suzuki et al. 2018).

Insulin *INS* is normally expressed in the yolk sac of both eutherians and marsupials (Ager et al. 2008b; Stringer et al. 2012a, b) but uniquely in marsupials *INS* is imprinted in the maternal liver and the mammary gland. Tammar *IGF2* is imprinted in the placenta, in the pouch young liver (but not adult liver), in the pouch young brain, and possibly in adult hypothalamus and mammary gland (Stringer et al. 2014) (Fig. 7).

These results show that genomic imprinting is not only important in growth regulation by the placenta but also for the survival of the young after birth, again suggesting an evolutionary the continuum between placentation and lactation in marsupials.

6 Conclusions

The relatively short-lived placenta of marsupials has functions similar to the phases of early placentation relying on the yolk sac as all other mammals. Where marsupials have diverged from other therian mammals is the choice of their reproductive strategy to deliver an altricial young that then transfers its dependency to postnatal

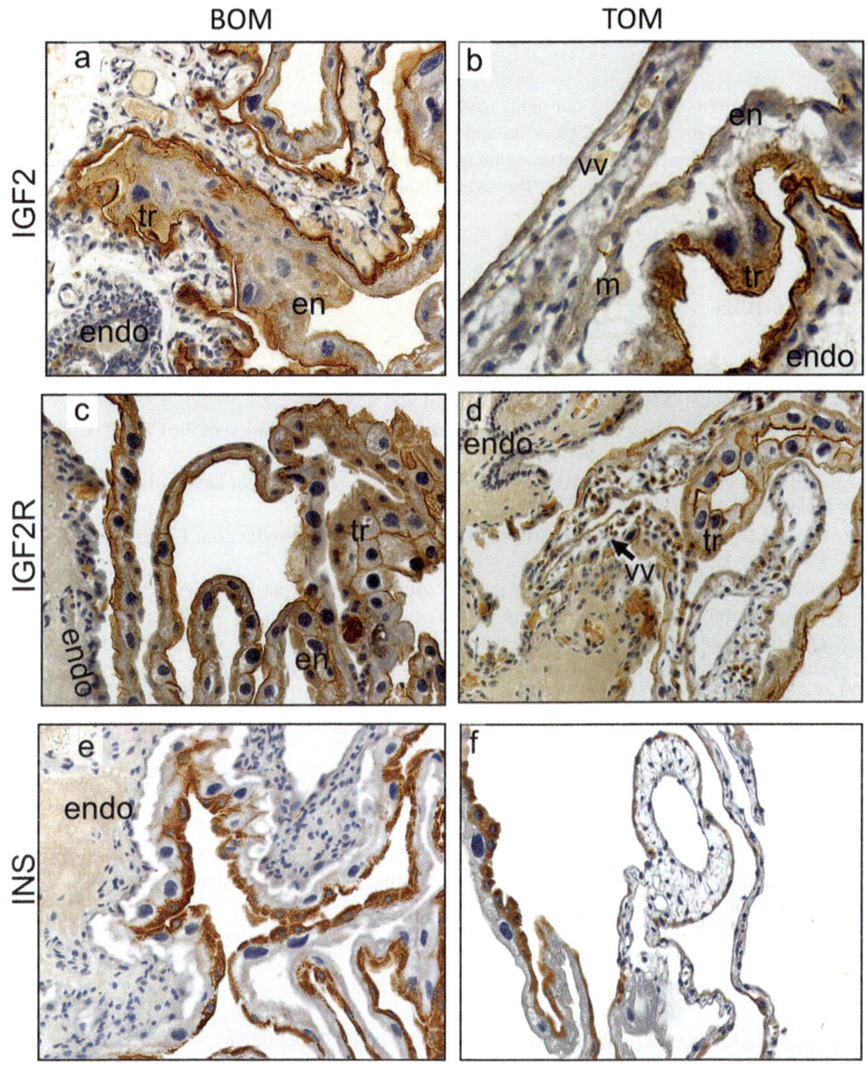

Fig. 7 Immunocytochemical detection of IGF2 (**a**, **b**), IGF2R (**c**, **d**), and INS (**e**, **f**) in the tammar BOM (**a**, **c**, **e**) and TOM (**b**, **d**, **f**). *endo* endometrium, *en* endoderm, *tr* trophoblast giant cells, *vv* vitelline vessels. (Data from Ager et al. (2007, 2008a) and Renfree et al. (2008))

stages, supported by the most sophisticated lactational physiology amongst all mammals. Marsupial milk does act as a de-facto placenta, in that the mammary gland produces an ever-changing milk supply, tailored precisely to the needs of the developing young, much as the placenta does in eutherian mammals. The delivery of a tiny young did not require major anatomical adaptations to the birth canal, and while the energetic "cost" of lactation is no doubt equivalent to the cost of a lengthy gestation, it probably has many advantages. This is simply an alternative reproductive strategy—marsupials are placental mammals but they are also lactational champions.

Acknowledgments We thank our many students and colleagues who have contributed to our new understanding of the marsupial placenta and all members of the Walgroup. We thank the late Dr Gordon Lyne for the bandicoot neonate images; Dr Stephen Frankenberg for the dunnart neonate images and Dr Brandon Menzies for the swamp wallaby neonate images.

References

Ager E, Suzuki S, Pask A, Shaw G, Ishino F, Renfree MB (2007) Insulin is imprinted in the placenta of the marsupial, Macropus eugenii. Dev Biol 309:317–328

Ager EI, Pask AJ, Gehring HM, Shaw G, Renfree MB (2008a) Evolution of the CDKN1C-KCNQ1 imprinted domain. BMC Evol Biol 8:163

Ager EI, Pask AJ, Shaw G, Renfree MB (2008b) Expression and protein localisation of IGF2 in the marsupial placenta. BMC Dev Biol 8:17

Amoroso EC (1952) Placentation, Marshall's physiology of reproduction. Longman, Green and Co., London, pp 127–311

Armstrong DL, McGowen MR, Weckle A, Pantham P, Caravas J, Agnew D, Benirschke K, Savage-Rumbaugh S, Nevo E, Kim CJ, Wagner GP, Romero R, Wildman DE (2017) The core transcriptome of mammalian placentas and the divergence of expression with placental shape. Placenta 57:71–78

Beckman DA, Brent RL, Lloyd JB (1996) Sources of amino acids for protein synthesis during early organogenesis in the rat. 4. Mechanisms before envelopment of the embryo by the yolk sac. Placenta 17:635–641

Buentjen I, Drews B, Frankenberg SR, Hildebrandt TB, Renfree MB, Menzies BR (2015) Characterisation of major histocompatibility complex class I genes at the fetal-maternal interface of marsupials. Immunogenetics 67:385–393

Burton GJ, Watson AL, Hempstock J, Skepper JN, Jauniaux E (2002) Uterine glands provide histiotrophic nutrition for the human fetus during the first trimester of pregnancy. J Clin Endocrinol Metab 87:2954–2959

Carter AM (2018) Recent advances in understanding evolution of the placenta: insights from transcriptomics. F1000Res 7:89

Carter AM (2020) The role of mammalian foetal membranes in early embryogenesis: lessons from marsupials. J Morphol:1–3

Carter AM, Enders AC (2016) Placentation in mammals: definitive placenta, yolk sac, and paraplacenta. Theriogenology 86:278–287

Cha J, Sun X, Bartos A, Fenelon J, Lefevre P, Daikoku T, Shaw G, Maxson R, Murphy BD, Renfree MB, Dey SK (2013) A new role for muscle segment homeobox genes in mammalian embryonic diapause. Open Biol 3:130035

Cha J, Fenelon J, Murphy BD, Shaw G, Renfree MB, Dey SK (2020) A role for Msx genes in mammalian embryonic diapause. Biosci Proc 10:44–51

Chavan AR, Griffith OW, Stadtmauer D, Maziarz J, Pavlicev M, Fishman R, Koren L, Romero R, Wagner GP (2018) Evolution of embryo implantation was enabled by the origin of decidual cells in eutherian mammals. BioRxiv. https://doi.org/10.1101/429571

Cheng Y, Wakefield M, Siddle HV, Coggill PC, Herbert CA, Beck S, Belov K, Eldridge MD (2009) Isolation and characterization of 10 MHC Class I-associated microsatellite loci in tammar wallaby (Macropus eugenii). Mol Ecol Resour 9:346–349

Cornelis G, Vernochet C, Carradec Q, Souquere S, Mulot B, Catzeflis F, Nilsson MA, Menzies BR, Renfree MB, Pierron G, Zeller U, Heidmann O, Dupressoir A, Heidmann T (2015) Retroviral envelope gene captures and syncytin exaptation for placentation in marsupials. Proc Natl Acad Sci U S A 112:E487–E496

Cruz YP, Selwood L (1997) Histological differences between gravid and non-gravid uteri in the dasyurid marsupial, Sminthopsis macroura (Spencer). J Reprod Fertil 111:319–325

Denker HW, Tyndale-Biscoe CH (1986) Embryo implantation and proteinase activities in a marsupial (Macropus eugenii). Histochemical patterns of proteinases in various gestational stages. Cell Tissue Res 246:279–291

Dudley JS, Murphy CR, Thompson MB, McAllan BM (2015) Desmoglein-2 during pregnancy and its role in the evolution of viviparity in a marsupial (Sminthopsis crassicaudata; Dasyuridae). J Morphol 276:261–272

Dudley JS, Murphy CR, Thompson MB, McAllan BM (2017) Epithelial cadherin disassociates from the lateral plasma membrane of uterine epithelial cells throughout pregnancy in a marsupial. J Anat 231:359–365

Enders AC, Carter AM (2006) Comparative placentation: some interesting modifications for histotrophic nutrition—a review. Placenta 27(Suppl A):S11–S16

Fenelon JC, Murphy BD (2019) New functions for old factors: the role of polyamines during the establishment of pregnancy. Reprod Fertil Dev 31:1228–1239

Fenelon JC, Shaw G, Frankenberg SR, Murphy BD, Renfree MB (2017) Embryo arrest and reactivation: potential candidates controlling embryonic diapause in the tammar wallaby and minkdagger. Biol Reprod 96:877–894

Flynn TT (1923) The yolk-sac and allantoic placenta in Perameles. Q J Microsc Sci 67:123–182

Flynn TT (1930) The uterine cycle of pregnancy and pseudo-pregnancy as it is in the diprotodont marsupial Bettongia cuniculus. Proc Linn Soc N S W 55:506–531

Freyer C (2003) The placenta in the original marsupial stem species, zoology. The Universiyt of Melbourne, Parkville

Freyer C, Renfree MB (2009) The mammalian yolk sac placenta. J Exp Zool B Mol Dev Evol 312:545–554

Freyer C, Zeller U, Renfree MB (2002) Ultrastructure of the placenta of the tammar wallaby, Macropus eugenii: comparison with the grey short-tailed opossum, Monodelphis domestica. J Anat 201:101–119

Freyer C, Zeller U, Renfree MB (2003) The marsupial placenta: a phylogenetic analysis. J Exp Zool A Comp Exp Biol 299:59–77

Freyer C, Zeller U, Renfree MB (2007) Placental function in two distantly related marsupials. Placenta 28:249–257

Griffith OW, Chavan AR, Protopapas S, Maziarz J, Romero R, Wagner GP (2017) Embryo implantation evolved from an ancestral inflammatory attachment reaction. Proc Natl Acad Sci U S A 114:E6566–E6575

Griffith OW, Chavan AR, Pavlicev M, Protopapas S, Callahan R, Maziarz J, Wagner GP (2019) Endometrial recognition of pregnancy occurs in the grey short-tailed opossum (Monodelphis domestica). Proc Biol Sci 286:20190691

Guernsey MW, Chuong EB, Cornelis G, Renfree MB, Baker JC (2017) Molecular conservation of marsupial and eutherian placentation and lactation. Elife 6

Gulbis B, Jauniaux E, Decuyper J, Thiry P, Jurkovic D, Campbell S (1994) Distribution of iron and iron-binding proteins in first-trimester human pregnancies. Obstet Gynecol 84:289–293

Gulbis B, Jauniaux E, Cotton F, Stordeur P (1998) Protein and enzyme patterns in the fluid cavities of the first trimester gestational sac: relevance to the absorptive role of secondary yolk sac. Mol Hum Reprod 4:857–862

Heap RB, Renfree MB, Burton RD (1980) Steroid metabolism in the yolk sac placenta and endometrium of the tammar wallaby, Macropus eugenii. J Endocrinol 87:339–349

Hill JP (1897) The placenation of Perameles. Q J Microsc Sci 40:385–446

Hill JP (1900) Contributions to the embryology of the Marsupialia. 2. On a further stage of placentation of Perameles. 3. On the foetal membrane of Macropus purma. Q J Microsc Sci 43:1–22

Hughes RL (1974) Morphological studies on implantation in marsupials. J Reprod Fertil 39:173–186

Hughes RL, Green RH (1998) Placentation and adaptations for survival at birth in the common wombat Vombatus ursinus. In: Wells RT, Pridmore PA (eds) Wombats. Sulley Beatty and Sons/ Royal Zoological Society of South Australia, Chipping Norton, NSW, pp 86–108

Jauniaux E, Gulbis B, Jurkovic D, Campbell S, Collins WP, Ooms HA (1994) Relationship between protein concentrations in embryological fluids and maternal serum and yolk sac size during human early pregnancy. Hum Reprod 9:161–166

Jones CJP, Jauniaux E (1995) Ultrastructure of the materno-embryonic interface in the first trimester of pregnancy. Micron 26:145–173

King B, Enders A (1993) Comparative development of the mammalian yolk sac. In: Nogales FF (ed) The human yolk sac and yolk sac tumors. Springer, Berlin, pp 1–32

Kojima T, Hinds LA, Muller WJ, O'Neill C, Tyndale-Biscoe CH (1993) Production and secretion of progesterone in vitro and presence of platelet activating factor (PAF) in early pregnancy of the marsupial, Macropus eugenii. Reprod Fertil Dev 5:15–25

Laird MK, Thompson MB, Murphy CR, McAllan BM (2014) Uterine epithelial cell changes during pregnancy in a marsupial (Sminthopsis crassicaudata; Dasyuridae). J Morphol 275:1081–1092

Laird MK, Dargan JR, Paterson L, Murphy CR, McAllan BM, Shaw G, Renfree MB, Thompson MB (2017a) Uterine molecular changes for non-invasive embryonic attachment in the marsupials Macropus eugenii (Macropodidae) and Trichosurus vulpecula (Phalangeridae). Mol Reprod Dev 84:1076–1085

Laird MK, McShea H, McAllan BM, Murphy CR, Thompson MB (2017b) Uterine remodelling during pregnancy and pseudopregnancy in the brushtail possum (Trichosurus vulpecula; Phalangeridae). J Anat 231:84–94

Laird MK, McShea H, Murphy CR, McAllan BM, Shaw G, Renfree MB, Thompson MB (2018) Non-invasive placentation in the marsupials Macropus eugenii (Macropodidae) and Trichosurus vulpecula (Phalangeridae) involves redistribution of uterine Desmoglein-2. Mol Reprod Dev 85:72–82

Lawton BR, Carone BR, Obergfell CJ, Ferreri GC, Gondolphi CM, Vandeberg JL, Imumorin I, O'Neill RJ, O'Neill MJ (2008) Genomic imprinting of IGF2 in marsupials is methylation dependent. BMC Genomics 9:205

Lefevre PL, Palin MF, Beaudry D, Dobias-Goff M, Desmarais JA, Llerena VE, Murphy BD (2011) Uterine signaling at the emergence of the embryo from obligate diapause. Am J Physiol Endocrinol Metab 300:E800–E808

Lloyd J, Brent R, Beckman D (1996) Sources of amino acids for protein synthesis during earlyorganogenesis in the rat. 3. Methionine incorporation. Placenta 17:629–634

Martin FC, Ang CS, Gardner DK, Renfree MB, Shaw G (2016) Uterine flushing proteome of the tammar wallaby after reactivation from diapause. Reproduction 152:491–505

Menzies BR, Pask AJ, Renfree MB (2011) Placental expression of pituitary hormones is an ancestral feature of therian mammals. Evodevo 2:16

Menzies BR, Hildebrandt TB, Renfree MB (2020) Unique reproductive strategy in the swamp wallaby. Proc Natl Acad Sci U S A 117:5938–5942

Merad Z, Wild AE (1992) The route of maternal IgM transport to the rabbit fetus. Placenta 13:291–304

Mossman HW (1937) Comparative morphogenesis of the fetal membranes and accessory uterine structures. Carnegie Inst. Washington Publ. 479. Contrib Embryol 26:129–246

Murphy BD, Fenelon J (2020) Models of embryonic diapause in Carnivora. Biosci Proc 10:85–100

Murphy BD, Jewgenow K, Renfree MB, Ulbrich SE (2000) Introduction to pregnancy in waiting: embryonic diapause in mammals. Biosci Proc 10:1–4

O'Neill MJ, Ingram RS, Vrana PB, Tilghman SM (2000) Allelic expression of IGF2 in marsupials and birds. Dev Genes Evol 210:18–20

Owen R (1834) On the generation of marsupial animals with a description of the impregnated uterus of the Kangaroo. Philos Trans R Soc Lond 124:333–364

Padykula HA, Taylor JM (1976) Ultrastructural evidence for loss of the trophoblastic layer in the chorioallantoic placenta of Australian bandicoots (Marsupialia: Peramelidae). Anat Rec 186:357–385

Padykula HA, Taylor JM (1982) Marsupial placentation and its evolutionary significance. J Reprod Fertil Suppl 31:95–104

Palis J, Yoder MC (2001) Yolk-sac hematopoiesis: the first blood cells of mouse and man. Exp Hematol 29:927–936

Parry LJ, Clark JM, Renfree MB (1997) Ultrastructural localization of relaxin in the corpus luteum of the pregnant and early lactating tammar wallaby, Macropus eugenii. Cell Tissue Res 290:615–622

Renfree MB (1972) Influence of the embryo on the marsupial uterus. Nature 240:475–477

Renfree MB (1973a) The composition of fetal fluids of the marsupial Macropus eugenii. Dev Biol 33:62–79

Renfree MB (1973b) Enzymes of intermediary metabolism in the fetal fluids of the tammar wallaby, Macropus eugenii. J Reprod Fertil 32:330–331

Renfree MB (1973c) Proteins in the uterine secretions of the marsupial Macropus eugenii. Dev Biol 32:41–49

Renfree MB (1983) Marsupial reproduction: the choice between placentation and lactation. In: Finn C (ed) Oxford reviews of reproductive biology. Oxford University Press, Oxford, pp 1–29

Renfree MB (1993) Diapause, pregnancy, and parturition in Australian marsupials. J Exp Zool 266:450–462

Renfree MB (2010) Review: Marsupials: placental mammals with a difference. Placenta 31(Suppl): S21–S26

Renfree MB, Blanden DR (2000) Progesterone and oestrogen receptors in the female genital tract throughout pregnancy in tammar wallabies. J Reprod Fertil 119:121–128

Renfree MB, Fenelon JC (2017) The enigma of embryonic diapause. Development 144:3199–3210

Renfree MB, Heap RB (1977) Steroid metabolism by the placenta, corpus luteum, and endometrium of the marsupial Macropus eugenii. Theriogenology 8:164

Renfree MB, McLaren A (1974) Foetal origin of transferrin in mouse amniotic fluid. Nature 252:159–161

Renfree MB (2000) Maternal recognition of pregnancy in marsupials. Rev Reprod 5:6–11

Renfree MB, Shaw G (2000) Diapause. Annu Rev Physiol 62:353–375

Renfree MB, Tyndale-Biscoe C (1973a) Transferrin variation between mother and fetus in the marsupial, Macropus eugenii. J Reprod Fertil 32:113–115

Renfree MB, Tyndale-Biscoe CH (1973b) Intrauterine development after diapause in the marsupial Macropus eugenii. Dev Biol 32:28–40

Renfree MB, Ager EI, Shaw G, Pask AJ (2008) Genomic imprinting in marsupial placentation. Reproduction 136:523–531

Renfree MB, Hore TA, Shaw G, Marshall Graves JA, Pask AJ (2009) Evolution of genomic imprinting: insights from marsupials and monotremes. Annu Rev Genomics Hum Genet 10:241–262

Renfree MB, Papenfuss AT, Deakin JE, Lindsay J, Heider T, Belov K, Rens W, Waters PD, Pharo EA, Shaw G (2011) Genome sequence of an Australian kangaroo, Macropus eugenii, provides insight into the evolution of mammalian reproduction and development. Genome Biol 12:1–26

Renfree MB, Suzuki S, Kaneko-Ishino T (2013) The origin and evolution of genomic imprinting and viviparity in mammals. Philos Trans R Soc Lond B Biol Sci 368:20120151

Roberts CT, Breed WG (1994) Embryonic-maternal cell interactions at implantation in the fat-tailed dunnart, a dasyurid marsupial. Anat Rec 240:59–76

Shaw G (1983) Effect of PGF-2 alpha on uterine activity, and concentrations of 13, 14-dihydro-15-keto-PGF-2 alpha in peripheral plasma during parturition in the tammar wallaby (Macropus eugenii). J Reprod Fertil 69:429–436

Shaw G (1996) The uterine environment in early pregnancy in the tammar wallaby. Reprod Fertil Dev 8:811–818

Shaw G, Renfree MB (2001) Fetal control of parturition in marsupials. Reprod Fertil Dev 13:653–659

Shaw G, Rose RW (1979) Delayed gestation in the potoroo, Potorous tridactylus (Kerr). Austr J Zool 27:901–912

Siddle HV, Deakin JE, Coggill P, Hart E, Cheng Y, Wong ES, Harrow J, Beck S, Belov K (2009) MHC-linked and un-linked class I genes in the wallaby. BMC Genomics 10:310

Siddle HV, Marzec J, Cheng Y, Jones M, Belov K (2010) MHC gene copy number variation in Tasmanian devils: implications for the spread of a contagious cancer. Proc Biol Sci 277:2001–2006

Smits G, Mungall AJ, Griffiths-Jones S, Smith P, Beury D, Matthews L, Rogers J, Pask AJ, Shaw G, VandeBerg JL, McCarrey JR, Renfree MB, Reik W, Dunham I (2008) Conservation of the H19 noncoding RNA and H19-IGF2 imprinting mechanism in therians. Nat Genet 40:971–976

Stringer JM, Suzuki S, Pask AJ, Shaw G, Renfree MB (2012a) Promoter-specific expression and imprint status of marsupial IGF2. PLoS One 7:e41690

Stringer JM, Suzuki S, Pask AJ, Shaw G, Renfree MB (2012b) Selected imprinting of INS in the marsupial. Epigenetics Chromatin 5:14

Stringer JM, Pask AJ, Shaw G, Renfree MB (2014) Post-natal imprinting: evidence from marsupials. Heredity (Edinb) 113:145–155

Suzuki S, Renfree MB, Pask AJ, Shaw G, Kobayashi S, Kohda T, Kaneko-Ishino T, Ishino F (2005) Genomic imprinting of IGF2, p57(KIP2) and PEG1/MEST in a marsupial, the tammar wallaby. Mech Dev 122:213–222

Suzuki S, Ono R, Narita T, Pask AJ, Shaw G, Wang C, Kohda T, Alsop AE, Marshall Graves JA, Kohara Y, Ishino F, Renfree MB, Kaneko-Ishino T (2007) Retrotransposon silencing by DNA methylation can drive mammalian genomic imprinting. PLoS Genet 3:e55

Suzuki S, Shaw G, Kaneko-Ishino T, Ishino F, Renfree MB (2011) The evolution of mammalian genomic imprinting was accompanied by the acquisition of novel CpG islands. Genome Biol Evol 3:1276–1283

Suzuki S, Shaw G, Renfree MB (2018) Identification of a novel antisense noncoding RNA, ALID, transcribed from the putative imprinting control region of marsupial IGF2R. Epigenetics Chromatin 11:55

Tyndale-Biscoe H, Renfree MB (1987) Reproductive physiology of marsupials. Cambridge University Press, Cambridge

Walker MT, Hughes RL (1981) Ultrastructural changes after diapause in the uterine glands, corpus luteum and blastocyst of the red-necked wallaby, Macropus rufogriseus banksianus. J Reprod Fertil Suppl 29:151–158

Wallace GI (1981) Uterine factors and delayed implantation in macropodid marsupials. J Reprod Fertil Suppl 29:173–181

Whittington CM, Griffith OW, Qi W, Thompson MB, Wilson AB (2015) Seahorse brood pouch transcriptome reveals common genes associated with vertebrate pregnancy. Mol Biol Evol 32:3114–3131

Whittington CM, O'Meally D, Laird MK, Belov K, Thompson MB, McAllan BM (2018) Transcriptomic changes in the pre-implantation uterus highlight histotrophic nutrition of the developing marsupial embryo. Sci Rep 8:2412

Young IR, Renfree MB (1979) The effects of corpus luteum removal during gestation on parturition in the tammar wallaby (Macorpus eugenii). J Reprod Fertil 56:249–254

Young IR, Renfree MB, Mesiano S, Shaw G, Jenkin G, Smith R (2011) The comparative physiology of parturition in mammals. In: Norris D, Lopez KH (eds) Hormones and reproduction of vertebrates, vol 5: mammals. Elsevier, pp 95–116

Zeller U, Freyer C (2002) Early ontogeny and placentation of the grey short-tailed opossum, Monodelphis domestica (Didelphidae: Marsupialia): contribution to the reconstruction of the marsupial morphotype. J Zool Syst Evol Res 39:137–158

The Early Stages of Implantation and Placentation in the Pig

Gregory A. Johnson, Fuller W. Bazer, and Heewon Seo

Contents

Abstract Pregnancy in pigs includes the events of conceptus (embryo/fetus and placental membranes) elongation, implantation, and placentation. Placentation in pigs is defined microscopically as epitheliochorial and macroscopically as diffuse. In general, placentation can be defined as the juxtapositioning of the endometrial/uterine microvasculature to the chorioallantoic/placental microvasculature to facilitate the transport of nutrients from the mother to the fetus to support fetal development and growth. Establishment of epitheliochorial placentation in the pig is achieved by: (1) the secretions of uterine glands prior to conceptus attachment to the uterus; (2) the development of extensive folding of the uterine–placental interface to maximize the surface area for movement of nutrients across this surface; (3) increased angiogenesis of the vasculature that delivers both uterine and placental blood and, with it, nutrients to this interface; (4) the minimization of connective tissue that lies between these blood vessels and the uterine and placental epithelia; (5) interdigitation of microvilli between the uterine and placental epithelia; and

G. A. Johnson (✉) · H. Seo
Department of Veterinary Integrative Biosciences, College of Veterinary Medicine and Biomedical Sciences, Texas A&M University, College Station, TX, USA
e-mail: gjohnson@cvm.tamu.edu; hseo@cvm.tamu.edu

F. W. Bazer
Department of Animal Science, Texas A&M University, College Station, TX, USA
e-mail: fbazer@cvm.tamu.edu

R. D. Geisert, T. E. Spencer (eds.), *Placentation in Mammals*, Advances in Anatomy, Embryology and Cell Biology 234, https://doi.org/10.1007/978-3-030-77360-1_5

(6) the secretions of the uterine glands, called histotroph, that accumulate in areolae for transport though the placenta to the fetus. Placentation in pigs is not achieved by invasive growth of the placenta into the uterus. In this chapter, we summarize current knowledge about the major events that occur during the early stages of implantation and placentation in the pig. We will focus on the microanatomy of porcine placentation that builds off the excellent histological work of Amoroso and others and provide a brief review of some of the key physiological, cellular, and molecular events that accompany the development of "implantation" in pigs.

Keywords Pig/Porcine · Pregnancy · Conceptus · Uterus · Implantation · Placentation

1 Introduction: Basic Definition for Epitheliochorial Placentation

Placentation is epitheliochorial in pigs, as uterine endometrial luminal epithelium (LE) remains intact throughout pregnancy. Porcine placental trophectoderm/chorion cells directly attach to the uterine LE, and these two epithelia serve as the conduit for maternal hemotrophic support for conceptus growth and development (Bjorkmann 1973). Amoroso stated that "In its histological structure the epitheliochorial placenta is exceeding simple. . . .with the pig as exhibiting structurally the simplest type of placenta" (Amoroso 1952). Epitheliochorial placentation is less intimate than other types of placentation because both the uterine LE and the chorionic epithelium remain intact. The result is a significant barrier to the transfer of nutrients from the uterine vasculature to the placental vasculature. A nutrient in a subepithelial uterine capillary must pass through the cytoplasm and basement membrane of the endothelial cell of the uterine capillary. It must then pass across the remaining stromal connective tissue. The nutrient then must pass through the basement membrane and cytoplasm of the uterine LE cell, through the cytoplasm and basement membrane of the chorionic epithelial cell, across the remaining allantoic connective tissue, and through the basement membrane and cytoplasm of the endothelial cell of the allantois (see Fig. 1). Although simplistic in microscopic structure, epitheliochorial placentation and the macroscopic classification of "diffuse" characteristic of pigs are not primitive and have evolved from a more invasive placental type, hemochorial/endotheliochorial, and discoid (Wildeman et al. 2006). Indeed, although it could be inferred that the noninvasive types of placentation are the least efficient because there are more physical barriers to limit the movement of nutrients from mother to fetus, the newborn pig is relatively mature compared to the newborn of a mouse or human. Epitheliochorial placentation has arisen evolutionarily in three distinct mammalian lineages and is present in pigs, horses/donkeys, camels, some whales and primates, and in multiple other species across different genera (Amoroso 1952; Wildeman et al. 2006).

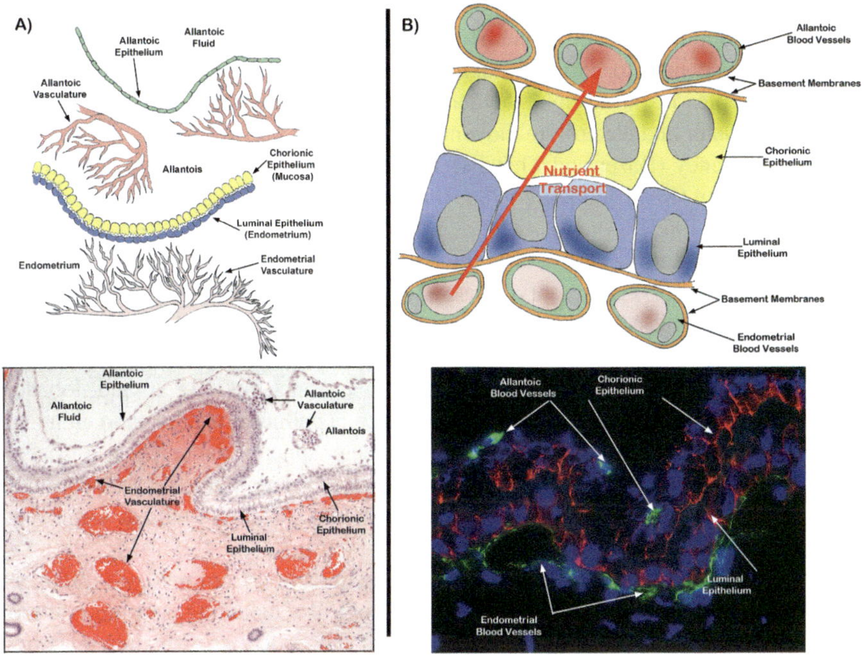

Fig. 1 Epitheliochorial placentation in the pig. (**a**) The upper panel is a cartoon depiction of porcine placentation at lower magnification, and the lower panel represents an H&E-stained thin section depicting porcine placentation on day 20 of gestation. (**b**) The upper panel is a cartoon depiction of porcine placentation at higher magnification, and the lower panel is a thin section depicting immunofluorescence-staining for epithelial cadherin (red color, width of field is 230 μm) and platelet endothelial cell adhesion molecule (green color). (Panels **a** and **b** are original drawings by Gregory A. Johnson. Panel **a** is adapted with permission from Bazer FW, Kim J, Song G, Ka H, Wu G, Johnson GA, and Vallet JL. 2013. Roles of selected nutrients in development of the porcine conceptus during pregnancy. In: Control of Pig Reproduction IX. A. Rodriquez-Martinez, N.M. Soede and W.L. Flowers Eds., Context Products Ltd., British Library Cataloguing in Publication Data, Leicestershire, United Kingdom, pp. 159–174)

2 Uterine Histoarchitecture and Early Conceptus Development

The pig uterine wall differentiates embryologically from the Mullerian ducts into the endometrial and myometrial compartments (Mossman 1937). The myometrium is composed primarily of smooth muscle cells and vascular elements, and bundles of smooth muscle fibers are organized into inner circular and outer longitudinal layers. In contrast, the endometrium is a complex epitheliomesenchymal organ consisting of a simple columnar LE that overlays a thick lamina propria subdivided into upper stratum compactum and lower stratum spongiosum regions composed of stromal cells, immune cells, and vasculature. Invaginating downward from the LE are long branched and coiled tubular glands that traverse the lamina propria and are referred to as the glandular epithelium (GE) (Cooke et al. 1998) (*see* Fig. 2a). These various cell types communicate in an autocrine, paracrine and/or, endocrine manner to coordinate responses to hormones and cytokines (Geisert et al. 1982a). Examples of these complex responses include increasing uterine blood flow (Ford et al. 1982), facilitating water and electrolyte movement (Veldius et al. 1980), maternal recognition of pregnancy (Bazer and Thatcher 1977), receptivity to conceptus attachment for implantation (Keys and King 1980; Burghardt et al. 1997), and uterine secretory activity (Bazer 1975).

By 26 h after fertilization within the oviduct, the one-cell porcine ovum/zygote cleaves to form a two-cell embryo that, 22–30 h later, enters the uterus at the 4- to 8-cell stage. The blastocyst forms when early embryonic cells differentiate into the embryonic disc, trophectoderm, and extraembryonic endoderm. The blastocoel forms and there is continued development to a conceptus (Bazer and Johnson 2014) (Fig. 2b). Conceptuses "hatch" from the zona pellucida and begin to elongate. At hatching, the conceptuses are 0.5–1 mm diameter spheres, then increase in size to 2–6 mm by day 10 of pregnancy. They then undergo a morphological transition to large spheres of 10–15 mm diameter, then to 15 mm by 50 mm tubular forms, and then to 1 mm by 100–200 mm filamentous forms on day 11 (*see* Fig. 2b). Pig conceptuses elongate at 30–45 mm/h during the transition from tubular to filamentous forms, primarily by proliferation, morphological remodeling, and migration of the trophectoderm and endoderm. However, the subsequent growth and elongation of the conceptus by day 15 into 800–1000 mm length filaments is supported by the hyperplasia of trophectoderm cells. An elongation zone of densely packed endoderm and trophectoderm cells extends from the inner cell mass (ICM) to the tip of the blastocyst on day 10. Then, through alterations in microfilaments and junctional complexes of trophectoderm cells and formation of filopodia by endodermal cells, there is further rapid elongation of the 100–200 mm long conceptus to a conceptus of 800–1000 mm in length by day 16 of pregnancy. Each conceptus within the litter eventually achieves maximum surface area for contact between the trophectoderm and uterine LE to facilitate uptake of nutrients from uterine LE and uterine GE, the numbers of which increase, coincidentally, with elongation of the conceptuses (Bazer and Johnson 2014) (*see* Fig. 2c).

3 Select Hormones and Cytokines That Prepare the Uterus for Placentation

Much is known about the localization of genes and their regulation by hormones within the uterus and conceptus during implantation and early placentation in pigs (Johnson et al. 2009; Bazer and Johnson 2014; Waclawik et al. 2017). Progesterone dominates the uterine environment during the establishment of pregnancy in pigs, but other factors are required to maintain a successful pregnancy. These include the secretion of estrogens from the conceptus (Bazer and Thatcher 1977; Geisert et al.

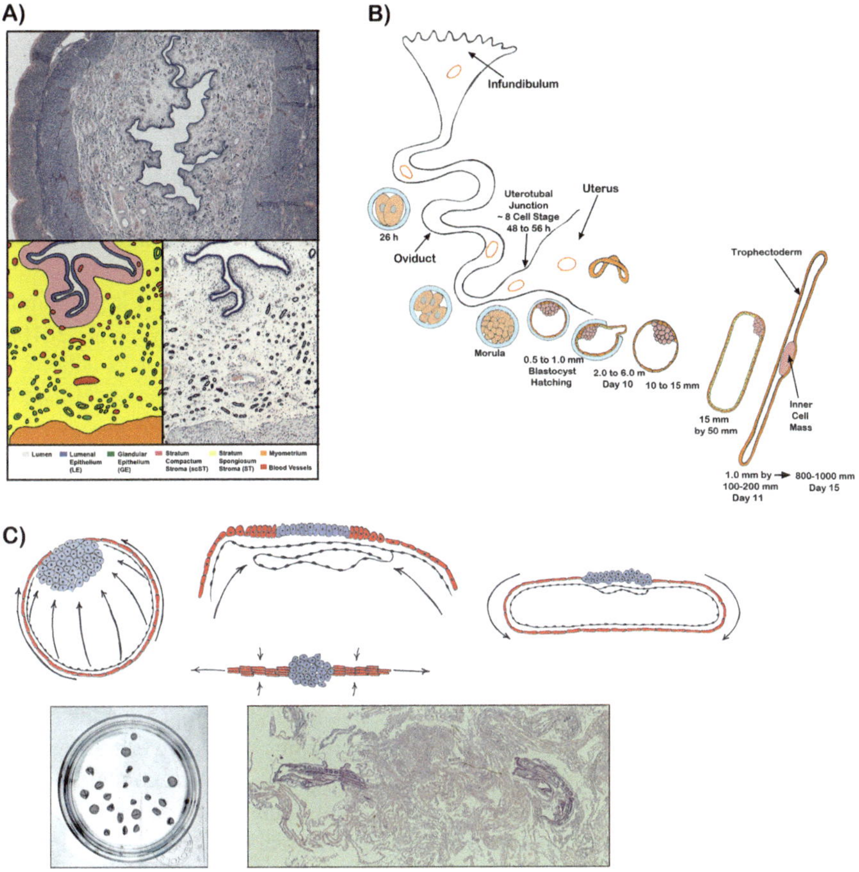

Fig. 2 (**a**) The top and bottom right panels show H&E-stained thin sections of the wall of the porcine uterus. Bottom left panel is a cartoon in which the different tissues and cell types within the uterus are color coded for identification. (**b**) Shows a cartoon depicting early development of porcine conceptuses. The embryos enter the oviduct at bout the eight-cell stage 48–56 h postfertilization, hatch from the zona pellucida, and increase in size to 10–15 mm disks by day 10, then transform to tubular and filamentous conceptuses by day 11 of gestation. (**c**) Top panel shows a cartoon depicting the trophectoderm cells (red color) of the round blastocyst (left panel) migrating inward towards the ICM to form the elongation zone (middle panels), then trophectoderm cells migrate outwards (right panel) to form the filamentous conceptus. Bottom panels show a uterine flushing from Day 10 of gestation (left panel) containing 22 round blastocysts, and H&E staining of two pig conceptuses (right panel) from a uterine flushing obtained on Day 15 of gestation. The cartoon in Panel **c** is an interpretation of data published by Geisert et al. (1982b). (Panels **b** and **c** are original drawings by Gregory A. Johnson. Panel **b** is adapted with permission from Bazer FW, Kim J, Song G, Ka H, Wu G, Johnson GA, and Vallet JL. 2013. Roles of selected nutrients in development of the porcine conceptus during pregnancy. In: Control of Pig Reproduction IX. A. Rodriquez-Martinez, N.M. Soede and W.L. Flowers Eds., Context Products Ltd., British Library Cataloguing in Publication Data, Leicestershire, United Kingdom, pp. 159–174. Panel **c** is adapted with permission from Bazer FW, Johnson GA (2014) Pig blastocyst-uterine interactions. Differentiation 87:52–65. Published on behalf of the International Society of Differentiation by Elsevier)

2005; Johnson et al. 2009), secretions from the uterine LE and GE, i.e., histotroph (Bazer 1975; Johnson et al. 2009), and cellular remodeling at the interface between the uterine LE and trophectoderm during implantation and early placentation (Burghardt et al. 2002; Johnson et al. 2014). These complex events are orchestrated through, among others, five important cell signaling pathways for endocrine and paracrine communication between the ovary, conceptus, and uterus. These five hormones and cytokines are progesterone, estrogen, prostaglandins, interleukin-1 beta (IL1B), and the interferons [interferon gamma (IFNG) and interferon delta (IFND)].

Progesterone is considered to be the hormone of pregnancy, and it profoundly influences the uterus and placenta throughout pregnancy (Bailey et al. 2010a). Progesterone binds its nuclear progesterone receptor (PGR) to exert its actions on tissues. The PGR is encoded by a single gene and has three isoforms (A, B, and C) that differ in their activities (Wetendorf and DeMayo 2014; Mulac-Jericevic and Conneely 2004). PRB is expressed in the uterine LE, GE, stromal fibroblasts, and myometrium through day 7 of the estrous cycle and pregnancy; however, prolonged exposure to progesterone downregulates PGR in the uterine LE by day 10 and in the majority of the uterine GE by day 12 of both the estrous cycle and pregnancy. PGR protein expression in uterine LE returns by day 17 of the estrous cycle, but is not expressed by the uterine LE between days 25 and 85 of gestation. Expression of PGR is maintained in the deep GE throughout the estrous cycle and early pregnancy and continues to decrease, with almost no expression in GE by day 50 of pregnancy. PGR is, however, expressed in the necks of the uterine GE that empty into the uterine lumen proximal to areolae, suggesting that these regions of GE are excretory ducts that are not involved in the synthesis of histotroph, but serve as conduits for the passage of histotroph into the lumens of areolae. In contrast to the epithelia, PGR protein expression is maintained in the uterine stroma and myometrium through day 85 of gestation (*see* Fig. 3a) (Geisert et al. 1994; Sukjumlong et al. 2005; Steinhauser et al. 2017). One of the major roles of progesterone is to stimulate the production and secretion of histotroph, a complex mixture of hormones, growth factors, nutrients, and other substances that are required for growth and development of the conceptus (Knight et al. 1974; Roberts and Bazer 1988; Bailey et al. 2010b). The consensus is that the role of progesterone in producing histotroph is mediated via PGR; however, PGR is not expressed in uterine LE or GE that secrete histotroph during the peri-implantation period (Geisert et al. 1994; Sukjumlong et al. 2005; Steinhauser et al. 2017). It is clear that progesterone regulation of gene expression in the endometrium during the peri-implantation period is complex. Induction of genes in uterine GE may require that progesterone downregulates PGR, thereby eliminating PGR-dependent inhibition of expression of progesterone-regulated genes. However, another explanation is that progesterone induces the expression of genes in uterine via regulation of one or more paracrine-acting factors (progestamedins) produced by the PGR-positive stromal cells (White et al. 2005).

Pig conceptuses secrete estrogens on days 11 and 12 of gestation, and the consensus since 1977 has been that pregnancy recognition in the pig is the result of secretion of estrogens by conceptuses. Evidence to support this idea includes:

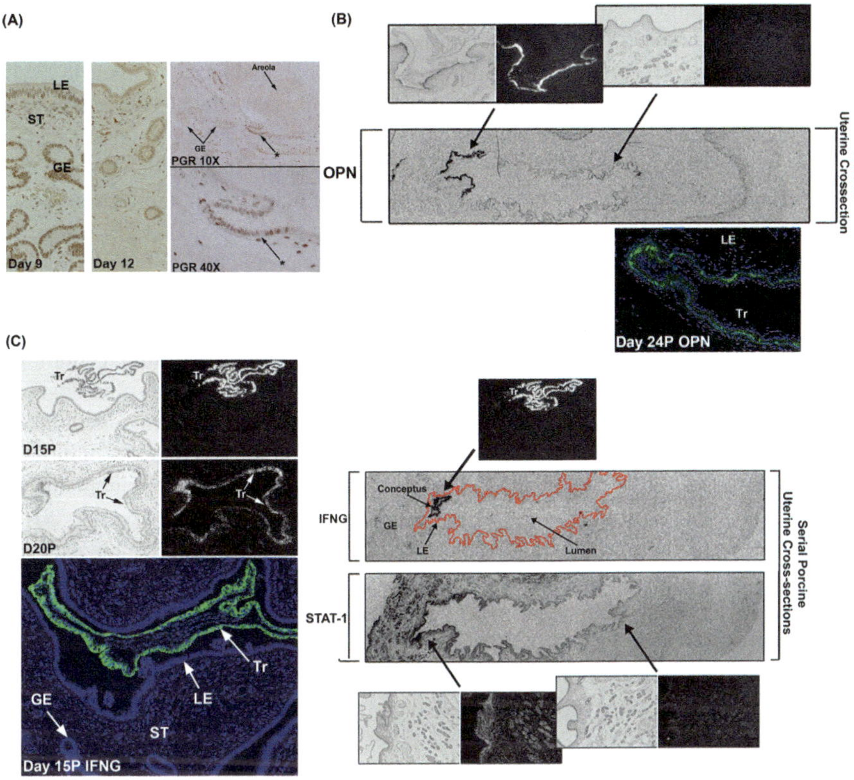

Fig. 3 (**a**) Progesterone (P4) downregulates progesterone receptor (PGR) in the uterine epithelia, but maintains expression in the necks of uterine glands. Immunohistochemistry for PGR in the endometrium of cyclic (left two panels) and pregnant (right two panels) pigs. (**b**) Conceptus estrogens (E2) induce osteopontin (OPN) in the uterine luminal epithelium (LE), during the peri-implantation period of pregnancy. Top panels show corresponding brightfield and darkfield images of in situ hybridization from a day 15 pregnant uterus probed with *OPN* cRNA. Middle panel shows in situ hybridization analysis of *OPN* mRNA in autoradiographic images (Biomax-MR; Kodak) showing an entire cross-section of the uterine wall from day 15 of pregnancy. Bottom panel shows immunofluorescence staining for OPN in the uterine LE on day 24 of pregnancy. (**c**) Interferon gamma (IFNG) upregulates the interferon-responsive gene (ISG) signal transducer and activator of transcription one (STAT1) in the uterine stroma (ST) and glandular epithelium (GE). Left panels show corresponding brightfield and darkfield images of in situ hybridization from day 15 and day 20 pregnant uteri probed with *IFNG* cRNA, and immunofluorescence staining for IFNG in the conceptus trophectoderm on day 15 of pregnancy. Right panels show in situ hybridization analysis of *IFNG* and *STAT1* mRNAs in autoradiographic images (Biomax-MR; Kodak) showing entire serial cross-sections of the uterine wall from day 15 of pregnancy. The luminal epithelium of tissue probed for *IFNG* mRNA has been outlined in red for histological reference. Corresponding brightfield and darkfield images from the same uterus probed with *IFNG* and *STAT1* cRNAs are also shown. Width of each field of autoradiographic images is 20 mm. Width of each field of brightfield and darkfield images is 940 μm. Width of each field of immunohistochemistry is 890 μm. (Panel **a** is adapted with permission from Steinhauser CB, Bazer FB, Burghardt RC, and Johnson GA (2017) Expression of Progesterone Receptor in the Porcine Uterus and Placenta throughout Gestation: Correlation with Expression of Uteroferrin and Osteopontin. Domestic Anim Endocrinol 58:1–11. Published by Elsevier. Panel **b** is adapted with permission from White FJ, Ross JW, Joyce MM, Geisert RD, Burghardt RC, Johnson GA (2005) Steroid regulation of cell specific secreted phosphoprotein 1 (osteopontin) expression in the pregnant porcine uterus. Biol Reprod

(1) the uterine endometrium secretes PGF during both the estrous cycle and pregnancy; (2) pig conceptuses secrete estrogens which appear to be antiluteolytic; (3) PGF is secreted toward the uterine vasculature in an endocrine manner in cyclic gilts to induce luteolysis; and (4) in pregnant gilts PGF is into the uterine lumen in an exocrine manner where it is sequestered from the corpora lutea (CL) and/or metabolized to prevent luteolysis (Bazer and Thatcher 1977). It should be noted that prostaglandin E2 (PGE2), as well as lysophosphatic acid (LPA), are also proposed to have roles in pregnancy recognition signaling. Expression of PGE2 synthase by the trophectoderm and the endometrium decreases the production of PGF in favor of PGE2, and this supports maintenance of CL (Ziecik et al. 2008; Waclawik et al. 2017). However, when estrogen synthesis by the conceptus was ablated by targeting the aromatase (CYP19A1) gene utilizing CRISPR/Cas9 genome editing technology, estrogen was observed to not be essential for preimplantation conceptus development, conceptus elongation, or early CL maintenance. Estrogen was essential for maintenance of pregnancy beyond 30 days and its ablation disrupted a number of biological pathways (Meyer et al. 2019). Conceptus estrogens also modulate uterine gene expression (Geisert et al. 1982c; Johnson et al. 2009; Waclawik et al. 2017). The importance of estrogen to porcine pregnancy is underscored by the fact that exposure of the pregnant uterus to estrogen on days 9 and 10, prior to when pig conceptuses normally secrete estrogens, results in degeneration of all pig conceptuses by day 15 (Ross et al. 2007). Figure 3b illustrates that the timing of estrogen secretion by the conceptus correlates with the induction of osteopontin [OPN, also known as secreted phosphoprotein 1 (SPP1)] expression in the uterine LE, and administration of exogenous estradiol to ovariectomized pigs induces OPN in the uterine LE (White et al. 2005). The upregulation of OPN within uterine LE in close proximity to the implanting conceptus implies paracrine regulation of genes by conceptus estrogens. It is likely that effects of estrogen on the uterus are restricted to regions near the conceptus due to the metabolic activity of trophectoderm. The endometrium of pigs can convert estradiol to estrone and then converts estrone to estrone sulfate, which is biologically inactive and present in high concentrations within the uterine lumen of pregnant pigs (Flood 1974). In contrast, the trophectoderm has sulfatase that can restore the biological activity of estrogen, allowing for estrogen to upregulate genes such as OPN in the LE immediately juxtaposed to the implanting conceptus. To date, only a limited number of estrogen-stimulated genes have been localized in the pig endometrium (reviewed in Johnson et al. 2009), but the number increases each year.

Fig. 3 (continued) 73:1294–1301. Published on behalf of the Society for the Study of Reproduction by Oxford University Press. Panel **c** is adapted with permission from Joyce MM, Burghardt RC, Geisert RD, Burghardt JR, Hooper RN, Ross JW, Ashworth MD, Johnson GA (2007) Pig conceptuses secrete estrogen and interferons to differentially regulate uterine STAT1 in a temporal and cell type-specific manner. Endocrinology 148: 4420–4431. Published on behalf of the Endocrine Society by Oxford University Press)

In addition to the important role in maintaining the lifespan of the CL, prostaglandins function as autocrine and paracrine regulators of conceptus development and endometrial functions during the peri-implantation period in pigs. Pig conceptuses and endometrium actively secrete PGE2 and PGF (Geisert et al. 1982a; Guthrie and Lewis 1986) and express their receptors (Waclawik et al. 2017). PGF, acting through its receptor (PTGFR), induces endometrial expression of vascular endothelial growth factor-A (VEGFA), biglycan, matrix metalloprotease 9 (MMP9), IL1A, and transforming growth factor B3 (TGFB3), suggesting that PGF is involved in angiogenesis and tissue remodeling during early pregnancy (Kaczynski et al. 2016). Elevated quantities of PGE2 in the uterine lumen during the peri-implantation period stimulate conceptus PTGER2 expression, which increases the synthesis and secretion of estrogen and enhances integrin-dependent trophoblast adhesion via an estrogen receptor-dependent mechanism and MEK/MAPK signaling (Waclawik et al. 2017). Several studies have proposed that PGE2 is involved in maternal recognition of pregnancy in the pig (Christenson et al. 1994; Ford and Christenson 1991; Gregoraszczuk and Michas 1999). Intrauterine infusion of PGE2 into cyclic gilts from days 7 to 23 delays CL regression and extends the length of the estrous cycle (Akinlosotu et al. 1986). PGE2, produced by the conceptuses and endometrium, can be directly luteoprotective due to an increase in the PGE2/PGF ratio. The conceptus produces about twice as much PGE2 as PGF to successfully protect luteal cells from regression (Gregoraszczuk and Michas 1999). Indeed, Meyer et al. (2019) demonstrated that PGE can protect the CL through day 24 in the absence of conceptus estrogen synthesis. However, when CRISPR/Cas9 gene editing was used to knock out prostaglandin-endoperoxide synthase 2 (PTGS2) during the period of conceptus elongation in pigs, the conceptuses elongated and pregnancy was maintained through day 35 of gestation (Pfeiffer et al. 2020). It is possible that conceptus production of estrogen and IL1B may compensate for the lack of PTGS2-derived prostaglandins as estrogen and IL1B2 increase endometrial PTGS2 expression and production of PGE2 (Waclawik et al. 2009a, b; Franczak et al. 2010).

As porcine conceptuses elongate and begin to secrete estrogens, they also secret large amounts of IL1B (Tuo et al. 1996; Ross et al. 2003). There is an initial and acute rise in IL1B on day 12 of pregnancy followed by a precipitous decline in the expression and secretion of IL1B by pig conceptuses by days 15–18 of gestation as the conceptuses become fully elongated filaments. The expression of receptors for IL1B increases in the uterus and conceptus in tandem with this transient expression of IL1B by pig conceptuses, between days 12 and 15 of pregnancy (Ross et al. 2003). This suggests that IL1B from the conceptus may bind IL1B receptors on conceptus and/or uterine tissues to modify interactions between the uterus and conceptus in an autocrine and/or paracrine fashion. It should be noted that there are two transcripts for IL1B expressed in the pig trophectoderm, with both genes in close proximity on chromosome 3 (Mathew et al. 2015). The classical IL1B is expressed in macrophages and endometrial tissue, but another form of IL1B, IL1B2, appears to be unique to the elongating pig conceptus and is expressed just prior to attachment of the conceptus to the uterine LE. The specific roles of conceptus IL1B in trophoblast elongation and uterine receptivity have not been determined, but when CRISPR/Cas9 gene editing was used to knock out IL1B2 during the period of conceptus elongation in pigs, the conceptuses failed to elongate. The expression of the majority of conceptus developmental genes was not altered in these conceptuses, but the conceptuses expressed less aromatase and secreted less estrogen than normal (Whyte et al. 2018).

Antiviral activity in the uterine flushings of pigs peaks on days 14 and 25 of gestation, and the media from cultured porcine conceptuses exhibits significant antiviral activity (Mirando et al. 1990). The cause for this phenomenon is that the trophectoderm of pigs secretes a Type II IFN gamma (IFNG, 75% of antiviral activity in pig conceptus secretory proteins) and a Type I IFN delta (IFND, 25% of antiviral activity in pig conceptus secretory proteins) during the peri-implantation period (La Bonnardière et al. 1991; Lefèvre et al. 1998a). There is a 567-fold increase in the abundance of IFNG mRNA in porcine conceptuses between days 13 and 20 of gestation, during the transition of conceptuses from spherical to day 14 filamentous forms (Cencič and LaBonnardiėre 2002; Joyce et al. 2007a) (*see* Fig. 3c). In contrast, RT-PCR analysis is required to detect IFND mRNA in the day 14 conceptuses of pigs (Joyce et al. 2007a, b). Although IFN tau (IFNT), a Type I IFN, is the pregnancy recognition signal in ruminants (Spencer et al. 2007a), IFNG and IFND do not appear to be antiluteolytic in pigs (Harney and Bazer 1989; Lefèvre et al. 1998b). However, paracrine effects for porcine conceptus IFNs are suggested by: (1) localization of IFN receptors on endometrial epithelial cells (Lefèvre et al. 1998b); (2) increased secretion of PGE2 (Harney and Bazer 1989); (3) endometrial expression of several known IFN-responsive genes (reviewed in Johnson et al. 2009); and (4) modulation of uterine stromal and GE gene expression by the IFNs in conceptus secretory protein preparations (Joyce et al. 2007a, b, 2008). Figure 3c illustrates that stromal induction of mRNA for *STAT1*, an IFN responsive gene, correlates with IFNG and IFND secretion by the conceptus. Indeed, intrauterine infusion of conceptus secretory proteins, which contain IFNG and IFND, into pseudopregnant pigs, increased *STAT1* as compared to intrauterine infusion of control proteins (Joyce et al. 2007a). Upregulation of *STAT1* within the uterine stroma and GE in close proximity to the implanting conceptus implies paracrine regulation of genes by conceptus IFNs. It is somewhat surprising that initial increases in stromal *STAT1* are restricted to sites of intimate association between the conceptus and uterus in pigs because the magnitude of IFNG production by pig conceptuses appears to be similar to IFNT synthesis and secretion by ovine conceptuses (Joyce et al. 2007a). *STAT1* increases universally in the stroma and GE of pregnant sheep without regard to conceptus location within the lumen, presumably due to the high levels of secretion of IFNT by conceptuses (Pontzer et al. 1988; Joyce et al. 2005a). One explanation for the more limited expression of *STAT1* in the pig uterus is that IFNG and IFND act synergistically to upregulate interferon responsive genes. Interactions between Type I and Type II IFNs to enhance their activities have been demonstrated previously (Decker et al. 1989). To date, only a limited number of IFN responsive genes have been localized in the pig endometrium and specific roles for these IFN responsive genes remain elusive (reviewed in Johnson et al. 2009).

4 Attachment of the Conceptus Trophectoderm to the Uterine LE (Implantation)

In strict terms, porcine conceptuses do not penetrate the uterine LE and invade into the uterine stroma; therefore, for pigs, the term "implantation" is a misnomer. Instead, pig conceptuses undergo a central type of implantation. Nevertheless, elongation and implantation are used to describe the initial stages of placental development in pigs. Although the duration of the preimplantation period is prolonged in the pig, and type of implantation differs from many other species, the initial stages of implantation/placentation are common across species and are characterized as the "Adhesion Cascade for Implantation" (Dantzer 1985; Guillomot 1995; Burghardt et al. 2002). This adhesion cascade begins with shedding of the zona pellucida, followed by elongation of the conceptus trophectoderm. The conceptus trophectoderm then orients itself to the uterine LE in a phase called "apposition," followed by adhesion of the apical surface of trophectoderm to the apical surface of uterine LE, and development of interdigitating microvilli between trophectoderm and uterine LE (see Fig. 4a). As this cascade concludes, adhesion seamlessly transitions to the progressive formation of epitheliochorial placentation that supports fetal-placental development throughout pregnancy (Bazer and Johnson 2014).

During the peri-implantation period of pregnancy in eutherian mammals, the conceptus trophectoderm and uterine LE develop adhesion competency in synchrony to initiate an adhesion cascade within a restricted period of pregnancy termed the "Window of Receptivity" (Fazleabas et al. 2004; Spencer et al. 2007b; Bazer et al. 2011). This window is defined by the actions of progesterone and estrogen to regulate locally produced cytokines and growth factors, cell surface glycoproteins and adhesion molecules, and extracellular matrix (ECM) proteins (Johnson et al. 2009). Progesterone initiates the adhesion cascade for implantation in pigs. Immediately prior to when the endometrium becomes receptive to implantation, just after day 10, progesterone downregulates the expression of PGR in the uterine LE (Geisert et al. 1994; Steinhauser et al. 2017). This downregulation of PGR is associated with a downregulation in the expression of the mucin 1 (MUC1), a component of the apical surface glycocalyx of the uterine LE that physically inhibits attachment of the conceptus trophectoderm to the uterine LE due to its extended carbohydrate configuration (Brayman et al. 2004). Indeed, when cyclic gilts are administered intramuscular injections of progesterone, there is a loss of MUC1 from the apical surface of the uterine LE (Bowen et al. 1996). It is accepted that in all mammals, firm attachment of the conceptus trophectoderm to the uterine LE requires a temporal loss of MUC1 at the apical surface of uterine LE cells (Brayman et al. 2004) that exposes low-affinity carbohydrate ligand-binding molecules including selectins and galectins and, perhaps heparan sulfate proteoglycan, heparin-binding EGF-like growth factors, cadherins, and CD44. These molecules are proposed to then contribute to the initial, if fragile, attachment of the conceptus trophectoderm to the uterine LE (Kimber et al. 1995; Kimber and Spanswick 2000; Spencer et al.

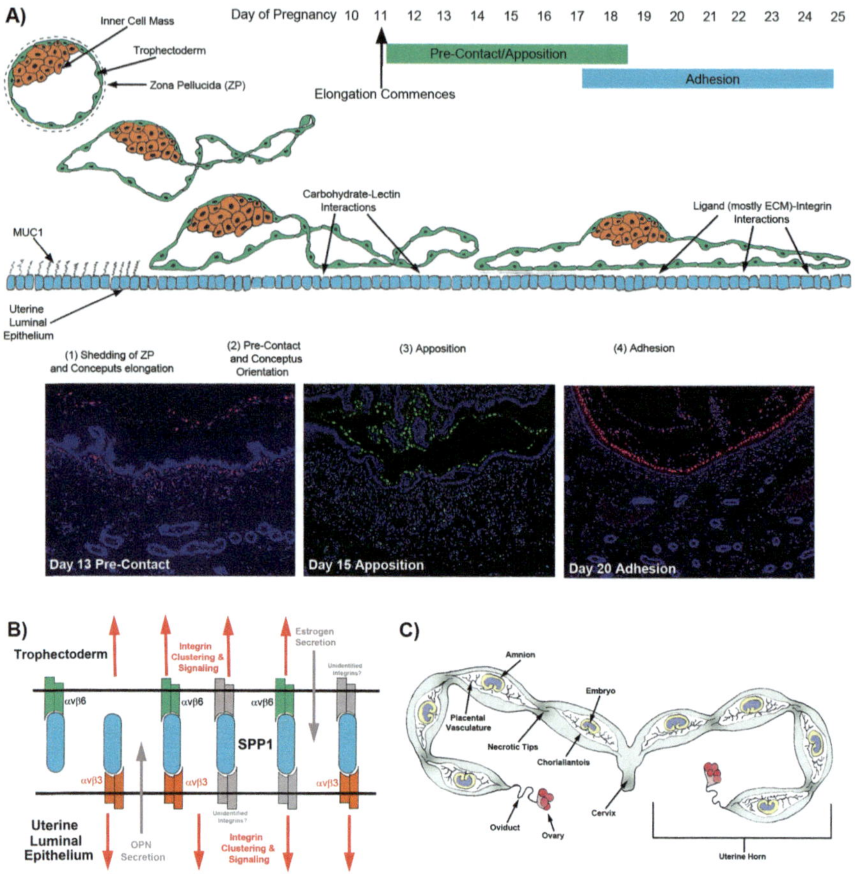

Fig. 4 (**a**) The initial stages of implantation are common across species and are characterized as the "Adhesion Cascade for Implantation." The phases of this adhesion cascade in pigs include: (1) elongation of the conceptus trophectoderm and shedding of the zona pellucida; (2) downregulation of MUC1 at the apical surface of uterine LE to expose potential, but not yet identified, low-affinity carbohydrate-lectin binding molecules to mediate precontact and conceptus trophectoderm orientation to the uterine LE; (3) low-affinity contacts are then replaced by a more stable and extensive repertoire of adhesive interactions between integrins and maternal ECM to mediate apposition of trophectoderm to LE; (4) integrin receptors expressed at the apical surface of uterine LE cells bind to Arg-Gly-Asp (RGD) and non RGD amino acid sequence-containing ECM molecules and bridge to another complement of potential integrin receptors expressed at the apical surface of conceptus trophectoderm cells to mediate conceptus trophectoderm adhesion; and (5) development of interdigitating microvilli between LE and trophectoderm (not illustrated in figure). Immunofluorescence staining for PCNA illustrates that the conceptus trophectoderm (Tr) proliferates, but the uterine luminal epithelium (LE) does not proliferate during the peri-implantation period of pigs. The left and right panels were immunolabeled with a goat anti-mouse IgG Alexa 594 secondary antibody. The middle panel was immunolabeled with a goat anti-mouse IgG Alexa 488 secondary antibody. (**b**) Shows a cartoon depicting results of in vitro experiments that have identified the αvβ6 integrin receptor on conceptus trophectoderm, and the αvβ3 integrin receptor on uterine LE as binding partners for OPN. OPN may bind individually to these receptors to act as a bridging ligand between these receptors. Alternatively, OPN may serve as a bridging ligand between one of these receptors and an as yet unidentified integrin receptor expressed on the opposing tissue. (**c**) Shows a cartoon depicting conceptus spacing within the

2004). Interactions between carbohydrates and lectins during the adhesion cascade of pigs have not been systematically investigated. However, it is likely that these carbohydrate ligands and their lectin receptors, expressed at the apical surfaces of the conceptus trophectoderm and uterine LE of pigs, undergo a series of attach-and-release events. This results in maximal apposition of the conceptus trophectoderm to the uterine LE, similar to the "rolling and tethering" that occurs when leukocytes adhere to the endothelium for extravasation out of the vasculature and into connective tissues (Kling et al. 1992). Indeed, rolling and tethering has been proposed for the initial attachment of human blastocysts to the uterine wall (Red-Horse et al. 2004). In support of this idea, other domestic farm species, including goats and sheep, express H-type-1 antigens and glycosylation-dependent glycam 1, respectively, at the interface between the conceptus trophectoderm and uterine LE during the attachment phase of implantation (Powell et al. 2000; Spencer et al. 1999). These low-affinity contacts are then stabilized by adhesion between an extensive repertoire of integrins and extracellular matrix (ECM) proteins. The binding of integrins to ECM proteins appears to be the major contributor to firm attachment of the conceptus trophectoderm to the uterine LE during implantation (Hynes 1987; Ruoslahti and Pierschbacher 1987; Aplin et al. 1994; Burghardt et al. 1997, 2002; Lessey 2002; Johnson et al. 2003, 2014). Integrins have been implicated in many cell adhesion cascades (Kling et al. 1992). They are transmembrane glycoprotein receptors composed of noncovalently linked α and β subunits. Integrin-ligand binding promotes cell-cell and cell-ECM adhesion, causes the cytoskeleton within the cells to reorganize and stabilize that adhesion, and transduces numerous intracellular signaling pathways (Giancotti and Ruoslahti 1999; Albelda and Buck 1990). There are 18 α- and 8 β-subunits that can dimerize to form 24 heterodimer combinations that then bind to numerous extracellular ligands including a variety of ECM proteins (Albelda and Buck 1990; Humphries et al. 2006; Gallant et al. 2005). Integrin receptors expressed at the apical surface of the uterine LE can bind to Gly-Arg-Gly-Asp-Ser (GRGDS) amino acid sequence containing ECM molecules and bridge to another complement of integrin receptors expressed at the apical surface of the conceptus trophectoderm (Johnson et al. 2014).

Seven integrin subunits have been shown to be expressed at the apical surface of both the conceptus trophectoderm and the uterine LE of pigs. These include alpha

Fig. 4 (continued) uterine horns and development of the allantoic and amniotic membranes (Panels **a**, **b**, and **c** are original drawings by Gregory A. Johnson). Panel **a** is adapted with permission from Geisert RD, Johnson GA and Burghardt RC. Implantation and establishment of pregnancy in the pig. In: Regulation of implantation and establishment of pregnancy in mammals: Tribute to 45 year anniversary of Roger V. Short's "Maternal Recognition of Pregnancy". Ed's R.D. Geisert and F.W. Bazer. 2015, pp. 137–164. Springer Sham. Panel **b** is adapted with permission from Erikson DW, Burghardt RC, Bayless KJ, Johnson GA (2009) Secreted phosphoprotein 1 (SPP1, osteopontin) binds to integrin alphavbeta6 on porcine trophectoderm cells and integrin alphavbeta3 on uterine luminal epithelial cells, and promotes trophectoderm cell adhesion and migration. Biol Reprod 81:814–825. Published on behalf of the Society for the Study of Reproduction by Oxford University Press

1 (α1), α3, α4, α5, αv, beta 1 (β1), β3, and β5. The expression of α4, α5, and β1 on the uterine LE increases during the peri-implantation period, and treatment with progesterone increases the expression of these integrins at the apical surface of uterine LE cells of cyclic pigs. Further, α4, α5, αv, β1, β3, and β5 are present at porcine implantation sites on days 12 through 15 of gestation (Bowen et al. 1996). These subunits have the potential to assemble into the integrin heterodimer receptors αvβ1, αvβ3, αvβ5, α4β1, and α5β1, and these integrin receptors may function in the adhesion cascade that adheres the conceptus trophectoderm to the uterine LE (Burghardt et al. 1997). Integrin activation through binding to a ligand results in dynamic macromolecular complexes, termed integrin adhesion complexes (IACs) that are composed of the heterodimeric transmembrane integrin receptors that connect ECM proteins to the actin cytoskeleton, along with a diverse array of cell signaling intermediates (Sastry and Burridge 2000; Wozniak et al. 2004; Larsen et al. 2006). Immunofluorescence staining has revealed that αv and β3 integrin subunits co-localize with an altered distribution of an intracellular signaling intermediate within IACs called talin within large aggregates at sites of attachment between the conceptus trophectoderm and uterine LE on Days 20 and 25 of gestation (Erikson et al. 2009; Frank et al. 2017). The size and nature of these aggregates are similar to the well-characterized IACs that form at the base of cultured cells when they attach to the ECM (Sastry and Burridge 2000; Burghardt et al. 2009). To date, five ligands capable of engaging integrin receptors to induce assembly of IACs have been characterized at sites of implantation in pigs. The inter-α-trypsin inhibitor heavy chain-like (IαIH4) protein contains a von Willebrand type A domain that is a recognition site for the αvβ3 integrin receptor (Geisert et al. 1998). The Latency Associated Peptide (LAP) of transforming growth factor beta (TGFB) binds to αvβ1, αvβ3, and αvβ5 (Massuto et al. 2009a). Fibronectin is capable of binding α4β1, α5β1, and αvβ3 (Bowen et al. 1996). Vitronectin is a major ligand for αvβ3 (Bowen et al. 1996). Finally, OPN is the most promiscuous of the ligands and interacts with αvβ1, αvβ3, αvβ5, and α4β1 (Johnson et al. 2003, 2014).

The IαIH4 is part of the kallikrein-kininogen-kinin protease system. Both IαIH4 protein expression and kallikrein enzymatic activity increase within the uterine environment during the peri-implantation period of pregnancy in pigs (Geisert et al. 1998; Vonnahme et al. 1999). The IαIH4 can not only bind to the αvβ3 integrin receptor but also interact with hyaluronic acid within the ECM, both of which could be involved in conceptus implantation. However, the primary role for IαIH4 during implantation in pigs may be to act in concert with bikunin to stabilize the uterine LE surface glycocalyx during conceptus attachment for implantation (Hettinger et al. 2001). The LAP associates with TGFB to form an inactive homodimer, called the small latent complex. The small latent complex remains in the cell until it is bound by latent TGFB-binding protein to form the large latent complex (LLC) that is then secreted into the ECM (Lawrence 1996). The secreted LAP is then cleaved from the LLC by proteases to release active TGFB (Jenkins et al. 2006). In pigs, TGFB1, TGFB2, and TGFB3, and their receptors, TGFBRI and TGFBRII, are expressed by the conceptus trophectoderm and by the uterine LE between days 10 and 14 of gestation. It has been demonstrated that TGFB acts through LAP to increase fibronectin synthesis and cell adhesion to fibronectin, and LAP was shown to initiate the formation of IACs in a porcine trophectoderm cell line (Jaeger et al. 2005). In addition, conceptuses failed to implant when LAP was infused into the uteri of pregnant pigs, suggesting that the infused LAP competed with the endogenous LLC for binding to integrins expressed on the conceptus trophectoderm (Massuto et al. 2009b). In support of this idea, IACs containing LAP, β1, β3, and β5 are present at the interface between the conceptus trophectoderm and uterine LE that may support attachment of the conceptus trophectoderm to the uterine LE during porcine implantation (Massuto et al. 2009a). Both fibronectin and vitronectin are present at sites of attachment of conceptus trophectoderm to uterine LE in pigs (Bowen et al. 1996), and oncofetal fibronectin (oFN), a glycosylation variant of fibronectin, is constitutively expressed by the porcine conceptus trophectoderm as well as the uterine LE throughout gestation (Tuo and Bazer 1996). Fibronectin and vitronectin are prototype cell adhesion proteins, and fibronectin recognizes 10 integrin receptors to generate different signals depending upon the specific integrin receptor involved (Johansson et al. 1997; Humphries et al. 2006). It is noteworthy that the human conceptus trophoblast produces oFN so precisely at sites of trophoblast contact with the endometrium that it has been referred to as "trophoblast glue" (Feinberg et al. 1994), and fibronectin is a strong candidate for mediating conceptus adhesion to the uterus in rodents (Armant 2005).

Of the ECM adhesion proteins expressed at the interface between the conceptus trophectoderm and the uterine LE during the peri-implantation period of pregnancy in pigs, OPN is the most extensively studied (Johnson et al. 2003, 2014) (see Fig. 4b). OPN is a member of the small integrin-binding ligand *N*-linked glycoprotein (SIBLING) family of ECM proteins that has multiple physiological functions (Denhardt and Guo 1993; Butler et al. 1996; Sodek et al. 2000), but with regard to implantation, focus has been on OPNs ability to bind integrins to mediate cell adhesion and cell migration (Senger et al. 1994). OPN is an abundant component of the intrauterine environment of pregnant humans, mice, rabbits, sheep, cattle, goats, and pigs (Johnson et al. 1999a, b; Garlow et al. 2002; Apparao et al. 2003; Kimmins et al. 2004; Mirkin et al. 2005; Joyce et al. 2005b; White et al. 2006). In pigs, conceptus estrogens induce OPN expression just prior to the initiation of implantation, beginning on day 13, in discrete regions of the uterine LE next to the conceptus trophectoderm. By day 20, OPN expression extends along the entire uterine LE when stable adhesion of the conceptus trophectoderm to uterine LE occurs, and OPN expression remains high at the uterine–placental interface throughout gestation (Garlow et al. 2002; White et al. 2005). In vitro affinity chromatography and immunoprecipitation experiments have shown that OPN binds the αvβ6 integrin heterodimer on porcine trophectoderm cells, and the αvβ3 integrin heterodimer on porcine uterine LE cells (Erikson et al. 2009). OPN binding, particularly to the αv integrin, promotes dose-dependent attachment of porcine trophectoderm and uterine LE cells and stimulates haptotactic trophectoderm cell migration, meaning that the cells migrated along a physical gradient of nonsoluble OPN (Erikson et al. 2009; Frank et al. 2017). Immunofluorescence staining for integrins at implantation sites of pigs revealed IACs containing these same integrins distributed in a pattern similar to that suggested by in vitro binding to OPN (Erikson et al. 2009; Frank et al. 2017). Immunofluorescent staining for the αv integrin subunit revealed large IACs at the junction between the apical surfaces of conceptus trophectoderm and uterine LE cells, suggesting these IACs facilitate conceptus trophectoderm attachment to the uterine LE for implantation. The β3 subunit, however, appeared in aggregates at the apical surface of uterine LE cells only, agreeing with affinity chromatography data indicating in vitro binding of αvβ3 to OPN on uterine LE cells (Erikson et al. 2009). Finally, OPN-coated microspheres were used to demonstrate co-localization of OPN, the αv integrin subunit, and talin to IACs at the apical domain of porcine trophectoderm cells (Erikson et al. 2009). Collectively, results indicate that OPN binds integrins to stimulate integrin-mediated IAC assembly within conceptus trophectoderm and uterine LE cells, attachment of conceptus trophectoderm to uterine LE cells, and migration of conceptus trophectoderm cells to promote conceptus implantation in pigs.

The physical changes that occur to the interface between the conceptus trophectoderm and uterine LE during the initial stages of epitheliochorial placentation in pigs have been eloquently described by Dantzer (1985). On all days examined, the glycocalyx at the surface of the uterine LE is thicker than the glycocalyx at the surface of the conceptus trophectoderm. On days 13 and 14, the uterine LE develops protrusions that become enclosed by caps of conceptus trophectoderm cells, and this unique configuration physically immobilizes the conceptus at the uterine LE surface. By day 14, there is close apposition between the apical membranes of conceptus trophectoderm and uterine LE cells, and microvilli form and interdigitate between these plasma membranes through days 15 and 16. The interface between the conceptus trophectoderm and uterine LE cells becomes increasingly complex as it functionally transitions from histotrophic to hemotrophic nutrient transport between days 15 and 20 of pregnancy. It is characterized by apical domes on the uterine LE cells that are closely apposed to the conceptus trophectoderm cells and provide long cytoplasmic extensions into a luminal space between the apical domes. Finally, adhesion transitions into placentation through ever-increasing development of interdigitating microvilli between the conceptus trophectoderm and uterine LE cells that extends into the peripheral zone by day 26 of gestation.

5 Folding of the Uterine–Placental Interface to Facilitate Hemotrophic Support of the Fetus

As pregnancy progresses, the uterine–placental interface, composes of the uterine (maternal) placenta and the fetal placenta, undergoes considerable morphological changes to progressively develop more complex folds that increase the surface area of contact between uterine LE and placental chorionic epithelium (CE) for exchange of nutrients, gases and waste products (Vallet and Freking 2007; Vallet et al. 2009). The interface between the maternal (uterine) placenta and fetal placenta of pigs begins to fold by day 25 of pregnancy, and these early folds increase in length through day 35, and again increase in length between days 50 and 60 of gestation (Seo et al. 2020) (see Figs. 5a and 6a, c). Placental and uterine capillaries lie immediately beneath these epithelia, minimizing the distance between maternal and fetal blood vessels (Dantzer and Leiser 1994). Therefore, the increase in folding at the maternal placental (uterine)–fetal placental (chorioallantoic) interface directly increases the surface area of contact between maternal and placental microvasculatures to optimize the potential for transport of nutrients from maternal to placental blood vessels for eventual utilization by the embryo/fetus. The small intestine utilizes a similar strategy to maximize exposure of the microvasculature for micronutrient transport (Hilton 1902; Simons 2013). Friess et al. (1980) described the microarchitecture of these folds at Days 30 and 58 of gestation. By day 30 of pregnancy, the chorioallantoic and endometrial surfaces interlock into folds composed of endometrial ridges and chorioallantoic troughs. The trophectoderm cells at the bottom of the chorioallantoic troughs are columnar and 40 μm in height, and the trophectoderm cells at the sides and tops of the troughs are 20 μm in height. The height of the uterine LE cells does not vary along the interface of the folds. The apical surfaces of the trophectoderm cells possess long microvilli which interdigitate with microvilli on the apical surfaces of the uterine LE cells. The subepithelial capillaries of both the chorioallantois and endometrium remain separated from the epithelial basal laminae by a limited, but easily visualized, layer of connective tissue (Friess et al. 1980). By day 58 of pregnancy, the height of the trophectoderm cells is 35 μm at the bottom of the chorioallantoic troughs, and these cells are high columnar and narrow in width. Along the sides and tops of the chorioallantoic troughs, the trophectoderm cells are 15 μm in height. The uterine LE cells opposite the base of the chorioallantoic troughs are 25 μm, whereas the uterine LE cells along the sides and at the top of the chorioallantoic troughs are 15 μm. The subepithelial capillaries indent into the trophectoderm cells on the sides and tops of the chorioallantoic troughs, and where the capillaries protrude into the trophectoderm cells, the trophectoderm cell height is reduced to 2 μm. In summary, the lateral sides and tops of the chorioallantoic ridges are designed for gaseous exchange, and the chorioallantoic barrier can often be less than 2 μm, whereas the base of the chorioallantoic troughs is designed for the transport of blood borne nutrients, i.e., hemotroph (Friess et al. 1980).

An emerging concept is that tissues respond to mechanical forces that coordinate morphogenesis, and it is hypothesized that mechanotransduction and mechanosensation at the interface between the uterus and placenta drives the morphological development of folding characteristic of the epitheliochorial placentation of pigs (Seo et al. 2020). Changes in the length of the placental folds, expression of mechanotransduction-implicated molecules in uterine and placental tissues, and changes in the size of subepithelial blood vessels were examined for days 20 through 60 of gestation in pigs (see Fig. 5). It was observed that: (1) the length of folds increased (2) OPN, talin, and focal adhesion kinase co-localized into aggregates at the uterine–placental interface; (3) filamin, actin-related protein 2, and F-actin were enriched at the tops of uterine folds extending into placental tissue (the bottom of the chorioallantoic troughs described in the preceding paragraph); (4) uterine stromal

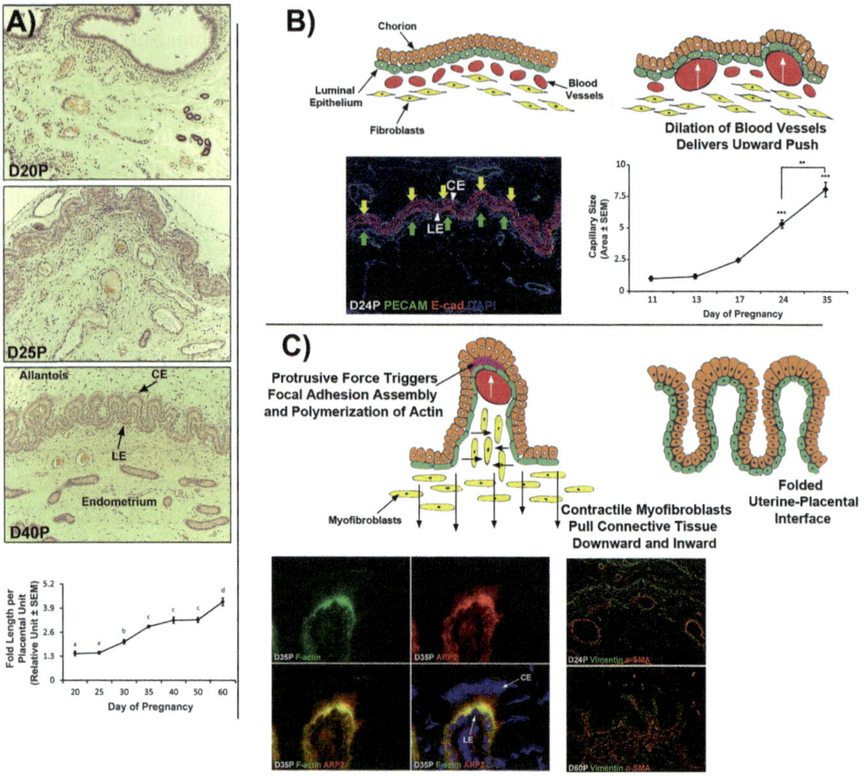

Fig. 5 (**a**) Representative paraffin-embedded H&E-stained thin sections of the porcine uterine–placental interface on days (D) 20, 25, and 40 of pregnancy (P). The interface between the maternal placenta (endometrium) and fetal placenta (chorioallantois) begins to fold between day 20 and day 25 of pregnancy. The bottom panel is a graphic illustration of morphometric analyses to quantify the length of placental folds on days 20 through 60 of pregnancy. Folding of the maternal placental-fetal placental epithelial bilayer significantly increases from day 30 through 60 of gestation. LE, luminal epithelium; CE, chorionic epithelium; Data are expressed as mean ± standard error of the mean (SEM). *, $P < 0.05$; ***, $P < 0.0001$. Width of field for microscopic images is 890 μm. (**b**) Upper panels show a cartoon depicting the dilation of subepithelial maternal placental (endometrial) blood vessels that increases blood flow to push upward on the interface between the endometrial LE and the chorioallantoic CE. The bottom left panel shows immunofluorescence microscopy for platelet/endothelial cell adhesion molecule 1 (PECAM-1), in uterine-placental tissues from day 24 of pregnancy. Subepithelial blood vessels are dilated from day 24 and day 60 of pregnancy. The right panel shows measurements of the mean area of subepithelial capillaries per unit endometrial tissue from day 11 to day 35 of pregnancy. The mean capillary area significantly increased between day 17 and day 24, and again between day 24 and day 35 of pregnancy. Data are expressed as mean ± standard error of the mean (SEM). **, $P < 0.01$; ***, $P < 0.0001$; E-cad, E-cadherin. Width of field for the microscopic image is 230 μm. (**c**) Upper panels depict protrusive forces at the sites of growing uterine capillaries that trigger IAC assembly and polymerization of actin between the LE and CE. These IACs serve to anchor the interface at the tops of the maternal placental folds and hold the maternal placenta and fetal placenta together. Maternal placental (endometrial) fibroblasts differentiate into contractile myofibroblasts. Because protrusive force continues to be applied to the placental interface at the tops of the maternal placental folds, and the tops of these folds are stabilized by the formation of IACs, contraction of the myofibroblasts does not pull the tops of the folds downward. Instead, the myofibroblast contraction pulls the sides of the folds inward and the tops of the fetal placental folds downward. The result is both a narrowing and a

fibroblasts acquired alpha smooth muscle actin; and (5) uterine blood vessels increased in size. These observations indicate that the lengthening of the folds is associated with polymerization of actin that coincides with IAC assembly, endometrial fibroblasts differentiate into myofibroblasts, and dilation of subepithelial blood vessels correlates with the development of the folds. It is proposed that dilation of subepithelial uterine blood vessels delivers increased blood flow that pushes upward on the interface between the uterine LE and the placental chorioallantois, protrusive forces from growing uterine blood vessels trigger IAC assembly and actin polymerization between the uterine LE and chorionic epithelium at the bottoms of the chorioallantoic troughs, and uterine fibroblasts differentiate into contractile myofibroblasts that pull the connective tissue downward and inward to sculpt folds at the uterine–placental interface (see Fig. 5b, c).

6 Areolae Provide for Histotrophic Support of the Fetus

In addition to having chorionic epithelium closely apposed to the uterine LE, there are specialized epithelial cells of the chorionic areolae at the openings of the mouths of uterine glands (see Fig. 6). These are tall columnar cells that have numerous vacuoles containing the secretions of uterine glands (histotroph or uterine milk). Indeed, the open space between the chorion and LE is filled with uterine milk (Dempsey et al. 1955). The blood vessels that supply the folds of the wall of the areolae form a ring towards the periphery and the areolar capillaries converge into one or two stem veins indicating facilitated external inflow of blood into the areola and outflow in a manner different from that of the inter-areolar regions of the placenta (Dantzer and Leiser 1993). This anatomy allows areolae to transport glandular secretions such as macromolecules, particularly proteins, by fluid-phase pinocytosis across the placenta and into the fetal-placental circulation.

After the apposition of the conceptus trophectoderm to the uterine LE at days 13–15 of gestation, by days 15–17, the chorioallantois immediately around each of

Fig. 5 (continued) lengthening of the folds at the interface between the maternal (uterine) placenta and fetal placenta of pigs. The lower left panels show co-staining for F-actin and actin-related protein 2 (ARP2) in uterine-placental tissues on day 35 of pregnancy. F-actin and ARP2 colocalized to the top of the maternal placental folds. The lower right panels show uterine-placental tissues from day 24 and day 60 of gestation that are double immunostained for vimentin and alpha smooth muscle actin (α-SMA). Vimentin, but not α-SMA, was evident in the stratum compactum of the endometrium of the maternal placenta on day 24, however immunostaining for α-SMA also increased between days 24 and 35 of gestation. Width of fields for microscopic images on the left is 230 μm. Width of fields for microscopic images on the right is 890 μm. (Panel **b** is an original drawing by Gregory A. Johnson. Figure is adapted with permission from Seo H, Li X, Wu G, Bazer FW, Burghardt RC, Bayless KJ and Johnson GA (2020) Mechanotransduction drives morphogenesis to develop folding at the uterine-placental interface of pigs. Placenta 90:62–70. Published on behalf of the International Federation of Placenta Associations by Elsevier)

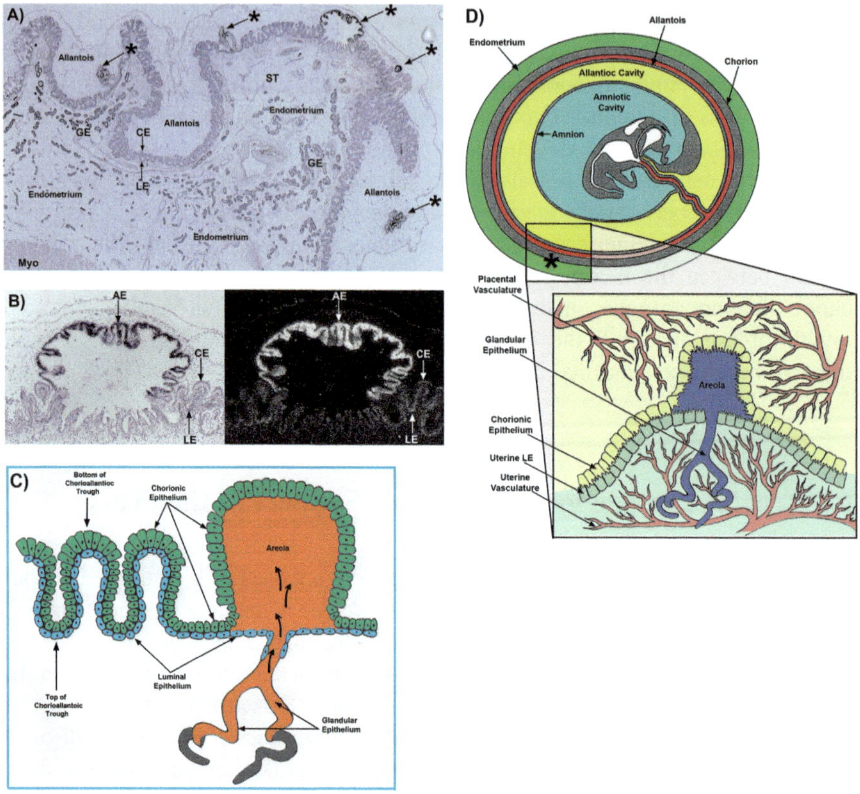

Fig. 6 (**a**) H&E-stained brightfield image of in situ hybridization analysis of Cathepsin L (*CTSL1*) mRNA, a gene that is expressed in the chorionic epithelium (CE) of areolae, within the chorioallantoic and endometrial tissues of a day-60 pregnant pig (the top-most areola in this panel and the areola in (**b**) are the same) (Song et al. 2010). Note the extensive folding interface between the chorioallantois and the endometrium and the intermittent areolae at this interface denoted by the astericks *. (**b**) Corresponding brightfield and darkfield images of a higher magnification of a placental areola from panel **a**. Width of field of panel **a** is 7.6 mm. *LE* luminal epithelium, *AE* areolar epithelium, *GE* glandular epithelium, *ST* stroma, *CE* chorionic epithelium, *Myo* myometrium. (**c**) Shows an illustration of the uterine placental interface of mature placentation in the pig illustrating folding for hemotrophic support of the fetus and an areola for histotrophic support of the fetus. The orange indicates the synthesis, secretion, and transport of histotroph by the glands and into the lumen of the areola. Note that the epithelial cells of the areola and the chorionic epithelial cells at the bottom of the chorioallantoic troughs are tall columnar. (**d**) Illustration depicting the uterine-placental microenvironment at a placental areola in relationship to the fetus. Nutrients and gases are transported from the maternal capillaries into the placental capillaries and then to the heart via the umbilical vein for distribution to all tissue of the fetal-placental unit. Macromolecules transported across the areolae also go to the heart via the umbilical circulation for utilization by various tissues and cells. (Panels **c** and **d** are original drawings by Gregory A. Johnson. Panels **a**, **b**, and **d** are adapted with permission from Song G, Bailey DW, Dunlap KA, Burghardt RC, Spencer TE, Bazer FW, Johnson GA (2010) Cathepsin B, Cathepsin L and Cystatin C in the Porcine Uterus and Placenta: Potential Roles in Endometrial/Placental Remodeling and in Fluid-Phase Transport of Proteins Secreted by Uterine Epithelia Across Placental Areolae and Neonatal Gut. Biol Reprod 82:854–864. Published on behalf of the Society for the Study of Reproduction by Oxford University Press)

the openings of the uterine glands begins to reach over the mouth of the gland(s) to develop a cavity that separates the uterine LE from the chorioallantois. These structures are termed the areolae. They are initially observed as small white circular discs with a prominent peripheral thickening of 1 mm in diameter (Friess et al. 1981), but quickly develop to cover the openings of the uterine gland(s). The cavity that forms receives the secretions of the uterine glands, and the columnar chorionic epithelial cells that line the placental border of this cavity form a seal between the uterine LE and the walls of the placental areola to prevent dissipation of histotroph into inter-areolar regions of the placenta. The allantoic vasculature that receives the histotroph is clearly discernable from the vasculature that supplies inter-areolar regions of the placenta (Leiser and Dantzer 1994). The endometrial vasculature that supplies the areola develops more slowly than the endometrial vasculature of inter-areolar regions presumably due a less intimate association with the trophectoderm. This prevents direct physical interaction between the trophectoderm and endometrium and decreases the influence of paracrine products that are secreted by the trophectoderm. As the placenta grows, areolar diameter increases and a stretching of the areolar capillary network leads to a progressively widening size. During the early stages of placentation, the placental surface of the areolae is flat, but as placentation progresses the flat surface becomes more complex with formation of ridges and papilla-like structures lined by a columnar chorionic epithelium (Amoroso 1952). The balloon shape of the areola implies that there is an interior pressure against the chorioallantoic surface of the areola delivered by the continuous accumulation of histotroph from the uterine glands. Indeed, the cavity of an areola is a small reservoir for the histotroph that is potentially secreted by the much larger uterine glands (Leiser and Dantzer 1994). There are some 2500 areolae distributed over the entire chorioallantois, and their number is correlated with fetal weight ($r = 0.65$). There are approximately 6 areolae per square centimeter of chorioallantois at mid-pregnancy, but approximately 2.5 areolae per square centimeter of chorioallantois by the end of gestation (Knight et al. 1977). The microanatomy of areolae facilitate the transport of glandular secretions such as macromolecules and proteins, including those for the transport of iron and vitamins, by fluid-phase pinocytosis across the placenta and into the fetal-placental circulation. Macromolecules and proteins are then transported to the heart via the fetal venous system for distribution to all fetal-placental tissues (Renegar et al. 1982; Ducsay et al. 1984; Roberts et al. 1986; Bazer et al. 1991). All nutrients transferred across the placenta may be cleared via the kidney and into the bladder from which they can enter the allantoic sac via the urachus for metabolism, degradation, or reuptake into the placental circulation for redistribution to affect development and function of fetal-placental tissues (Bazer 1989).

7 Conclusion

Figure 7 summarizes some of the major events during implantation and placentation in the pig. A primary function of the placenta is the transplacental exchange of gases, micronutrients (amino acids, glucose), and macromolecules (proteins), as well as the production of hormones, cytokines, and other regulatory molecules that affect growth and development of the conceptus throughout gestation. In all mammals, placentation initiates with implantation that includes specialized cell adhesion and cell migration leading to attachment of conceptus trophectoderm to the uterine LE. Once attached, maternal and fetal blood must be brought into close apposition to allow for transplacental exchange of molecules while maintaining separation of the maternal and fetal circulatory systems. Endometrial and placental tissues are remodeled to achieve areas of reduced interhaemal distance regardless of whether the placenta is epitheliochorial, synepitheliochorial, endotheliochorial, or hemochorial. Extensive remodeling to form chorionic (placental) ridges and corresponding endometrial invaginations results in folding that further increases the area of uterine-placental association. Indentation of the uterine LE and conceptus trophectoderm by underlying capillaries further reduces the distance between maternal and fetal blood, providing a short diffusion distance across the placenta.

Placentation is epitheliochorial in pigs as uterine LE remains intact throughout pregnancy. Porcine conceptus trophectoderm directly attaches to the uterine LE, and

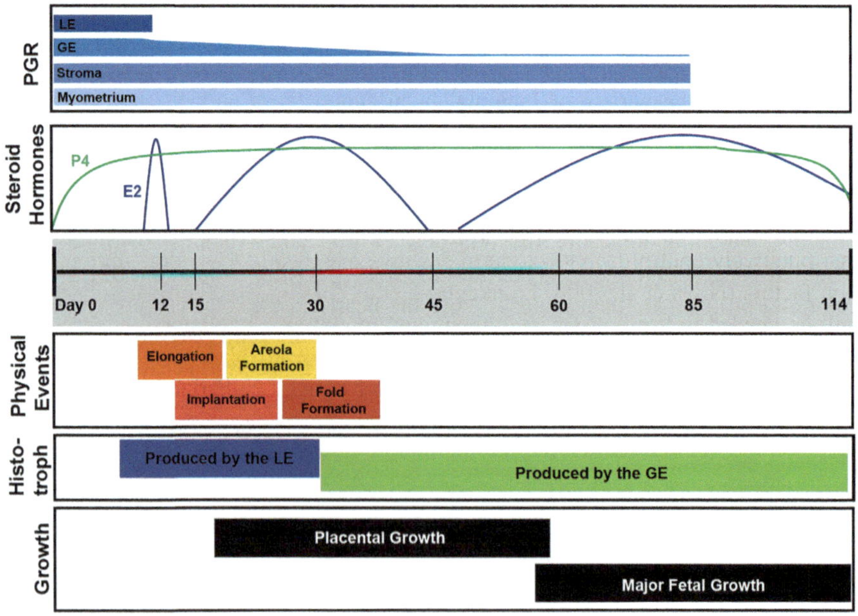

Fig. 7 Major events during implantation and placentation in pigs. (Figure concept by Dr. Chelsie Steinhauser)

these epithelia serve as the conduit for maternal hematotrophic and histotrophic support for conceptus growth and development. There is progressive interdigitation of microvilli on conceptus trophectoderm and uterine LE. The interface between the maternal (uterine) placenta and fetal placenta of pigs begins to fold by day 25 of pregnancy, and these early folds increase in length through day 35, and again increase in length between days 50 and 60 of gestation to eventually cover the entire placenta, except at the openings of uterine glands. Here the conceptus trophectoderm never fuses with uterine LE, rather it forms a pocket referred to as an areola. Secretions from superficial GE and deep uterine glandular GE, as well as molecules representing selective transudation from maternal serum into uterine GE and LE, are absorbed and transported across the chorioallantoic placenta by fluid-phase pinocytosis for release into the fetal circulation.

References

Akinlosotu BA, Diehl JR, Gimenez T (1986) Sparing effects of intrauterine treatment with prostaglandin E2 on luteal function in cycling gilts. Prostaglandins 32:291–299

Albelda SM, Buck CA (1990) Integrins and other cell adhesion molecules. FASEB J 4:2868–2880

Amoroso E (1952) Placentation. In: Parkes A (ed) Marshall's physiology of reproduction. Little Brown & Co.Amoroso reference, Boston, MA, pp 127–311

Aplin JD, Seif MW, Graham RA, Hey NA, Behzad F, Campbell S (1994) The endometrial cell surface and implantation. Expression of the polymorphic mucin MUC-1 and adhesion molecules during the endometrial cycle. Ann N Y Acad Sci 30:103–121

Apparao KBC, Illera MJ, Beyler SA, Olson GE, Osteen KG, Corjay MH, Boggess K, Lessey BA (2003) Regulated expression of osteopontin in the peri-implantation rabbit uterus. Biol Reprod 68:1484–1490

Armant DR (2005) Blastocysts don't go it alone. Extrinsic signals fine-tune the intrinsic developmental program of trophoblast cells. Dev Biol 280:260–280

Bailey DW, Dunlap KA, Frank JW, Erikson DW, White BG, Bazer FW, Burghardt RC, Johnson GA (2010a) Effects of long-term progesterone on developmental and functional aspects of porcine uterine epithelia and vasculature: progesterone alone does not support development of uterine glands comparable to that of pregnancy. Reproduction 140:583–594

Bailey DW, Dunlap KA, Erikson DW, Patel AK, Bazer FW, Burghardt RC, Johnson GA (2010b) Effects of long-term progesterone exposure on porcine uterine gene expression: progesterone alone does not induce secreted phosphoprotein 1 (osteopontin) in glandular epithelium. Reproduction 140:595–604

Bazer FW (1975) Uterine protein secretions: relationship to development of the conceptus. J Anim Sci 41:1376–1382

Bazer FW (1989) Allantoic fluid: regulation of volume and composition. In: Brace RA (ed) Fetal and neonatal body fluids. Perinatology Press, Cornell, NY, pp 135–157

Bazer FW, Johnson GA (2014) Pig blastocyst-uterine interactions. Differentiation 87:52–65

Bazer FW, Thatcher WW (1977) Theory of maternal recognition of pregnancy in swine based on estrogen controlled endocrine versus exocrine secretion of prostaglandin F2α by the uterine endometrium. Prostaglandins 14:397–400

Bazer FW, Worthington-White D, Fliss MFV, Gross S (1991) Uteroferrin: a progesterone-induced hematopoietic growth factor of uterine origin. J Exp Hematol 19:910–915

Bazer FW, Spencer TE, Johnson GA, Burghardt RC (2011) Uterine receptivity to implantation of blastocysts in mammals. Front Biosci S3:745–767

Bazer FW, Kim J, Song G, Ka H, Wu G, Johnson GA, Vallet JL (2013) Roles of selected nutrients in development of the porcine conceptus during pregnancy. In: Rodriquez-Martinez A, Soede NM, Flowers WL (eds) Control of pig reproduction IX. Context Products Ltd, British Library Cataloguing in Publication Data, Leicestershire, pp 159–174

Bjorkmann N (1973) Fine structure of the fetal-maternal area of exchange in the epitheliochorial and endotheliochorial types of placentation. Acat Anat 61(Suppl):1–22

Bowen JA, Bazer FW, Burghardt RC (1996) Spatial and temporal analysis of integrin and Muc-1 expression in porcine uterine epithelium and trophectoderm *in vivo*. Biol Reprod 55:1098–1106

Brayman M, Thathiah A, Carson DD (2004) MUC1: a multifunctional cell surface component of reproductive tissue epithelia. Reprod Biol Endocrinol 2:4–12

Burghardt RC, Bowen JA, Newton GR, Bazer FW (1997) Extracellular matrix and the implantation cascade in pigs. J Reprod Fertil 52:151–164

Burghardt RC, Johnson GA, Jaeger LA, Ka H, Garlow JE, Spencer TE, Bazer FW (2002) Integrins and extracellular matrix proteins at the maternal/fetal interface in domestic animals. Cells Tissues Organs 172:202–217

Burghardt RC, Burghardt JR, Taylor JD II, Reeder AT, Nguyen BT, Spencer TE, Johnson GA (2009) Enhanced focal adhesion assembly reflects increased mechanosensation and mechanotransduction along the maternal/conceptus interface during pregnancy in sheep. Reproduction 137:583–593

Butler WT, Ridall AL, McKee MD (1996) Osteopontin. In: Principles of bone biology. Academic Press, Inc, New York, pp 167–181

Cencič A, LaBonnardière C (2002) Trophoblastic interferon-gamma: current knowledge and possible role(s) in early pig pregnancy. Vet Res 33:139–157

Christenson LK, Farley DB, Anderson LH, Ford SP (1994) Luteal maintenance during early pregnancy in the pig: role for prostaglandin E2. Prostaglandins 47:61–75

Cooke PS, Buchanan DL, Kurita T, Lubahn DB, Cunha GR (1998) Stromal-epithelial cell communication in the female reproductive tract (Chap. 17). In: Bazer FW (ed) The endocrinology of pregnancy. Ahumana Press, Totowa, NJ, pp 491–506

Dantzer V (1985) Electron microscopy of the initial stages of placentation in the pig. Anat Embryol 172:281–293

Dantzer V, Leiser R (1993) Microvascularization of regular and irregular areolae of the areola-gland submunit of the porcine placenta: structural and functional aspects. Anat Embryol (Berl) 188:257–267

Dantzer V, Leiser R (1994) Initial vascularisation in the pig placenta: I. Demonstration of nonglandular areas by histology and corrosion casts. Anat Rec 238:177–190

Decker T, Lew DJ, Cheng YS, Levy DE, Darnell JEJ (1989) Interactions of alpha- and gamma-interferon in the transcriptional regulation of the gene encoding a guanylate-binding protein. EMBO J 8:2009–2014

Dempsey FW, Wislocki B, Amoroso EC (1955) Electron microscopy of the pig's placenta, with especial reference to the cell membranes of the endometrium and chorion. Am J Anat 91:65–101

Denhardt DT, Guo X (1993) Osteopontin: a protein with diverse functions. FASEB J 7:1475–1482

Ducsay CA, Buhi WC, Bazer FW, Roberts RM, Combs GE (1984) Role of uteroferrin in placental iron transport: effect of maternal iron treatment on fetal iron and uteroferrin content and neonatal hemoglobin. J Anim Sci 62:1303–1308

Erikson DW, Burghardt RC, Bayless KJ, Johnson GA (2009) Secreted phosphoprotein 1 (SPP1, osteopontin) binds to integrin alphavbeta6 on porcine trophectoderm cells and integrin alphavbeta3 on uterine luminal epithelial cells, and promotes trophectoderm cell adhesion and migration. Biol Reprod 81:814–825

Fazleabas AT, Kim JJ, Strakova Z (2004) Implantation: embryonic signals and the modulation of the uterine environment—a review. Placenta 25(Suppl A):S26–S31

Feinberg RF, Kliman HJ, Wang CL (1994) Transforming growth factor-beta stimulates trophoblast oncofetal fibronectin synthesis in vitro: implications for trophoblast implantation in vivo. J Clin Endocrinol Metab 78:1241–1248

Flood PF (1974) Steroid-metabolizing enzymes in the early pig conceptus and in the related endometrium. J Endocrinol 63:413–414

Ford SP, Christenson LK (1991) Direct effects of oestradiol-17 beta and prostaglandin E-2 in protecting pig corpora lutea from a luteolytic dose of prostaglandin F-2 alpha. J Reprod Fertil 93:203–209

Ford SP, Christenson RK, Ford JJ (1982) Uterine blood flow and uterine arterial, venous and luminal concentrations of oestrogens on days 11, 13, and 15 after oestrus in pregnant and non-pregnant sows. J Repord Fertil 64:185–190

Franczak A, Zmijewska A, Kurowicka B, Wojciechowicz B, Kotwica G (2010) Interleukin 1-β-induced synthesis and secretion of prostaglandin E_2 in the porcine uterus during various periods of pregnancy and the estrous cycle. J Physiol Pharmacol 61:733–742

Frank JW, Seo H, Burghardt RC, Bayless KJ, Johnson GA (2017) ITGAV (alpha V integrins) bind SPP1 (osteopontin) to support trophoblast cell adhesion. Reproduction 153:695–706

Friess AE, Sinowatz F, Skolek-Winnisch R, Trautner W (1980) The placenta of the pig. I. Finestructural changes of the placental barrier during pregnancy. Anat Embryol 158:179–191

Friess AE, Sinowatz F, Skolek-Winnisch R, Trautner W (1981) The placenta of the pig. II. Finestructural changes of the placental barrier during pregnancy. Anat Embryol 163:43–53

Gallant ND, Michael KE, García AJ (2005) Cell adhesion strengthening: contributions of adhesive area, integrin binding, and focal adhesion assembly. Mol Biol Cell 16:4329–4340

Garlow JE, Ka H, Johnson GA, Burghardt RC, Jaeger LA, Bazer FW (2002) Analysis of osteopontin at the maternal-placental interface in pigs. Biol Reprod 66:718–725

Geisert RD, Renegar RH, Thatcher WW, Roberts RM, Bazer FW (1982a) Establishment of pregnancy in the pig: I. Interrelationships between preimplantation development of the pig blastocyst and uterine endometrial secretions. Biol Reprod 27:925–939

Geisert RD, Brookbank JW, Roberts RM, Bazer FW (1982b) Establishment of pregnancy in the pig. II. Cellular remodelling of the porcine blastocysts during elongation of day 12 of pregnancy. Biol Reprod 27:941–955

Geisert RD, Thatcher WW, Roberts RM, Bazer FW (1982c) Establishment of pregnancy in the pig: III. Endometrial secretory response to estradiol valerate administered on day 11 of the estrous cycle. Biol Reprod 27:957–965

Geisert RD, Pratt TN, Bazer FW, Mayes JS, Watson GH (1994) Immunocytochemical localization and changes in endometrial progestin receptor protein during the porcine oestrous cycle and early pregnancy. Reprod Fertil Dev 6:749–760

Geisert RD, Yelich JV, Pratt T, Pomp D (1998) Expression of an inter-α-trypsin inhibitor heavy chain-like protein in the pig endometrium during the oestrous cycle and early pregnancy. J Reprod Fert 114:35–43

Geisert RD, Ross JW, Ashworth MD, White FJ, Johnson GA, Hu J, DeSilva U (2005) Maternal recognition of pregnancy signal or endocrine disruptor: the two faces of oestrogen during establishment of pregnancy in the pig. In: Kraeling R, Ashworth C (eds) Control of pig reproduction VII. Nottingham University Press, Nottingham, pp 131–145

Giancotti FG, Ruoslahti E (1999) Integrin signaling. Science 285:1028–1032

Gregoraszczuk EL, Michas N (1999) Progesterone and estradiol secretion by porcine luteal cells is influenced by individual and combined treatment with prostaglandins E2 and F2alpha throughout the estrus cycle. Prostaglandins Other Lipid Mediat 57:231–241

Guillomot M (1995) Cellular interactions during implantation in domestic ruminants. J Reprod Fertil Suppl 49:39–51

Guthrie HD, Lewis GS (1986) Production of prostaglandin F2α and estrogen by embryonal membranes and endometrium and metabolism of prostaglandin F2α by embryonal membranes, endometrium and lung from gilts. Domest Anim Endocrinol 3:185–198

Harney JP, Bazer FW (1989) Effect of porcine conceptus secretory proteins on interestrous interval and uterine secretion of prostaglandins. Biol Reprod 41:277–284

Hettinger AM, Allen MR, Zhang BR, Goad DW, Malayer JR, Geisert RD (2001) Presence of the acute phase protein, bikunin, in the endometrium of gilts during estrous cycle and early pregnancy. Biol Reprod 65:507–513

Hilton WA (1902) The morphology and development of intestinal folds and villi in vertebrates. Am J Anat 1:459–504

Humphries JD, Byron A, Humphries MJ (2006) Integrin ligands at a glance. J Cell Sci 119:3901–3903

Hynes RO (1987) Integrins: a family of cell surface receptors. Cell 48:549–554

Jaeger LA, Spiegel AK, Ing NH, Johnson GA, Bazer FW, Burghardt RC (2005) Functional effects of transforming growth factor beta (TGFβ) on adhesive properties of porcine trophectoderm. Endocrinology 146:3933–3942

Jenkins RG, Su X, Su G, Scotton CJ, Camerer E, Laurent GJ, Davis GE, Chambers RC, Matthay MA, Sheppard D (2006) Ligation of protease-activated receptor 1 enhances alph(v)beta6 integrin-dependent TGF-beta activation and promotes acute lung injury. J Clin Invest 116:1601–1614

Johansson S, Svineng G, Wennerberg K, Armulik A, Lohikangas L (1997) Fibronectin-integrin interactions. Front Biosci 2:d126–d146

Johnson GA, Spencer TE, Burghardt RC, Bazer FW (1999a) Ovine osteopontin. I. Cloning and expression of mRNA in the uterus during the peri-implantation period. Biol Reprod 61:884–891

Johnson GA, Burghardt RC, Spencer TE, Newton GR, Ott TL, Bazer FW (1999b) Ovine osteopontin II. Osteopontin and alpha(v)beta(3) integrin expression in the uterus and conceptus during the peri-implantation period. Biol Reprod 61:892–899

Johnson GA, Burghardt RC, Bazer FW, Spencer TE (2003) Minireview: osteopontin: roles in implantation and placentation. Biol Reprod 69:1458–1471

Johnson GA, Bazer FW, Burghardt RC, Spencer TE, Wu G, Bayless KJ (2009) Conceptus-uterus interactions in pigs: endometrial gene expression in response to estrogens and interferons from conceptuses. Soc Reprod Fertil Suppl 66:321–332

Johnson GA, Burghardt RC, Bazer FW (2014) Osteopontin: a leading candidate adhesion molecule for implantation in pigs and sheep. J Anim Sci Biotechnol 5:56–70

Joyce MM, White FJ, Burghardt RC, Muñiz JJ, Spencer TE, Bazer FW, Johnson GA (2005a) Interferon stimulated gene 15 conjugates to endometrial cytosolic proteins and is expressed at the uterine-placental interface throughout pregnancy in sheep. Endocrinology 146:675–684

Joyce MM, Gonzalez JF, Lewis S, Woldesenbet S, Burghardt RC, Newton GR, Johnson GA (2005b) Caprine uterine and placental osteopontin expression is distinct among epitheliochorial implanting species. Placenta 26:160–170

Joyce MM, Burghardt RC, Geisert RD, Burghardt JR, Hooper RN, Ross JW, Ashworth MD, Johnson GA (2007a) Pig conceptuses secrete estrogen and interferons to differentially regulate uterine STAT1 in a temporal and cell type-specific manner. Endocrinology 148:4420–4431

Joyce MM, Burghardt JR, Burghardt RC, Hooper RN, Jaeger LA, Spencer TE, Bazer FW, Johnson GA (2007b) Pig conceptuses increase uterine interferon regulatory factor-1 (IRF-1), but restrict expression to stroma through estrogen-induced IRF-2 in luminal epithelium. Biol Reprod 77:292–302

Joyce MM, Burghardt JR, Burghardt RC, Hooper RN, Bazer FW, Johnson GA (2008) Uterine major histocompatibility class I molecules and beta 2 microglobulin are regulated by progesterone and conceptus interferons during pig pregnancy. J Immunol 181:2494–2505

Kaczynski P, Kowalewski MP, Waclawik A (2016) Prostaglandin F2alpha promotes angiogenesis and embryo-maternal interactions during implantation. Reproduction 151:539–552

Keys JL, King GL (1980) Microscopic examination of porcine conceptus-maternal interface between days 10 and 19 of pregnancy. Am J Anat 188:221–238

Kimber SJ, Spanswick C (2000) Blastocyst implantation: the adhesion cascade. Semin Cell Dev Biol 11:77–92

Kimber SJ, Illingworth IM, Glasser SR (1995) Expression of carbohydrate antigens in the rat uterus during early pregnancy and after ovariectomy and steroid replacement. J Reprod Fertil 103:75–87

Kimmins S, Lim HC, MacLaren LA (2004) Immuohistochemical localization of integrin alpha v and beta 3 and osteopontin suggests that they do not interact during embryo implantation in ruminants. Reprod Biol Endocrinol 2:19–32

Kling D, Fingerle J, Harlan JM (1992) Inhibition of leukocyte extravasation with a monoclonal antibody to CD18 during formation of experimental intimal thickening in rabbit carotid arteries. Arterioscler Thromb 12:997–1007

Knight JW, Bazer FW, Wallace HD (1974) Effect of progesterone induced increase in uterine secretory activity on development of the porcine conceptus. J Anim Sci 39:743–746

Knight JW, Bazer FW, Thatcher WW, Franke DE, Wallace HD (1977) Conceptus development in intact and unilaterally hysterectomized ovariectomized gilts: interrelations among hormonal status, placental development, fetal fluids and fetal growth. J Anim Sci 44:620–637

La Bonnardière C, Martinat-Botté F, Terqui M, Lefèvre F, Zouari K, Martal J, Bazer FW (1991) Production of two species of interferon by Large White and Meishan pig conceptuses during the peri-attachment period. J Reprod Fertil 91:469–478

Larsen M, Artym VV, Green JA, Yamada KM (2006) The matrix reorganized: extracellular matrix remodeling and integrin signaling. Curr Opin Cell Biol 18:463–471

Lawrence DA (1996) Transforming growth factor-β: a general review. Eur Cytokine Netw 7:363–374

Lefèvre F, Martinat-Botté F, Locatelli A, De Niu P, Terqui M, La Bonnardière C (1998a) Intrauterine infusion of high doses of pig trophoblast interferons has no antiluteolytic effect in cyclic gilts. Biol Reprod 58:1026–1031

Lefèvre F, Guillomot M, D'Andrea S, Battegay S, La Bonnardière C (1998b) Interferon-delta: the first member of a novel type I interferon family. Biochimie 80:779–788

Leiser R, Dantzer V (1994) Initial vascularization in the pig placenta: II. Demonstration of gland and areola-gland subunits by histology and corrosion casts. Anat Rec 238:326–334

Lessey BA (2002) Adhesion molecules and implantation. J Reprod Immunol 55:101–112

Massuto DA, Kneese EC, Johnson GA, Hooper NH, Burghardt RC, Ing NH, Jaeger LA (2009a) Transforming growth factor beta (TGFB) signaling is activated during porcine implantation: proposed role for latency associated peptide-integrins at the conceptus-maternal interface. Reproduction 139:465–478

Massuto DA, Hooper RN, Kneese EC, Johnson GA, Ing NH, Weeks BR, Jaeger LA (2009b) Intrauterine infusion of latency associated peptide (LAP) during early porcine pregnancy affects conceptus elongation and placental size. Biol Reprod 82:534–542

Mathew DJ, Newsom EM, Guyton JM, Tuggle CK, Geisert RD, Lucy MC (2015) Activation of the transcription factor nuclear factor-kappa B in uterine luminal epithelial cells by interleukin 1 beat 2L, a novel interleukin 1 expressed by the elongating pig conceptus. Biol Reprod 107:1–13

Meyer AE, Pfeiffer CA, Brooks KE, Spate LD, Benne JA, Cecil R, Samuel MS, Murphy CN, Behura S, McLean MK, Ciernia LA, Smith MF, Whitworth KM, Wells KD, Spencer TE, Prather RS, Geisert RD (2019) New perspective on conceptus estrogens in maternal recognition and pregnancy establishment in the pig. Biol Reprod 101:148–161

Mirando MA, Harney JP, Beers S, Pontzer CH, Torres BA, Johnson HM, Bazer FW (1990) Onset of secretion of proteins with antiviral activity by pig conceptuses. J Repord Fertil 88:197–203

Mirkin S, Arslan M, Churikov D, Corica A, Diaz JI, Williams S, Bocca S, Oehninger S (2005) In search of candidate genes critically expressed in the human endometrium during the window of implantation. Hum Reprod 20:2104–2117

Mossman HW (1937) Comparative morphogenesis of the foetal membranes and accessory uterine structures. Contrib Embryol 26:129–246

Mulac-Jericevic B, Conneely OM (2004) Reproductive tissue selective actions of progesterone receptors. Reproduction 128:139–146

Pfeiffer CA, Meyer AE, Brooks KE, Chen PR, Milano-Foster J, Spate LD, Benne JA, Cecil RF, Samuel MS, Ciernia LA, Spinka CM, Smith MF, Wells KD, Spencer TE, Prather RS, Geisert RD (2020) Ablation of conceptus PTGS2 expression does not alter early conceptus development and establishment of pregnancy in the pig. Biol Reprod 102(2):475–488

Pontzer CH, Torres BA, Vallet JL, Bazer FW, Johnson HM (1988) Antiviral activity of the pregnancy recognition hormone ovine trophoblast protein-1. Biochem Biophys Res Commun 152:801–807

Powell JK, Glasser SR, Woldesenbet S, Burghardt RE, Newton GR (2000) Expression of carbohydrate antigens in the goat uterus during early pregnancy and on steroid-treated polarized uterine epithelial cells *in vitro*. Biol Reprod 62:277–284

Red-Horse K, Zhou Y, Genbacev O, Prakobphol A, Foulk R, McMaster M, Fisher SJ (2004) Trophoblast differentiation during embryo implantation and formation of the maternal-fetal interface. J Clin Invest 114:744–754

Renegar RH, Bazer FW, Roberts RM (1982) Placental transport and distribution of uteroferrin in the fetal pig. Biol Reprod 27:1247–1260

Roberts RM, Bazer FW (1988) The functions of uterine secretions. J Reprod Fertil 82:875–892

Roberts RM, Raub TJ, Bazer FW (1986) Role of uteroferrin in transplacental iron transport in the pig. Fed Proc 45:2513–2518

Ross JW, Ashworth MD, Hurst AG, Malayer JR, Geisert RD (2003) Analysis and characterization of differential gene expression during rapid trophoblastic elongation in the pig using suppression subtractive hybridization. Reprod Biol Endocrinol 1:23

Ross JW, Ashworth MD, White FJ, Johnson GA, Aoubi PJ, DeSilva UE, Whitworth KM, Green JA, Prather RS, Geisert RD (2007) Premature estrogen exposure alters endometrial gene expression to disrupt pregnancy in the pig. Endocrinology 148:4761–4773

Ruoslahti E, Pierschbacher MD (1987) New perspectives in cell adhesion: RGD and integrins. Science 238:491–497

Sastry SK, Burridge K (2000) Focal adhesions: a nexus for intracellular signaling and cytoskeletal dynamics. Exp Cell Res 261:25–36

Senger DR, Perruzzi CA, Papadopoulos-Sergiou A, Van de Water L (1994) Adhesive properties of osteopontin: regulation by a naturally occurring thrombin-cleavage in close proximity to the GRGDS cell-binding domain. Mol Biol Cell 5:565–574

Seo H, Li X, Wu G, Bazer FW, Burghardt RC, Bayless KJ, Johnson GA (2020) Mechanotransduction drives morphogenesis to develop folding at the uterine-placental interface of pigs. Placenta 90:62–70

Simons BD (2013) Development. Getting your gut into shape. Science 342:203–204

Sodek J, Ganss B, McKee MD (2000) Osteopontin. Crit Rev Oral Biol Med 11:279–303

Song G, Bailey DW, Dunlap KA, Burghardt RC, Spencer TE, Bazer FW, Johnson GA (2010) Cathepsin B, cathepsin L and cystatin C in the porcine uterus and placenta: potential roles in endometrial/placental remodeling and in fluid-phase transport of proteins secreted by uterine epithelia across placental areolae and neonatal gut. Biol Reprod 82:854–864

Spencer TE, Bartol FF, Bazer FW, Johnson GA, Joyce MM (1999) Identification and characterization of glycosylation dependent cell adhesion molecule 1 (GlyCAM-1) expression in the ovine uterus. Biol Reprod 60:241–250

Spencer TE, Johnson GA, Bazer FW, Burghardt RC (2004) Implantation mechanisms: insights from the sheep. Reproduction 128:656–668

Spencer TE, Johnson GA, Bazer FW, Burghardt RC, Palmarini M (2007a) Pregnancy recognition and conceptus implantation in domestic ruminants: roles of progesterone, interferons and endogenous retroviruses. Reprod Fertil Dev 19:65–78

Spencer TE, Johnson GA, Bazer FW, Burghardt RC (2007b) Fetal-maternal interactions during the establishment of pregnancy in ruminants. Reproduction 64:379–396

Steinhauser CB, Bazer FB, Burghardt RC, Johnson GA (2017) Expression of progesterone receptor in the porcine uterus and placenta throughout gestation: correlation with expression of uteroferrin and osteopontin. Domestic Anim Endocrinol 58:1–11

Sukjumlong S, Dalin A-M, Sahlin L, Persson E (2005) Immunohistochemical studies on the progesterone receptor (PR) in the sow uterus during the oestrous cycle and in inseminated sows at oestrus and early pregnancy. Reproduction 129:349–359
Tuo WB, Bazer FW (1996) Expression of oncofetal fibronectin in porcine conceptuses and uterus throughout gestation. Reprod Fertil Dev 8:1207–1213
Tuo W, Harney JP, Bazer FW (1996) Developmentally regulated expression of interleukin-1b by peri-implantation conceptuses in swine. J Reprod Immunol 31:185–198
Vallet JL, Freking BA (2007) Differences in placental structure during gestation associated with large and small pig fetuses. J Anim Sci 85:3267–3275
Vallet JL, Miles JR, Freking BA (2009) Development of the pig placenta. Soc Reprod Fertil Suppl 66:265–279
Veldius JP, Klase P, Hammond JM (1980) Characteristic of changes in volume, osmolarity and electrolyte composition of porcine fetal fluids during gestation. Biol Reprod 22:1168–1180
Vonnahme KA, Malayer JR, Spivey HO, Ford SP, Clutter A, Geisert RD (1999) Detection of kallikrein gene expression and enzymatic activity in porcine endometrium during the estrous cycle and early pregnancy. Biol Reprod 61:1235–1241
Waclawik A, Jabbour HN, Blitek A, Ziecik AJ (2009a) Estradiol-17beta, prostaglandin E2 (PGE2), and the PGE2 receptor are involved in PGE2 positive feedback loop in the porcine endometrium. Endocrinology 150:3823–3832
Waclawik A, Blitek A, Kaczmarek MM, Kiewisz J, Ziecik AJ (2009b) Antiluteolytic mechanisms and the establishment of pregnancy in the pig. Soc Reprod Fertil Suppl 66:307–320
Waclawik A, Kaczemarek MM, Blitek A, Kaczynski P, Ziecik AJ (2017) Embryo-maternal dialogue during pregnancy establishment and implantation in the pig. Mol Reprod Dev 84:842–855
Wetendorf M, DeMayo FJ (2014) Progesterone receptor signaling in the initiation of pregnancy and preservation of a healthy uterus. Int J Dev Biol 58:95–106
White FJ, Ross JW, Joyce MM, Geisert RD, Burghardt RC, Johnson GA (2005) Steroid regulation of cell specific secreted phosphoprotein 1 (osteopontin) expression in the pregnant porcine uterus. Biol Reprod 73:1294–1301
White FJ, Burghardt RC, Hu J, Joyce MM, Spencer TE, Johnson GA (2006) Secreted phosphoprotein 1 (osteopontin) is expressed by stromal macrophages in cyclic and pregnant endometrium of mice, but is induced by estrogen in luminal epithelium during conceptus attachment for implantation. Reproduction 132:919–929
Whyte JJ, Meyer AE, Spate LCD, Benne JA, Cecil R, Samuel MS, Murphy CN, Prather RS, Geisert RD (2018) Inactivation of porcine interleukin-1β results in failure of rapid conceptus elongation. Proc Natl Acad Sci U S A 115:307–312
Wildeman DE, Chen C, Erez O, Grossman LI, Goodman M, Romero R (2006) Evolution of the mammalian placenta revealed by phylogenetic analysis. PNAS 103:3203–3208
Wozniak MA, Modzelewska K, Kwong L, Keely PJ (2004) Focal adhesion regulation of cell behavior. Biochim Biophys Acta 1692:103–119
Ziecik AJ, Waclawik BM (2008) Conceptus signals for establishment and maintenance of pregnancy in pigs – lipid singnaling system. Exper Clin Endocrinol Diabetes 116:443–449

Placentation in Equids

Douglas F. Antczak and W. R. (Twink) Allen

Contents

Abstract This chapter focuses on the early stages of placental development in horses and their relatives in the genus *Equus* and highlights unique features of equid reproductive biology. The equine placenta is classified as a noninvasive, epitheliochorial type. However, equids have evolved a minor component of invasive

---

Dedicated to the memory of Professor W. R. (Twink) Allen, FRCVS, ScD, CBE, a pioneer in equine reproductive biology.

D. F. Antczak (✉)
Department of Microbiology and Immunology, College of Veterinary Medicine, Baker Institute for Animal Health, Cornell University, Ithaca, NY, USA
e-mail: doug.antczak@cornell.edu; dfa1@cornell.edu

W. R. (Twink) Allen *(deceased)*
Sharjah Equine Hospital, Sharjah, United Arab Emirates

Robinson College, University of Cambridge, Cambridge, UK

The Paul Mellon Laboratory of Equine Reproduction, 'Brunswick', Newmarket, Suffolk, UK

R. D. Geisert, T. E. Spencer (eds.), *Placentation in Mammals*, Advances in Anatomy, Embryology and Cell Biology 234, https://doi.org/10.1007/978-3-030-77360-1_6

trophoblast, the chorionic girdle and endometrial cups, which links the equine placenta with the highly invasive hemochorial placentae of rodents and, particularly, with the primate placenta. Two types of fetus-to-mother signaling in equine pregnancy are mediated by the invasive equine trophoblast cells. First, endocrinological signaling mediated by equine chorionic gonadotrophin (eCG) drives maternal progesterone production to support the equine conceptus between days 40 and 100 of gestation. Only in primates and equids does the placenta produce a gonadotrophin, but the evolutionary paths taken by these two groups of mammals to produce this placental signal were very different. Second, florid expression of paternal major histocompatibility complex (MHC) class I molecules by invading chorionic girdle cells stimulates strong maternal anti-fetal antibody responses that may play a role in the development of immunological tolerance that protects the conceptus from destruction by the maternal immune system. In humans, invasive extravillous trophoblasts also express MHC class I molecules, but the loci involved, and their likely function, are different from those of the horse. Comparison of the cellular and molecular events in these disparate species provides outstanding examples of convergent evolution and co-option in mammalian pregnancy and highlights how studies of the equine placenta have produced new insights into reproductive strategies.

Keywords Horse · Placenta · Trophoblast · Chorionic gonadotrophin · Immunology · Endometrial cups

Abbreviations

CL	Corpus luteum
eCG	Equine chorionic gonadotrophin
MHC	Major histocompatibility complex

1 Introduction

It is a great pleasure and an honor to contribute to this volume dedicated to the memory of Professor Emmanuel Ciprian Amoroso (Amo). One of the authors (WRA) knew Amo well, while the other (DFA) met him on only a few occasions, but even in a short meeting Amo left a lasting impression. Here we have attempted to present some of the fascinating aspects of placentation in the horse and related species of the genus *Equus*. Many of the advances of the past six decades in our understanding of reproduction in equids have been built on the foundation in Amo's classic chapter on placentation (Amoroso 1958).

Several aspects of embryo development and placentation in the mare and other female equids are unique to the genus *Equus* and are of great scientific puzzlement. And they are made all the more interesting by virtue of the fact that each of the physically and karyotypically diverse members of the genus, from the ancient Przewalski's Horse of Mongolia ($2n = 66$) to all the breeds of domestic horse ($2n = 64$), then to the many breeds of domestic donkey ($2n = 62$), through the Kulan and Kiang at $2n = 60$ and 58, respectively, to the Somali Wild Ass at $2n = 56$, followed by a further reduction in chromosome number in the beautiful Grevy's zebra in Northern Kenya at $2n = 46$, then to the Grant's or Common zebra throughout Central and Southern Africa at $2n = 44$, and with a final decline to Mrs. Hartman's Mountain zebra in the Cape Colony of South Africa at only $2n = 32$ (Fig. 1), can not only interbreed with all the other member species to conceive and produce viable, although sterile, hybrid offspring (Gray 1972), but can also carry to term extraspecific pregnancies created by embryo transfer (Fig. 2) (Allen and Short 1997). The mule (female horse X male donkey), which has been bred and used successfully as a pack animal for thousands of years past, may reasonably be called man's first successful attempt at genetic engineering (Short 1975; Clutton-Brock 1992), and these sterile hybrids can also serve as surrogate mothers for horse or donkey embryos (Fig. 2) (Davies et al. 1985).

This unusual ability of so many phenotypically and karyotypically diverse species to interbreed successfully and carry both hybrid and extra-specific offspring safely to term is made all the more surprising by a unique feature of early equine placentation. Namely, the endometrial cup reaction (Cole and Goss 1943; Clegg et al. 1954; Allen 1975b), which involves active invasion of the maternal endometrium by specialized trophoblast cells of the chorionic girdle region of the developing conceptus that stimulate strong humoral and cell-mediated responses from the maternal immune system which appear not to be in any way detrimental to the establishment or progress of pregnancy. The physical, hormonal, and immunological intricacies of this unique and puzzling component of equine placentation form a significant proportion of the present chapter. How and why do female equids produce gonadotrophin-secreting endometrial cups in early pregnancy? The endometrial cups held great fascination for Amo (Amoroso 1958), but at that time the cups were considered to be maternal, not fetal, in origin.

Fig. 1 Selection of the equine species. (**a**) Przewalski's Ancient Horse of Mongolia, $2n = 66$; (**b**) Shire Horse and Welsh Pony domestic horses, $2n = 64$; (**c**) Domestic donkeys, $2n = 62$; (**d**) Somali Wild Ass, $2n = 56$; (**e**) Grevy's zebra, $2n = 46$; (**f**) Common or Grant's zebra, $2n = 44$

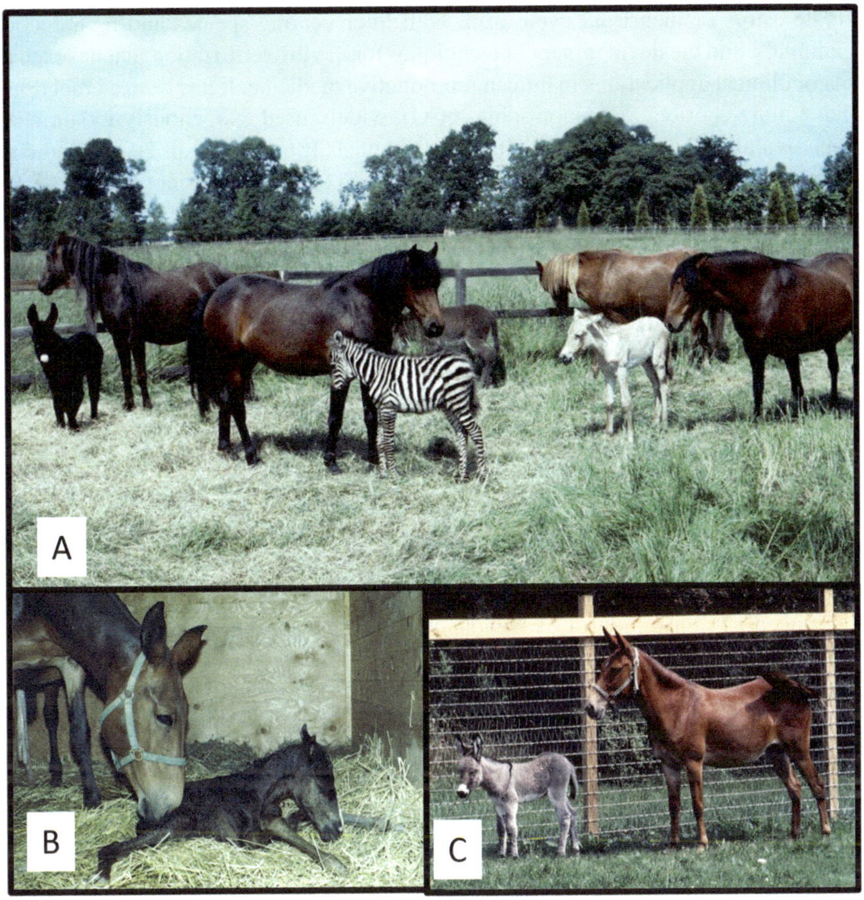

Fig. 2 The plasticity of equine reproduction. (**a**) Extraspecies equids created by transfer of Przewalski's horse ($2n = 66$), donkey ($2n = 62$), and Grant's zebra ($2n = 44$) embryos to domestic pony ($2n = 64$) recipient mares. (**b** and **c**) Horse and donkey foals born to sterile female mules after embryo transfer

2 Early Embryonic Development

Amoroso paid scant attention to the earliest stages of placental development and, indeed, very little was known about this topic at the time of his writing. Over the past 50 years, there has been an explosion of theory and practical knowledge about fertilization, zygotic diploid genome activation, epigenetic reprogramming, morula and blastocyst formation, and the induction of the trophectoderm, the earliest cell layer to develop (Piliszek and Madeja 2018; Posfai et al. 2019; Gerri et al. 2020; Toyooka 2020). Many of these advances were facilitated by methods of superovulation that enabled production of large numbers of fertilized or unfertilized ova in a single estrus or menstrual cycle from both litter bearing species and monotocous mammals, and the development of techniques for in vitro fertilization that have led to major clinical applications in human reproductive medicine. It has been a cruel irony that equine chorionic gonadotrophin (eCG), widely used exogenously to stimulate superovulation in many species, including humans (Sommer et al. 2007; Medicine 2008; Takeo and Nakagata 2015; Naranjo Chacón et al. 2020), is not efficacious for this purpose in the horse.

The ineffectiveness of superovulation in the horse has an anatomical cause that is now well understood. In female equids, the fibrosity and general toughness of the tunica albuginea that envelops the exterior of each ovary physically prevents ovulation of the large (4–5 cm) estrus follicle through this exterior surface as occurs in all other mammalian species. Instead, the follicle must migrate through the ovarian stroma in order to ovulate through the ventrally situated tunica-free "ovulation fossa" (Fig. 3a). This, in turn, means that the relatively small fimbria of the oviduct in the mare only has to surround the ovulation fossa to "collect" the ovulated oocyte and pass it on down the oviduct. However, this tough and impenetrable tunica albuginea in the mare imposes the practical disadvantage in modern-day horse breeding of effective prevention of the induction of superovulation in donor mares in embryo transfer programs. Even if multiple follicles are stimulated to develop by prolonged administration of exogenous gonadotrophins (Allen and Wilsher 2020), the growing follicles compete with one another to migrate towards the ovulation fossa with the result that they luteinize before they are able to ovulate. All manner of studies have been undertaken to attempt to induce multiple ovulations in embryo transfer donor mares, all of which have proved disappointingly negative. Even a recent advance in using recombinant-derived equine FSH and LH to successfully stimulate follicle growth in anestrous and cycling mares (Roser and Meyers-Brown 2019) has not been able to induce superovulation that results in repeatable recovery of multiple, viable horse embryos. Thus, fertile superovulation of embryo transfer donor mares, as is achieved routinely in other domestic animal species and women (Devroey et al. 2004; Fonseca et al. 2016; Mikkola et al. 2019; Hansen 2020), remains a sought-after pipedream in the mare.

In addition to the relative ineffectiveness of superovulation in mares, a second problem has further impeded progress in research on early horse embryo development; that is, the difficulty in producing embryos by in vitro fertilization (IVF). Despite attempts by equine embryologists in many laboratories around the world, only two successful births of foals after IVF have been reported (Palmer et al. 1991). The reasons for this failure are not fully understood but may relate to unresolved problems with in vitro capacitation of equine spermatozoa (Leemans et al. 2016). The persistent work of equine reproductive specialists has paid off in another way—through the development of highly successful techniques for ovum pickup (OPU), in vitro oocyte maturation, and intracytoplasmic sperm injection (ICSI) (Stout 2020). These techniques have fostered widespread use of Assisted Reproduction Technology (ART) in clinical equine reproduction in the breeds that allow such procedures. This has included successful freezing of embryos and routine long-distance shipping prior to embryo transfer (Hinrichs 2018; Morris 2018; Carnevale et al. 2019; Squires

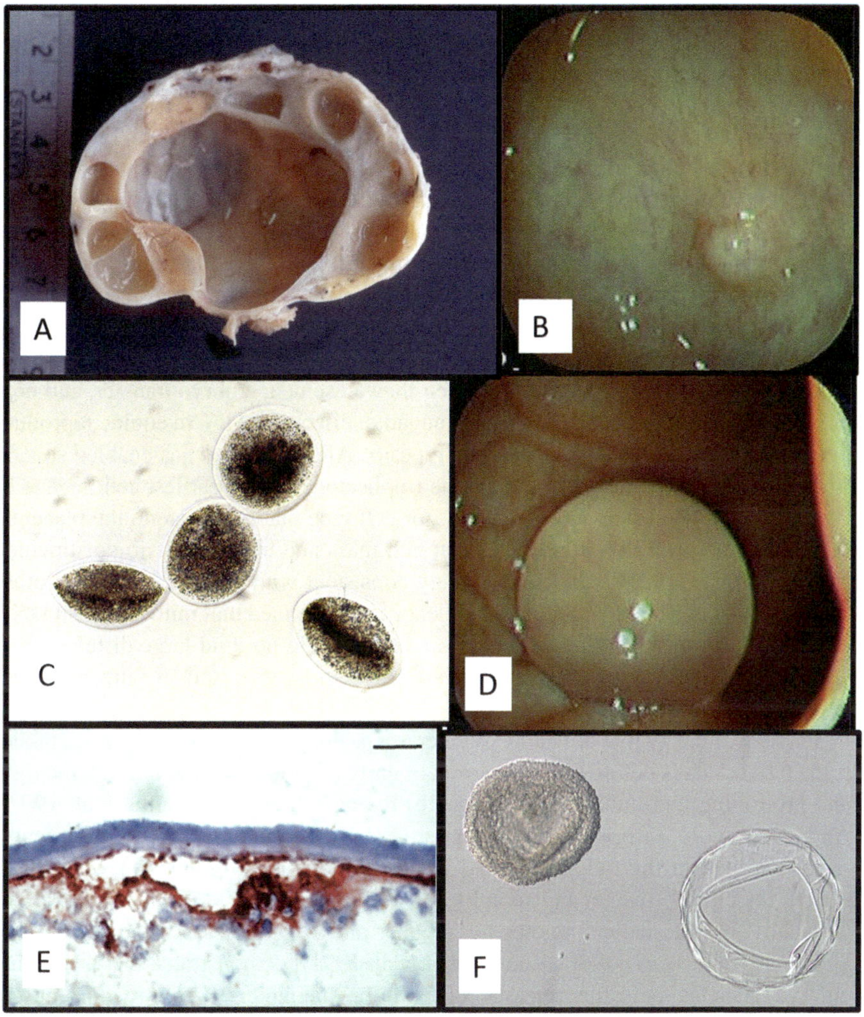

Fig. 3 Images and stages of early equine pregnancy. (**a**) A sectioned ovary from an estrus mare showing a preovulatory follicle protruding towards the ovulation fossa. (**b**) The tight and protruding uterotubal papilla at the tip of a mare's uterine horn, as viewed using video-endoscopy. (**c**) Four degenerate oocytes flushed from a cycling mare's oviduct, viewed through a dissecting microscope. (**d**) A spherical, unattached embryo in the uterus of a mare at day 13 after ovulation, viewed using video-endoscopy. (**e**) Cryostat section of the embryonic acellular capsule surrounding the underlying trophectoderm of a day 8 blastocyst, labeled with anti-trophoblast antibody 102.1 (Photograph kindly supplied by Dr. Julio Oriol). (**f**) A collapsed blastocyst capsule (right) after its removal from a day 8 embryo

2020). And in a surprising twist, the procedure of cloning horses by somatic cell nuclear transfer (SCNT) was found to be quite effective in producing viable cloned offspring (Gambini and Maserati 2017). This has led to the production of hundreds of cloned horses that are performing at high levels in several disciplines, particularly in the sport of polo. Newer methods of using mesenchymal stem cells as SCNT donor cells holds promise of even greater cloning efficiencies in the future (Olivera et al. 2018).

Embryos fertilized with cryopreserved sperm showed downregulation of genes involved in oxidative phosphorylation, DNA replication, and the immune system at 8, 10, and 12 days post ovulation (Ortiz-Rodriguez et al. 2019). That study suggests that some forms of ART may have detrimental effects on embryo and placental development. However, another recent study of foals and placentas showed no significant differences between naturally conceived, embryo transfer, and ICSI offspring, except for length of the foal's upper hind limbs (Valenzuela et al. 2018). And in fact, the widespread commercial use over several decades of artificial insemination with fresh, cooled, or frozen-thawed semen, embryo transfer, and now even somatic cloning, suggests that the negative effects of ART in equine reproduction may be minor. The use of various types of ART in horses has enabled several studies of the development of the equine trophectoderm, the earliest cell lineage of the mammalian embryo, and the progenitor cell type that gives rise to the placenta. Transcriptome profiling of isolated inner cell mass and trophectoderm documented changes in gene expression that is largely consistent with results obtained in other species (Iqbal et al. 2014). Stout and colleagues determined that mitochondrial DNA replication begins at blastocyst formation, but they did not find large differences in the characteristics of mitochondrial DNA between in vivo and in vitro produced embryos (Hendriks et al. 2019).

Passage of the zygote down the oviduct is a prolonged process in the mare, taking some 6.0–6.5 days before the late morula or early blastocyst passes through the tight and protruding uterotubal papilla (Fig. 3b) to enter the uterus (Battut et al. 1997). And even more surprising and apparently equine-specific, unfertilized oocytes remain lodged in the oviduct around the ampullary–isthmus junction where they slowly degenerate over many months and even years (Fig. 3c) (van Niekerk and Gerneke 1966). Excellent studies by Webber and colleagues at Washington State University in the early 1990s [see (Weber et al. 1991)] demonstrated that the equine embryo secretes pulsatile releases of both prostaglandin F2α (smooth muscle contractant) and PGE2 (smooth muscle relaxant) from day 4 after ovulation, and these stimulate the necessary peristaltic contractions of the oviducal smooth muscle to propel the embryo on down the oviduct and through the uterotubal papilla into the uterus on day 6, while also moving on any degenerating oocytes from previous estrus cycles. Thus, even as early as day 4 after ovulation, the equine embryo has become the driver of its own onward passage, and this self-driven mobility continues unabated after the embryo enters the uterus. During passage through the oviduct, the equine embryo influences innate immune response genes in oviductal tissue, with marked upregulation of interferon-associated genes (Smits et al. 2016).

3 The Embryonic Capsule and the Migrating Equine Embryo

Once in the uterus, the steadily enlarging and still spherical embryo continues to move throughout the entire uterine lumen over the next 10 days, driven by its continuing pulsatile releases of PGF2α and PGE2 to stimulate peristaltic contractions of the myometrium (Fig. 3d) (Ginther 2014; Stout et al. 2000). Essential to this unusual process of continuing intrauterine movement is maintenance of the embryo's spherical shape by formation of the so-called blastocyst capsule (Oriol 1994) soon after the embryo has entered the uterus on day 6. This translucent proteinaceous membrane (Fig. 3e) is secreted by the trophoblast cells (Oriol et al. 1993), and it completely envelops the embryo (Fig. 3f) until around day 22 when it begins to fragment and peel off the outer surface of the conceptus. It is presumed that the capsule plays the central role in preventing elongation of the equine conceptus around days 12–14 after ovulation, as occurs in the conceptuses of other large domestic animal species (Geisert et al. 2017), as it is in direct contact with the lumenal epithelium of the endometrium during the embryo's movement around the uterine lumen. Furthermore, the estrogens secreted by the young, mobile, equine embryo have been shown not to influence the lifespan or productivity of the cyclical corpus luteum (CL) (Wilsher and Allen 2011). Therefore, it seems reasonable to suspect that some sort of physical, rather than biochemical, interaction between the capsule and the endometrium is involved in initiating the so-called "maternal recognition of pregnancy signal" in the mare which suppresses the release of PGF2α from the endometrium during days 14–16 after ovulation that would normally occur to induce luteolysis and a return to estrus in the cycling mare (Stabenfeldt et al. 1974; Allen 1981). However, the identity and mode of action of the equine "maternal recognition of pregnancy signal" remains unproven and is a matter of considerable conjecture in the equine reproduction research community at the present time (Klohonatz et al. 2015; Klein 2016; Smits et al. 2018).

4 The Invasive Component of the Equine Placenta

Current theory suggests that invasive placentation, such as found in primates and rodents, developed earlier than the diffuse forms of placenta characteristic of the ungulates, which are now considered to be derived from more invasive types (Wildman et al. 2006; Carter and Mess 2007; Carter and Enders 2013). Amoroso classified the placenta of the horse as belonging to the least-invasive epitheliochorial type, along with that of the pig (Amoroso 1958). And in fact, the histological appearance of the placentae proper of these two species is remarkably similar. Both have a microcotyledonary structure and six tissue layers separating fetal and maternal circulations. It is at this interface between mother and fetus that the main housekeeping functions of the placenta take place: the provision of nutrients and

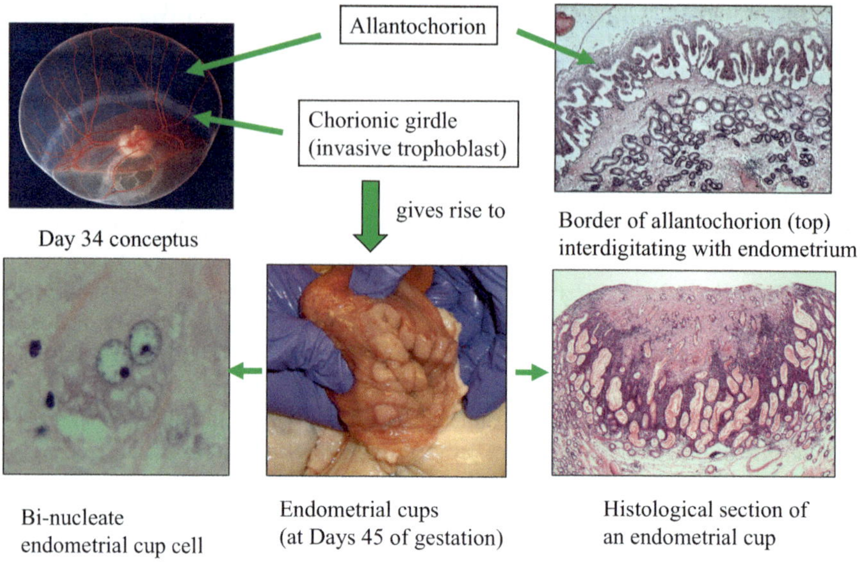

Fig. 4 Selected tissues critical to early placental development in the horse. The prominent feature of the days 30–36 equine conceptus is the chorionic girdle trophoblast (upper left image). The girdle cells invade the endometrium to form the endometrial cups. The cups are comprised of large, binucleate, terminally differentiated trophoblast cells (lower left) that secrete eCG, which can be detected in maternal blood between days 40 and 120 of gestation. The invading trophoblast cells attract numerous maternal lymphocytes around the edge of the cups (lower right). The noninvasive trophoblast of the allantochorion forms villi that interdigitate with corresponding crypts in the maternal endometrium (upper right) to comprise the maternal–fetal interface of the placenta proper

oxygen and the removal of waste (Carter and Enders 2013). However, the horse placenta has an additional component of invasive trophoblast, comprised of the chorionic girdle and endometrial cups (Fig. 4). This invasive trophoblast functions in endocrinological and immunological signaling between the conceptus and mother, as will be described below.

4.1 *Development of the Chorionic Girdle and Endometrial Cups*

Around day 16–17 after ovulation, the still-spherical and steadily expanding conceptus becomes too large to continue its intrauterine movement and is held stationary at the base of one or other of the uterine horns by an astounding increase in the general tonicity of the myometrium—the uterus literally clamps down upon the wandering conceptus, the precise stimulus for which also remains unknown. Once

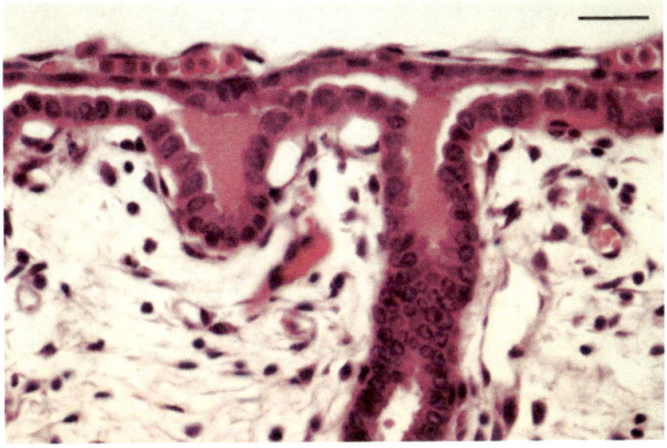

Fig. 5 Equine histiotroph. Histological section of the endometrium–trophoblast junction in a mare at day 29 of gestation showing histiotroph ("uterine milk") being exuded from the mouths of endometrial glands for uptake by the overlying allantochorion. Scale bar = 45 μm

held in place in this manner, continued steady expansion of the conceptus occurs over the next 25 days, fueled by the uptake of histiotroph secreted by the endometrial glands for absorption by its outermost layer of trophoblast cells, which lies in close contact with, but is not yet physically attached to the endometrium (Fig. 5). Amoroso referred to this histiotroph as "uterine milk" (Amoroso 1958). It is now recognized that the developing human conceptus benefits from nutrients provided by histiotroph during the first trimester, in a similar fashion to the situation in ungulates (Burton et al. 2002; Filant and Spencer 2014; Jones et al. 2015).

During this prolonged period of "detachment," the embryonic yolk sac regresses steadily and is replaced by the well-vascularized allantois, an outpouching of the hindgut that fuses with the outermost chorion to create the allantochorion that will begin to form the definitive equine placenta from day 40 onwards. However, before that commencing maternofetal attachment takes place, the curious and equine-unique endometrial cup reaction begins with development of the chorionic girdle. This starts around day 25 when a 5–10 mm wide "strip" of the trophoblast at the junction of the enlarging allantois and regressing yolk sac starts to become "ruffled" in appearance due to an increase in multiplication rate of the trophoblast cells only in that region (Fig. 6a–c). Over the next 10 days, continued rapid multiplication and elongation of these specialized trophoblast cells causes them to pile up upon one another to form a discrete band-like thickening, termed the chorionic girdle (Allen et al. 1973) which encircles the entire conceptus at the allantois–yolk sac junction. Then, between days 36 and 38, the entire thickened girdle peels off the fetal membranes and attaches itself to the opposing endometrium (Fig. 7a) where the now-binucleate (Wooding et al. 2001) trophoblast cells vigorously disrupt and destroy the lumenal epithelium before passing down the lumenae of the endometrial glands, dislodging the lining epithelium as they do so (Fig. 7b). Initially, they remain contained within the basement membrane of the glands, but soon break through to stream out into the endometrial stroma (Fig. 7c) where they enlarge further, round up, and lose their invasive properties to become closely

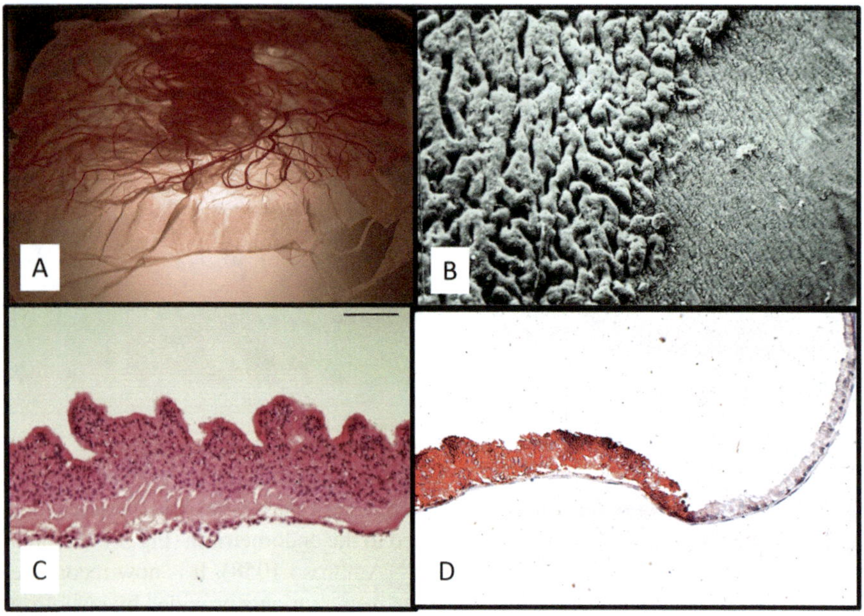

Fig. 6 Images of the chorionic girdle trophoblast. (**a**) Petri dish containing a non-surgically recovered horse conceptus at day 34 of gestation showing the discrete, thickened chorionic girdle at the junction of the enlarging allantois (above) and the regressing yolk sac (below). (**b**) A scanning electron micrograph of the 34-day chorionic girdle, showing piled-up invasive trophoblast cells on the left and the smooth surface of the expanding allantochorion membrane on the right. (**c**) Formalin-fixed section of chorionic girdle from day 33 of gestation showing rapidly dividing and piled-up girdle trophoblast cells. (Scale bar = 50 μm). (**d**) Cryostat section of day 33 chorionic girdle (left side) and chorion (right side) trophoblast labeled with antibody to equine major histocompatibility complex class I molecule. The invasive chorionic girdle is strongly labeled, indicating high expression of MHC class I, while the chorion is negative

packed together in the endometrial stroma (Fig. 7d), thereby forming a series of discrete, endometrial plaques of variable size, the so-called endometrial cups, arranged in a circle around the still—spherical conceptus at the base of the gravid uterine horn (Fig. 8a, b). The width or diameter of these endometrial cup plaques is dependent upon the width of the progenitor chorionic girdle, which can vary greatly between the different equine species (Allen 1982), and their variation in length originates from the degree of folding of the endometrium at the time of initial invasion of the chorionic girdle—numerous small cups develop only on the tips of the folds of very undulating endometrium whereas longer, unbroken strips of endometrial cup form when the endometrium is less folded and flatter at the time of girdle invasion (Allen 1975b). The cups persist for only about 60 to 80 days and have usually disappeared from the uterus by day 120 of the 330-day gestation of the mare.

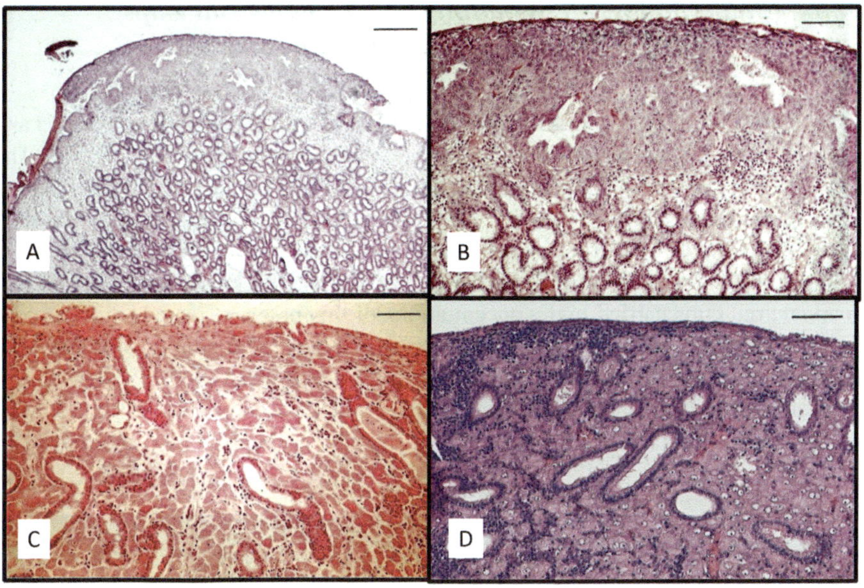

Fig. 7 Histology of endometrial cup formation. (**a**) Low power section of a day 38 chorionic girdle having just attached itself to the maternal endometrium (Scale bar = 250 μm). (**b**) Section of 38-day chorionic girdle attached to the endometrium. The lumenal epithelium of the endometrium has been obliterated and the large, binucleate girdle cells are seen passing down the endometrial glands as they dislodge and obliterate the glandular epithelium (Scale bar = 170 μm). (**c**) Developing endometrial cup at day 40 of gestation showing the enlarging binucleate chorionic girdle cells streaming out into the endometrial stroma having broken through the walls of the endometrial glands. Appreciable numbers of leucocytes are already beginning to accumulate around the invading trophoblast cells (Scale bar = 170 μm). (**d**) A newly formed endometrial cup at day 42 of gestation. The enlarged fetal cup cells are now becoming tightly packed together in the endometrial stroma and the number of accumulated leucocytes is beginning to diminish (Scale bar = 150 μm). H&E staining of formalin-fixed tissues

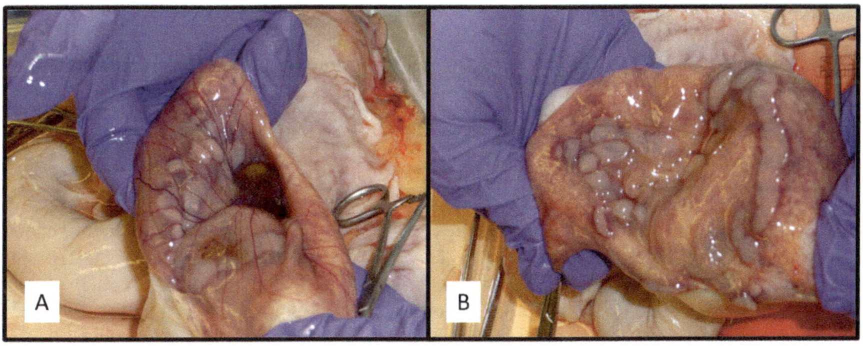

Fig. 8 Endometrial cups in situ at 42 days of gestation. (**a**) Everted uterus showing allantochorionic membranes overlying the endometrium in an area with numerous endometrial cups. (**b**) The same specimen with the allantochorion removed, allowing better visualization of a typical pattern of endometrial cups showing small, individual cups and longer strips of cup tissue. The variation in cup size results from the gross morphological variation in the endometrial surface when the girdle cells invade into a flattened or folded area of the endometrium

4.2 *Endocrinological Signaling: The Endometrial Cups and Equine Chorionic Gonadotrophin*

The chorionic girdle was first described in 1897 by Cossar Ewart (Ewart 1897) and accompanied by an astonishingly accurate drawing (compare Fig. 9a, b). The endometrial cups were first reported not long afterward (Schauder 1912). A function was attributed to the cups in the 1940s, when Cole and colleagues described the gonad-stimulating properties present in the serum of pregnant mares between days 40 and 120 of gestation (Cole and Hart 1930), and later demonstrated that the endometrial cups carried high concentrations of this pregnant mare serum gonadotrophin (PMSG), now known as equine chorionic gonadotrophin (eCG), and were thus the likely source of this active hormone (Cole and Goss 1943). By that time, it was already known that the human placenta was the source of a substance with similar gonadotrophic actions, now called human chorionic gonadotrophin (hCG)

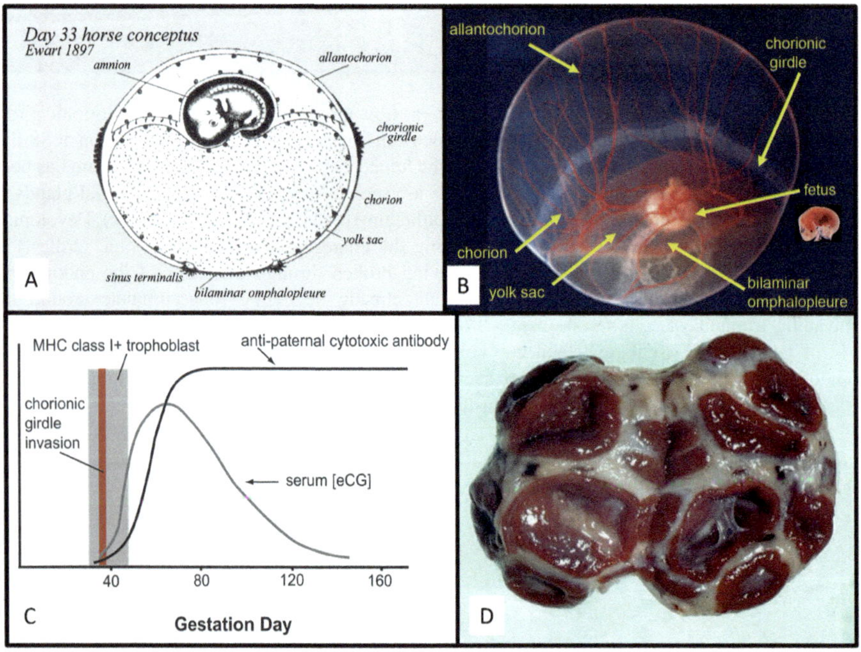

Fig. 9 Components of the days 32–36 equine conceptus and consequences of equine trophoblast invasion. (**a**) 1897 drawing of day 33 equine conceptus by Ewart, who discovered the chorionic girdle (Ewart 1897). (**b**) Day 33 horse conceptus recovered non-surgically with tissues labelled per Ewart (image courtesy of Dr. Julio Oriol). (**c**) Temporal plot of elements of early equine pregnancy, showing duration of MHC class I expression by the invasive trophoblast of the chorionic girdle cells and early endometrial cups and stylized curves of appearance and duration of maternal antibody to paternal MHC class I antigens and eCG in maternal blood. (**d**) Sectioned ovary from a mare at 102 days of gestation showing multiple accessory corpora lutea formed from eCG-induced secondary ovulations and luteinization of unruptured follicles after day 40

[reviewed by (Cole 2010)], but the link between the chorionic girdle and endometrial cups was not appreciated, and the origin of the endometrial cups was a mystery that eluded even Amoroso, who repeated past suppositions that the cups were of maternal, and not fetal, origin (Amoroso 1958). The connection between the chorionic girdle trophoblasts and the endometrial cups was not made until the early 1970s, when one of the present authors, W. R. Allen, and his colleagues described how the endometrial cups are formed by invasion of the endometrium by the chorionic girdle cells (Allen and Moor 1972; Allen et al. 1973; Hamilton et al. 1973). This work explained the origin of the endometrial cups and it also created a framework for understanding their function.

Only two groups of mammals—primates and equids—produce a placental gonadotrophin. The primary function of both eCG and hCG is to promote a maternal hormonal environment that will maintain pregnancy. In the mare, this is accomplished by stimulating the maternal ovarian follicle growth and luteinization that supplies progesterone early in gestation. The target receptors of eCG are thus on maternal, not fetal, tissues. This may explain why the only placental cells that produce eCG are the mature endometrial cup trophoblasts that have embedded themselves within maternal tissue and which are separated entirely from the developing placenta and fetus. Levels of this high molecular weight glycoprotein hormone (76,000) (Papkoff 1981), which expresses both FSH-like and LH-like biological activities (Evans and Irvine 1975), peak in maternal blood at around 65–75 days of gestation and thereafter decline steadily to finally disappear completely between 120 and 150 days (Fig. 9c). Simultaneously with the onset of eCG secretion by the developing endometrial cups around day 38–40 of gestation, the first of what becomes a series of eCG-induced secondary or accessory corpora lutea develops in the maternal ovaries, either from normal ovulation or by luteinization of an unruptured mature follicle which has been stimulated to grow by continuing 10 to 12 day releases of FSH from the maternal pituitary gland (Evans and Irvine 1975; Urwin and Allen 1982). These regular, seasonally driven FSH releases continue beyond day 40 and result in one or more secondary ovulations or luteinizations on each occasion, thereby building a "crop" of secondary corpora lutea in the maternal ovaries (Fig. 9d) that maintain the pregnancy state until around day 90–120 when the slowly developing allantochorion has now enlarged to occupy and attach to the endometrium throughout the whole of the uterine lumen and is secreting sufficient progestagens to take over this role entirely (Squires and Ginther 1975).

The genome-level tools and assays for the horse that have become widely available following the publication of the equine genome reference sequence (Wade et al. 2009) have facilitated the elucidation of molecular pathways controlling differentiation of the chorionic girdle and endometrial cups. This is a very dramatic and dynamic process that results in up and downregulation of hundreds of genes over a very short period of time as the chorionic girdle forms and as its cells alter their behavior (Brosnahan et al. 2012; Read et al. 2018b). de Mestre and colleagues determined the expression patterns in developing chorionic girdle cells of several transcription factors known to function in mouse and human trophoblast. That study suggested that glial cells missing homologue 1 (GCM-1) might function to regulate the expression of the eCG beta chain gene (*CGB*) (de Mestre et al. 2009). Subsequent in vitro experiments confirmed that GCM-1 binds to site 1 in the *LHB* promoter and also identified other GCM-1 binding partners likely to be involved in the control of eCG expression (Read et al. 2018a). One such factor is Bone Morphogenetic Protein (BMP) 4, a member of the TGF-beta superfamily, which was shown to act in paracrine fashion to promote terminal differentiation of chorionic girdle cells into binucleate endometrial cup cells through the SMAD1/5 pathway (Cabrera-Sharp et al. 2014).

Although there are many similarities between hCG and eCG, there are also substantial differences (Bousfield et al. 1996). eCG and hCG are both alpha-beta heterodimers that combine a beta chain with placenta-specific expression with an alpha chain that is common to follicle stimulating hormone (FSH), luteinizing hormone (LH), and thyroid stimulating hormone (TSH). However, the placenta-specific beta chains evolved by distinct mechanisms. The hCG beta chain is encoded by a gene that arose through duplication and subsequent modification of the human LH gene (Policastro et al. 1983). The major modification is the addition of a COOH-terminal extension that functions by prolonging the half-life of the molecule in the circulation.

In horse, the eCG and eLH molecules are encoded by a single gene, the primordial equine LH gene (Sherman et al. 1992). Surprisingly, the equine LH/CG gene has a COOH-terminal extension of approximately the same length as the human hCG gene, but there is virtually no homology between the two COOH extensions (Sherman et al. 1992). The equine eCG and eLH molecules differ in carbohydrate content, but not in amino acid sequence (Bousfield et al. 1985; Matsui et al. 1991). Many more functions have been attributed to human hCG than have been yet identified for eCG, including several immune modulating activities (Schumacher and Zenclussen 2019; Gridelet et al. 2020). Much more research has been conducted on hCG compared to eCG, but it is also possible that hCG has more functions because it exists as a separate gene that has evolved on its own. The continued evolution of eCG, on the other hand, may have been restricted because it is encoded by the eLH gene, and thus must retain the essential functional properties of LH.

Although commonly referred to as hormones, eCG and its primate counterpart, hCG, function more like pheromones than conventional hormones. Produced by highly specialized cells of the fetal-derived placenta, they convey strong molecular signals from offspring to mother prenatally to help secure the transfer of life from one generation to the next.

4.3 Immunological Signaling: The Endometrial Cup Reaction

The second type of signaling mediated by the chorionic girdle and endometrial cups is the strong stimulation of the mother's immune system by paternally derived fetal antigens encoded by the major histocompatibility complex (MHC), reviewed by Noronha and Antczak 2010. The highly polymorphic MHC class I and class II molecules constitute the primary barrier to tissue and organ transplantation (Alegre et al. 2016; Rock et al. 2016; DeWolf and Sykes 2017; Nakamura et al. 2019), and thus they also represent potential immune targets for maternal anti-fetal immune reactions that, in theory, could damage or destroy the conceptus if expressed on trophoblast cells. Because the MHC genomic region is so polymorphic in mammals, most pregnancies are MHC disparate; that is, the fetal MHC antigens inherited from the father are usually different from those of the mother and therefore potentially antigenic. Soon after the fetal origin of the endometrial cups was discovered, it was hypothesized that the striking accumulations of maternal mononuclear cells observed surrounding and infiltrating the endometrial cups (Fig. 10) represented a maternal cell-mediated immune response against the conceptus (Allen 1975a). Thereafter, early studies of the equine leukocyte antigen (ELA) genetic system (ELA is the species-specific name for the MHC of the horse) reported that virtually all mares carrying MHC incompatible conceptuses developed cytotoxic antibodies

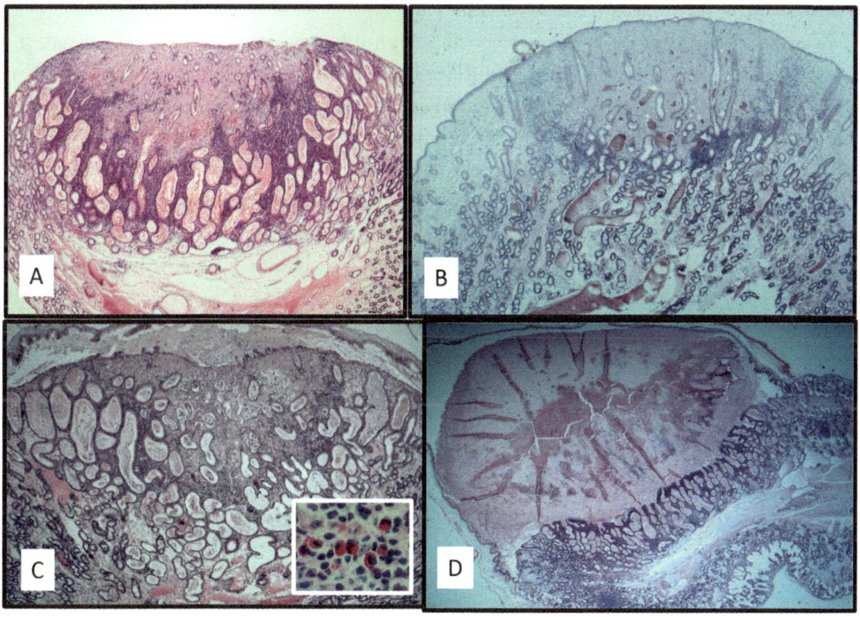

Fig. 10 Histology of the leukocyte response to the endometrial cups. (**a**) Early endometrial cup showing healthy binucleate trophoblasts surrounded by dense accumulations of maternal lymphocytes and dilated endometrial glands and lymphatics. (**b**) Day 60 endometrial cup showing reduced numbers of lymphocytes surrounding the cup cells. (**c**) Day 80 endometrial cup with increased numbers of invading lymphocytes accompanied by eosinophils (red cells in inset image). Note also the dilated endometrial glands and the mixture of sloughed cup cells and gland secretion on the uterine surface of the cups. (**d**) Endometrial cup in the final stages of its lifespan at 110 days of gestation showing advanced degeneration of the cup cells and the accumulated eCG-rich exocrine endometrial gland secretion beneath the allantochorion on the lumenal surface of the cup

against the foreign MHC antigens of the mating stallions (Bright et al. 1978; Antczak et al. 1982). Furthermore, the antibodies appeared shortly after the invasion of the endometrium by the chorionic girdle cells, that is, by day 45 of gestation (Fig. 9c) (Kydd et al. 1982; Antczak et al. 1984). This suggested that the stimulus for maternal antibody production and the accumulation of maternal lymphocytes around and within the endometrial cups might be paternal MHC antigens expressed by the cup cells.

Experiments designed to detect expression of MHC antigens on endometrial cups cells using immunohistochemical assays produced surprising results. First, mature endometrial cup trophoblast cells appeared to be devoid of MHC antigens (Fig. 11), and second, high levels of MHC class I antigens were detected on chorionic girdle trophoblasts recovered at days 32 to 36 of gestation (Fig. 6d) (Donaldson et al. 1990). No MHC class I expression was detected on trophoblast from 10 to 25 day embryos and conceptuses. Similar to findings in other species, no MHC class II antigen expression was detected on any equine trophoblast population. By testing timed pregnancies between days 35 and 45 of gestation, it was discovered that MHC class I antigen expression decreases rapidly after the chorionic girdle cells invade the endometrium (Donaldson et al. 1992).

Formal proof that the chorionic girdle trophoblast cells are the source of paternal MHC class I antigens that stimulate the mother's antibody response in early equine pregnancy was provided by experimental ectopic transplants of chorionic girdle cells recovered nonsurgically at day 34 of gestation and transplanted to the vulval mucosa of unrelated, nonpregnant recipient mares. In the parlance of transplantation biology, these trophoblast transplants were fully allogeneic. The donor trophoblast cells were of a different MHC type to the recipient mares and thus, not surprisingly, they stimulated the same types of cytotoxic antibodies observed in normal horse pregnancy. The ectopically transplanted girdle cells also developed into terminally differentiated endometrial cup cells and attracted lymphocytes around them, as in

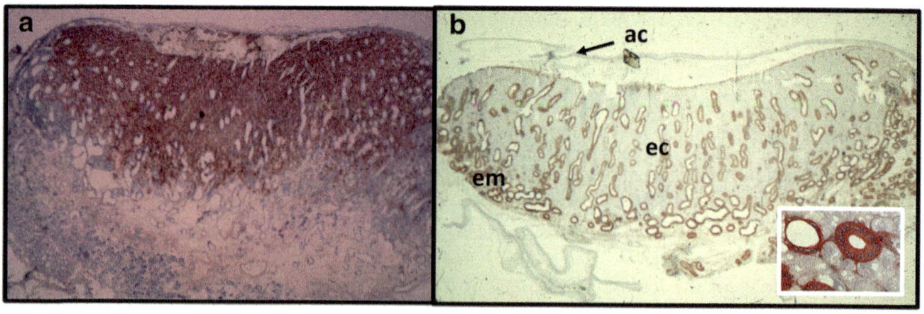

Fig. 11 Downregulation of major histocompatibility complex class I molecule expression by mature endometrial cups cells. Cryostat sections of day 60 endometrial cups labeled with monoclonal antibodies. (**a**) Antibody 71.8 (anti-horse trophoblast) showing strong labeling of endometrial cup trophoblasts but not maternal cells. (**b**) Antibody 116.1 (anti-horse MHC class I antigen) fails to label the endometrial cup trophoblasts but strongly labels maternal endometrial gland cells. Inset shows lack of MHC class I expression on endometrial cup cells in higher power). Labels: *em* endometrium, *ec* endometrial cup, *ac* allantochorion

the uterus during normal pregnancy (Adams and Antczak 2001). Surprisingly, the ectopically transplanted trophoblast survived in the vulva of recipient mares for approximately the same duration as the endometrial cups in normal horse pregnancy (de Mestre et al. 2011). Furthermore, repeated ectopic transplants of the same antigenic character induced secondary antibody responses in the recipients (Brosnahan et al. 2016), providing conclusive evidence that invasive horse trophoblast cells can induce both primary and anamnestic antibody responses while apparently avoiding immunological destruction by these responses. It was of considerable interest to find that the transplanted invasive trophoblasts produced sufficient eCG to be detectable in recipient serum. Furthermore, this eCG promoted the formation of corpora lutea in the ovaries of the recipient mares which prevented them from displaying estrus signs for the duration of the lifespan of the transplants (de Mestre et al. 2011).

The maternal cytotoxic antibodies generated against paternal MHC class I antigens expressed by the chorionic girdle cells in equine pregnancy represent the most dramatic example of a maternal antigen-specific, humoral immune response to trophoblast yet described in any species (Antczak 1989). This may be because equine trophoblast expresses classical MHC class I antigens, molecules which are highly antigenic within a species because of their high degree of polymorphism. The invasive human extravillous trophoblast, on the other hand, expresses pauci-polymorphic MHC class I molecules of the HLA-C, −G, and E loci, which are much less antigenic (Apps et al. 2009). Equally remarkable is the cell-mediated response to the equine conceptus. At the time of the initial invasion of the chorionic girdle cells at day 36–38 and for a few days afterwards, significant numbers of maternal leucocytes accumulate around the perimeter of the developing endometrial cup (Fig. 10a) (Allen 1975a; Enders and Liu 1991). Subsequent immunohistochemical studies using monoclonal antibodies to equine lymphocyte subsets indicated that the majority of the leukocytes are thymus-derived lymphocytes, or T-cells; that is, they express the CD3 molecule carried by all T-cells and either CD4 (T-helper cell subset marker) or CD8 (cytotoxic T-cell subset marker), with CD4+ T-cells predominating (Grunig et al. 1995). A few B-cells were also identified (Grunig et al. 1995), but no markers for macrophages, or for other types of leukocytes such as natural killer cells, were used. This same investigation documented striking changes in the density of leukocytes around the endometrial cups at different stages of their life cycle. Large numbers of T-cells were observed in the early stages of cup development (Fig. 10a) and in the late phase of deterioration and death of the endometrial cup trophoblasts, and at this end stage, eosinophils were also often seen (Fig. 10c). In contrast, at the peak of eCG production around day 60, fewer lymphocytes were found around or within the cups (Fig. 10b) (Grunig et al. 1995).

A plausible hypothesis for the reduction in lymphocyte numbers in the middle stage of endometrial cup development is that eCG, which peaks in secretion at that time, may have a local immunosuppressive effect, similar to the immunoregulatory actions of hCG (Schumacher 2017). However, to our knowledge only a single study addressed this issue and no evidence was reported for immune modulation by placental extracts containing eCG (Lea and Bolton 1991). Another possibility is that the maternal T-cell response is negatively regulated by galectin-1 which has been shown to be expressed on the endometrial cup trophoblast cells (Wilsher et al. 2019). Cultured chorionic girdle trophoblast cells, but not fetal fibroblasts, were shown to secrete factors that can inhibit lymphocyte proliferation, but the molecules involved were not identified (Flaminio and Antczak 2005). More recently, a family of equine pregnancy-specific glycoproteins (PSGs) expressed by invasive and non-invasive trophoblast has been described (Aleksic et al. 2016). In other species, PSGs have been postulated to have immunoregulatory properties, and this might be the case for equine PSGs, too (Kammerer et al. 2020).

Whatever mechanism results in reduction of the lymphocyte response, it appears to be a temporary one. In the late stage of endometrial cup life, coincident with falling eCG concentrations in maternal blood, increasing numbers of maternal lymphocytes, macrophages, and eosinophils again begin to accumulate in the endometrial stroma surrounding each endometrial cup (Fig. 10c). This marks the start of degeneration and death of the large trophoblast cells at the lumenal surface of the cup, presumably as a consequence of a relative lack of blood vessels in that region following their destruction during the original girdle cell invasion at days 38 to 40. And as the eCG levels in maternal blood decline, the accumulated leucocytes begin to invade into the body of the endometrial cup (Fig. 10c).

Despite this appearance of T-cell–mediated destruction of the endometrial cups, several findings suggest a more complicated process. First, in mares carrying MHC-matched compatible pregnancies, no maternal antibody response is detected (Antczak et al. 1984). This is expected if the target antigens are encoded by the MHC. Surprisingly, however, no reduction in the local lymphocyte response to the endometrial cups was noted in pregnancies from such MHC-matched matings (Allen et al. 1984). Furthermore, in repeated pregnancies derived from MHC-incompatible matings that were established using a stallion homozygous for the MHC as semen donor, there was no demonstrable reduction in lifespan of the endometrial cups and no detectable increase in the lymphocyte response to the endometrial cups, as would be predicted if the demise of the endometrial cup trophoblasts is caused by an antigen-specific T-cell response to paternal MHC antigens (Adams et al. 2007).

Thus, although the antibody response to paternal MHC class I antigens follows the expected, conventional rules of immunological specificity and memory, there is no evidence for any deleterious effects of this cytotoxic antibody response on the survival of the invasive trophoblast or the success of the pregnancy. It is possible that the equine conceptus escapes antibody-mediated destruction because the MHC class I antigens are expressed only on the invasive trophoblast of the chorionic girdle and early endometrial cups, and not on any other trophoblast cells. On the other hand, for the T-cell–mediated response, there is less evidence for either antigen specificity or memory. One possible explanation for the observed findings is that the conventional T-cell response to MHC class I antigens is subject to strong immune regulation. Current research in pregnancy immunology has highlighted the importance of regulatory T-cells (Tregs) in promoting the survival of the conceptus (Rowe et al. 2013; Salvany-Celades et al. 2019). There is evidence that equine T-regs (Robbin et al. 2011) are enriched in lymphocyte populations isolated from endometrial cup tissue (de Mestre et al. 2010) and possibly at other sites in the endometrium during pregnancy (Fedorka et al. 2019). Equine Tregs can also be detected in circulating blood, and low levels of these regulatory cells has been associated with early pregnancy loss (Aurich et al. 2014). A distinct mechanism of T-cell exhaustion has been demonstrated in mice and proposed as a critical component of the maternal immune system adjustments that prevent immunological destruction of the conceptus (Kinder et al. 2020). Although this mechanism has not been demonstrated in horse, a profound pregnancy-associated decrease in the capacity of CD8+ T-cells from pregnant mares to mount in vitro cytotoxic responses to foreign MHC class I antigens has been shown to occur in MHC disparate (Baker et al. 1999) and MHC compatible pregnancies (Noronha and Antczak 2012). The strong maternal antibody responses to paternal MHC class I antigens in equine pregnancy coupled with decreased cytotoxic T-cell responses has been termed "Split Immunological Tolerance to Trophoblast" (de Mestre et al. 2010). This concept of split tolerance fits the observations well, but there is no doubt that much more remains to be learned about the T-cell response to the equine invasive trophoblast.

Most studies of immune system cells in the equine endometrium have focused on conventional T-cells, which appear to be the dominant lymphocyte type (Grunig et al. 1995; Meeusen et al. 2001). In contrast, in humans and rodents, a specialized type of leukocyte called the uterine natural killer (uNK) cell is very prevalent in the pregnant endometrium (Croy et al. 2002; Moffett and Colucci 2014). There have been few investigations of equine NK cells or their receptors (Noronha et al. 2012a; Futas et al. 2020), and even fewer that studied NK cell distributions in the endometrium. However, the limited information available suggests that NK cells are enriched in the endometrium in the pregnant compared to the nonpregnant state (Tachibana et al. 2013), and higher in lymphocytes isolated from the endometrial cups compared to those extracted from peripheral blood obtained at the same time from individual mares (Noronha et al. 2012b). No information on the function of equine uNK cells is available. There is also very little data on other minor subpopulations of equine lymphocytes, so this remains a fertile area for future research.

Another feature of the endometrial cup reaction is blockage of the outlets of the endometrial glands during the initial invasion process. They remain active, however, so that their lumenae become increasingly distended with accumulated exocrine secretion. This begins to be liberated with commencing death and sloughing of the large cup cells at the lumenal surface of the cups (Fig. 10d). The inspissated secretion and admixed dead cup tissue accumulates underneath the allantochorion covering the surface of the cup, to form a "blob" of exocrine glurp on the surface of each cup which contains very high concentrations of eCG (Clegg et al. 1954). Take up of eCG into the maternal bloodstream occurs via a nest of lymph vessels which form at the base of each cup. And following the eventual death and desquamation of the necrotic cups and their accumulated secretion from day 100 of gestation onwards, the desquamated lumps of eCG-rich material in the dorsal region of the gravid uterine horn may indent into the underlying allantochorion to create a series of pendulous "allantochorionic pouches" (Clegg et al. 1954) that hang into the allantoic cavity (Fig. 12).

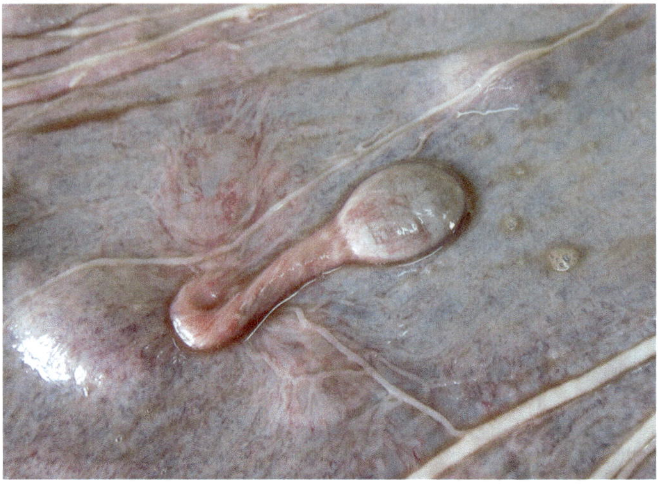

Fig. 12 Inner surface of the allantochorion of a term placenta showing an allantochorionic pouch that had hung into the allantoic cavity

5 Co-Option of Immune System Molecules by the Equine Placenta

The placenta evolved late in the evolutionary process, long after the vertebrate body plan was established (Carter and Mess 2007). The vertebrate gene set was already in place when the placenta evolved and it is therefore not surprising that the placenta expresses relatively few organ or tissue-specific genes (Rawn and Cross 2008). Molecular studies have identified a set of 115 genes that are expressed in the placentae of many, or perhaps even all, mammals and may thus have important conserved functions (Armstrong et al. 2017). Most of those genes are not specific to the placenta; rather, they have been co-opted, or borrowed, from other cell types to serve specific functions of the placenta. Some of those genes are involved in regulation of the immune system (Armstrong et al. 2017). However, the placenta is also a focus for rapid, continuing evolution in mammals, with great variation in form between species that reflects underlying patterns of gene expression (Wildman 2011). Furthermore, many genes active in the placenta have expression patterns that are restricted to particular tissues of the placenta and/or to certain developmental stages. Therefore, it is difficult to make generalizations about many genes found to be expressed in the placenta of any single species. The horse provides some interesting examples of restricted expression patterns of co-opted genes that affect immune function.

Galectins are a family of lectins that carry conserved carbohydrate recognition domains. Many of the members of this family have immune modulatory function and have been associated with the placentae of rodents (Blois et al. 2007) and humans (Than et al. 2009). As described earlier, the outermost trophoblast layer of the equine conceptus remains closely applied, but not physically attached, to the

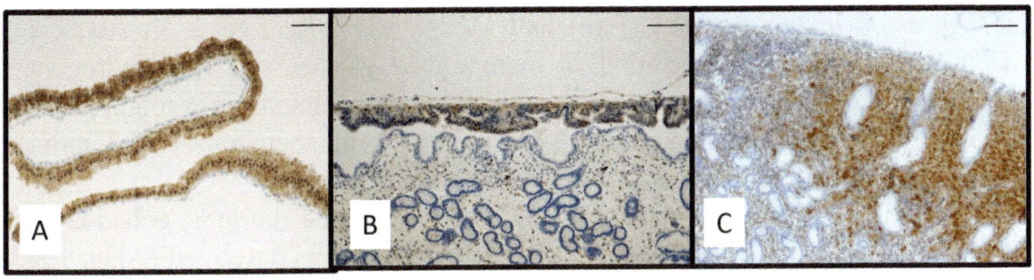

Fig. 13 Expression pattern of galectin-1 in equine trophoblast. (**a**) Trophectoderm of a day 18 conceptus. Note the lack of labeling of the underlying endoderm cell layer (Scale bar = 45 μm). (**b**) Day 32 chorionic girdle overlying the endometrium, showing greatly reduced labeling compared to day 18 (Scale bar = 200 μm). (**c**) Endometrial cup cells at day 40 of gestation, showing strong reexpression (Scale bar = 150 μm)

maternal endometrium between 20 and 40 days of gestation. These trophoblast cells stain strongly for galectin-1 between days 10 and 20 when the blastocyst capsule still surrounds the embryo (Fig. 13a). This galectin-1 staining disappears between days 20 and 30 after the capsule has been shed, only to appear again strongly thereafter in just the invasive trophoblast of the developing chorionic girdle (Fig. 13b) and the resulting eCG-secreting binucleate endometrial cup cells after the girdle has invaded the maternal endometrium at days 36 to 38 (Fig. 13c) (Wilsher et al. 2019). It is interesting to speculate if an important role of the blastocyst capsule, as well as maintaining the spherical shape and hence ability of the conceptus to move around the uterine lumen to distribute widely its maternal recognition of pregnancy signal between days 6 and 16 after ovulation, might be to prevent the galectin-1 secreting trophoblast cells from attempting to invade the maternal endometrium.

Interleukin 22 (IL-22) is an unusual cytokine. It is produced by a variety of immune system cells, particularly in inflammatory environments, but its receptor is expressed not on leukocytes, but on non-hematopoietic cells, typically epithelial cells. Engagement of IL-22 with its receptor can promote tissue repair and/or production of antimicrobial peptides (Rutz et al. 2013). A screen of transcriptional activity in invasive chorionic girdle cells revealed that equine IL-22 is expressed by the chorionic girdle trophoblast and by early endometrial cup cells (Brosnahan et al. 2012); this is the only known example of expression of IL-22 by a non-immune cell type. Because the lumenal epithelial and endometrial gland cells express the IL-22 receptor, it has been suggested that the purpose of IL-22 expression by invasive trophoblast may be to promote reepithelialization of the endometrium in the areas where the chorionic girdle has invaded (Brosnahan et al. 2012).

6 Epigenetic Regulation in the Equine Placenta

Genomic imprinting is a form of epigenetic regulation that results in monoallelic expression determined by parent-of-origin. About 200 imprinted genes have been described in humans and mice (Tucci et al. 2019), and many of the imprinted genes are expressed in the placenta (Frost and Moore 2010; Renfree et al. 2013). The discovery of genomic imprinting in manipulated gynogenetic and androgenetic mouse embryos suggested that paternally expressed imprinted genes might predominate in the formation and function of the extraembryonic membranes (McGrath and Solter 1984; Barton et al. 1984).

Much earlier, an unusual pattern of expression of equine chorionic gonadotrophin (eCG) in the sera of pregnant equids and their hybrids was described (Allen 1969). Mares carrying horse pregnancies had high levels of serum eCG, as did Jenny donkeys carrying interspecies hinny pregnancies. On the other hand, Jenny donkeys carrying normal donkey pregnancies and mares carrying interspecies mule pregnancies had very low levels of eCG. In these four types of pregnancy, the level of eCG in maternal blood appeared to be determined by the paternal genome; that is, conceptuses sired by horses produced high levels of eCG while conceptuses sired by donkeys produced low levels. This expression pattern is consistent with a hypothesized cause by imprinted genes. Transcriptome profiling of horse, donkey, mule, and hinny chorionic girdle trophoblast identified a large number of imprinted genes in this tissue, with a predominance of paternally expressed genes (Wang et al. 2013). It is of considerable interest to speculate whether the demonstrated differences in gene expression in the placentae of mule and hinnies is translated into phenotypic differences between the reciprocal hybrids in adult life (Fig. 14). A second result from this same dataset showed conclusively that X-chromosome inactivation, which is a specialized form of epigenetically determined monoallelic expression, was random in both embryonic and extraembryonic tissues (Wang et al. 2012). This result demonstrated that the selective paternal X inactivation in mouse extraembryonic tissues (Takagi and Sasaki 1975; Huynh and Lee 2001) does not apply to all mammals. Additional imprinted genes of the horse remain to be discovered. For example, recently the paternally expressed retroelement *RTL1* was identified and characterized in the equine placenta and associated with promoting angiogenesis in

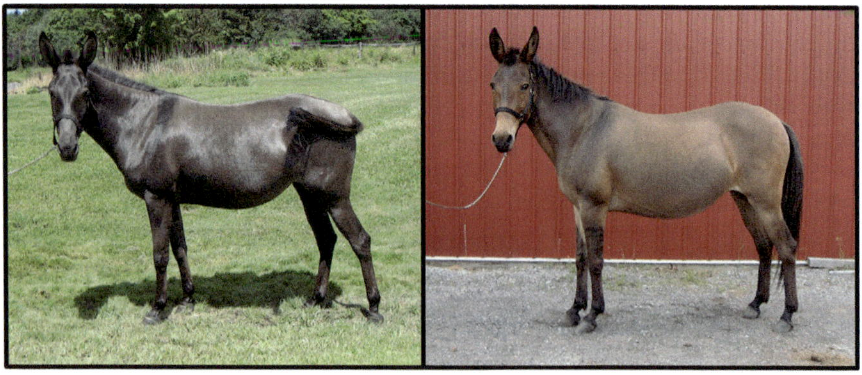

Fig. 14 A mule (female horse X male donkey) and hinny (female donkey X male horse). Which is which?

the chorioallantois (Dini et al. 2021). Epigenetic regulation remains a rich area for future research in equids.

7 Development of the Equine Placenta Proper

Placentation proper in the mare really begins from day 40 onwards with a microvillous contact between the trophoblast and lumenal epithelium of the endometrium, combined with a commencing protrusion of blunt villi of trophoblast into accommodating crypts in the endometrial surface (Fig. 15a) (Samuel et al. 1975, 1976). This interdigitation process continues steadily thereafter in parallel with enlargement and elongation of the conceptus to occupy the entire lumen of the uterus by around

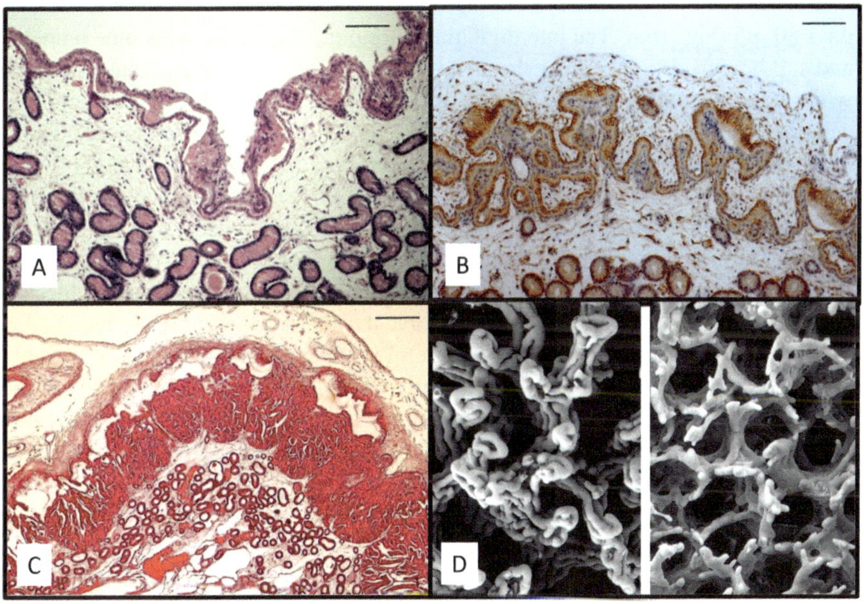

Fig. 15 Formation of the Equine Placenta Proper. (**a**) Histological section showing commencing interdigitation of the allantochorion with the endometrium at Day 46 of gestation (Scale bar = 150 μm). (**b**) Section of the allantochorion–endometrium border at day 68 of gestation stained with an eGF-receptor antiserum and illustrating specialized absorptive "areolae" for the uptake of endometrial histiotroph (Scale bar = 200 μm). (**c**) Low power section of a horse placenta at day 309 of gestation showing the multibranched and sub-branched microcotyledons that maximize the surface area contact between trophoblast and maternal endometrium. Areolae for the absorption of endometrial gland histiotroph are seen above the microcotyledons. (**d**) Scanning electron micrograph (SEM) images of equine microcotyledons at day 309 of gestation. (Left side) Trophoblast-denuded microcotyledons reveal the dense accumulation of fetal capillaries within these placental subunits. (Right side) Maternal endometrium showing the crypts into which the microcotyledons are normally inserted to bring fetal and maternal blood supplies into intimate physical contact throughout the entire epitheliochorial placenta

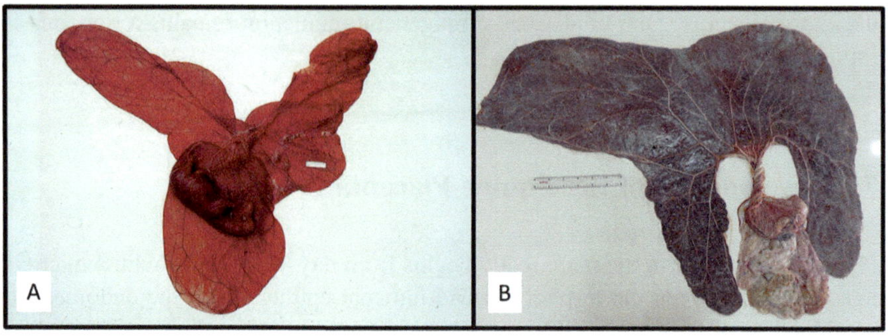

Fig. 16 The equine placenta expands to fill the space available within the uterus. (**a**) Day 93 horse conceptus that has now expanded to occupy the entire lumen of the uterus. (**b**) A horse term placenta laid out to demonstrate its considerable surface area which maximizes maternofetal circulatory interchange

days 80–85 (Fig. 16a). The interdigitating chorionic villi begin to become branched and sub-branched to maximize the area of contact between fetal and maternal tissue while at the same time leaving areas of apparently specialized absorptive trophoblast cells called "areolae" overlying the mouths of the still very active endometrial glands to enable the continued uptake of histiotroph (Fig. 15b) (Allen et al. 2017). In this way, the growing equine fetus benefits from both hemotrophic and histiotrophic forms of nutrition throughout the remainder of gestation (Wilsher and Allen 2012).

The continuing branching of the interdigitating chorionic villi eventually leads to the formation of discrete, so-called microcotyledons (Fig. 15c) between which the areolae continue their uptake of histiotroph throughout the entire gestation. And during the second half of pregnancy, each microvillus is, in effect, a mass of tightly coiled fetal capillaries interdigitated with an equivalently coiled mass of maternal capillaries (Fig. 15d), with as little as 4–6 μm of maternal and fetal tissues separating the two blood supplies (Abd-Elnaeim et al. 2006). And at term in Thoroughbred mares, use of the technique of stereology has demonstrated that the total microscopic area of feto-maternal contact across the entire placenta exceeds a whopping 50 square meters (Wilsher and Allen 2003) (Fig. 16b)!

The very large quantities of estrogens secreted by the fetal gonad-placenta cooperation during the second half of gestation are likely a major driver of the development and complicated interdigitation of maternal and fetal capillaries at the feto-maternal interface to facilitate maternal-fetal exchange of nutrients and waste products. Surgical removal of the gonads from four horse fetuses between days 197 and 251 of gestation (Pashen and Allen 1979) resulted in an immediate drop in maternal serum estrogen concentrations to baseline values. All four mares remained pregnant until spontaneous parturition at term, but three of the foals died during an abnormal parturition when each mare simply "stopped pushing" early in the process of expelling the foal through the birth canal, presumably due to a lack of prosta-glandin synthesis and storage in the myometrium to effect the necessary contractions

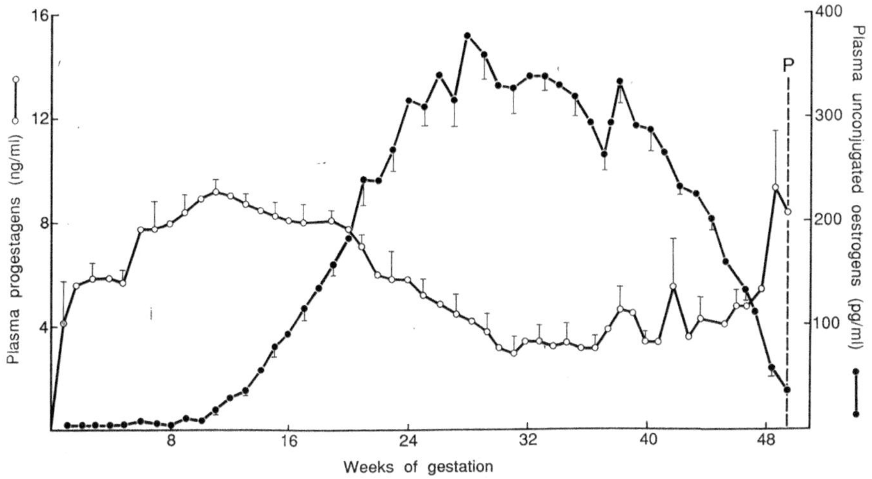

Fig. 17 Conjugated estrogen and progestagen profiles in the sera of five mares throughout pregnancy (courtesy of Robert Pashen, PhD thesis, University of Cambridge, 1978)

during the birth process (Pashen and Allen 1979). The fourth foal was born alive with assistance but it was very small, perhaps as a result of reduced placental interface resulting from the lack of placental estrogen synthesis, and it remained stunted as it matured into adult life (Pashen and Allen 1979).

Actually, the placenta secretes very little native progesterone per se, but it does produce a range of closely related and biologically active progestagens, prominent amongst which is 5 alpha dihydroprogesterone (Daels et al. 1998; Scholtz et al. 2014). And to accompany this copious supply of non-ovarian progestagens, the placenta also begins to secrete increasing quantities of estrogens from around days 80–90 of gestation which include both the common estrogens, estradiol-17β and estrone, and the two equine-unique and biologically active estrogens, equilin and equilenin (Bhavnani et al. 1969, 1971). Levels of all four estrogens in maternal blood and urine rise steeply to microgram per milliliter concentrations between days 200 and 280 of gestation before declining again steadily towards term (Fig. 17). Of particular interest, and another apparently equine-unique phenomenon, is the androstenedione and other carbon-19 precursors for this enormous production of pregnancy estrogens that are secreted by the interstitial cells of the fetal gonads, both testes and ovaries. These become tremendously enlarged between 150 and 280 days of gestation when they occupy over half the total volume of the fetal abdomen (Fig. 18a) (Pashen and Allen 1979). Each consists of a tightly packed mass of large, epithelioid, interstitial cells secreting the C-19 estrogen precursors (Fig. 18b) (Pashen et al. 1982) and they regress steadily during the final few weeks of gestation to become small and more gonad-like in appearance by the time of parturition.

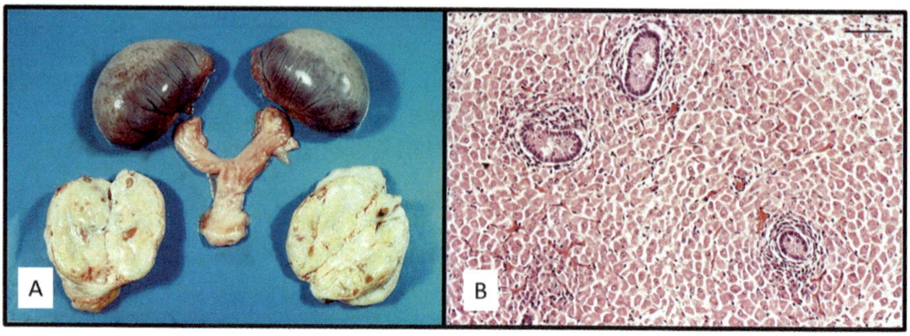

Fig. 18 Enlargement of fetal ovaries during equine pregnancy. (**a**) The very enlarged fetal ovaries still attached to the undeveloped fetal uterus and the inactive maternal ovaries from a mare at 36 weeks of gestation. (**b**) Histological section of a fetal testis at 239 days of gestation showing the enlarged and tightly packed estrogen precursor secreting interstitial cells (Scale bar = 150 μm)

8 Early Pregnancy Loss

Placental development has a direct relationship to fetal health and reproductive success. Advances in reproductive medicine such as the use of artificial insemination, transrectal ultrasonography, and prostaglandins have made improvements in breeding efficiency, but early pregnancy loss in horses persists. It has been known for several decades that pregnancy rates in mares are negatively correlated with maternal age and age-related degeneration of the endometrium (Woods et al. 1987; Ball 1988). The fertility rate per estrus cycle has been estimated to be in the range of 60–65% (Allen et al. 2007), and little is known about the reasons for pregnancy failure before day 14, when the equine conceptus can now reliably be detected using transrectal ultrasonography. Pregnancy loss between days 14 and 65 hovers around 7% (Hanlon et al. 2012; de Mestre et al. 2019). Pregnancy loss due to chromosomal disruption is common in human pregnancy and increases with maternal age (Vitez et al. 2019), but evidence for aneuploidy in domestic species is scarce. A recent study of failing conceptuses recovered non-surgically from privately owned mares used a whole genome genotyping array to discover that approximately 20% of these miscarriages had detectable chromosomal aneuploidy, while such genomic abnormalities were not found in placentae from any of the normal pregnancies that were examined as controls (Shilton et al. 2020). This high level of aneuploidy is consistent with results obtained using in vitro produced horse embryos (Rambags et al. 2005; Rizzo et al. 2019). It is possible that some embryos are lost before day 14 due to genetic abnormalities and that some additional pregnancy loss after day 14 may be due to small genetic rearrangements that were not detected by Shilton and colleagues. It may be that reproductive efficiency in the mare is reaching a point where little additional progress can be expected in the near future.

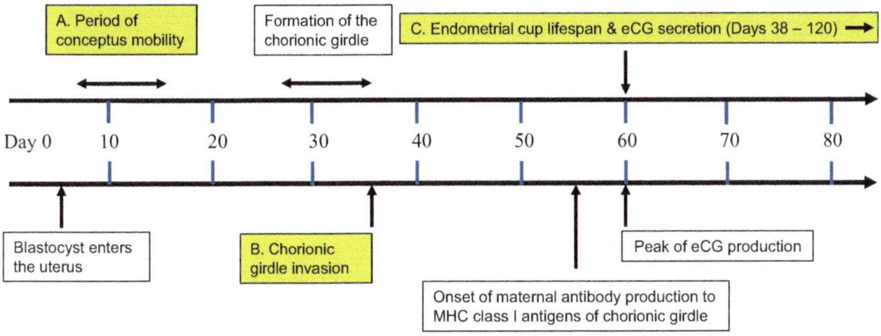

Fig. 19 Timeline of endocrinological and immunological fetus-to-mother signaling in early equine pregnancy. (**a**) Early conceptus migration between the uterine horns is the likely signal for maintenance of the primary corpus luteum (maternal recognition of pregnancy). (**b**) MHC class I antigen expression on invading chorionic girdle trophoblasts stimulates strong cytotoxic maternal anti-fetal antibody responses. (**c**) eCG secretion by the endometrial cup cells promotes secondary ovulations and continued progesterone production between days 40 and 100 to support the developing fetal-placental unit

9 Conclusion

In Fig. 19, we summarize the critical endocrinological and immunological signaling events in early equine pregnancy and placental development that, when correctly orchestrated, lead to the miracle of birth of a healthy foal (Fig. 20). The extensive, diffuse, epitheliochorial placenta of female equids, combined with the intriguing endometrial cup reaction with its immunological and endocrinological connotations, followed by the combined involvement of the fetal gonads and placenta in the production of relatively enormous quantities of biologically active estrogens involved in vascularization of the placental interface during the second half of gestation, singles out the genus *Equus* for special interest in and study of the whole process of placentation. And how that placentation phenomenon can continue to function normally and effectively in the face of both interspecies hybrid and transferred extraspecies pregnancies remains a fascinating area of investigation.

Acknowledgments The authors thank the many members of our laboratories (students, advanced trainees, and scientific and horse care staff) whose efforts over many years made major contributions to the information provided in this chapter.

Fig. 20 The finished product of equine reproduction. A mare and her young foal

References

Abd-Elnaeim MM, Leiser R, Wilsher S, Allen WR (2006) Structural and haemovascular aspects of placental growth throughout gestation in young and aged mares. Placenta 27:1103–1113

Adams AP, Antczak DF (2001) Ectopic transplantation of equine invasive trophoblast. Biol Reprod 64:753–763

Adams AP, Oriol JG, Campbell RE, Oppenheim YC, Allen WR, Antczak DF (2007) The effect of skin allografting on the equine endometrial cup reaction. Theriogenology 68:237–247

Alegre ML, Lakkis FG, Morelli AE (2016) Antigen presentation in transplantation. Trends Immunol 37:831–843

Aleksic D, Blaschke L, Missbach S, Hanske J, Weiss W, Handler J, Zimmermann W, Cabrera-Sharp V, Read JE, De Mestre AM, O'riordan R, Moore T, Kammerer R (2016) Convergent evolution of pregnancy-specific glycoproteins in human and horse. Reproduction 152:171–184

Allen WR (1969) Factors influencing pregnant mare serum gonadotrophin production. Nature 223:64–65

Allen WR (1975a) Immunological aspects of the equine endometrial cup reaction. In: Edwards RG, Howe CWS, Johnson MH (eds) Immunobiology of trophoblast. Cambridge University Press, Cambridge

Allen WR (1975b) The influence of fetal genotype upon endometrial cup development and PMSG and progestagen production in equids. J Reprod Fertil Suppl 23:405–413

Allen WR (1981) Use of prostaglandins for synchronization of oestrus and treatment of prolonged dioestrus in mares. Acta Vet Scand Suppl 77:227–239

Allen WR (1982) Immunological aspects of the endometrial cup reaction and the effect of xenogeneic pregnancy in horses and donkeys. J Reprod Fertil Suppl 31:57–94

Allen WR, Moor RM (1972) The origin of the equine endometrial cups. I. Production of PMSG by fetal trophoblast cells. J Reprod Fertil 29:313–316

Allen WR, Short RV (1997) Interspecific and extraspecific pregnancies in equids: anything goes. J Hered 88:384–392
Allen WRT, Wilsher S (2020) Historical aspects of equine embryo transfer. J Equine Vet Sci 89:102987
Allen WR, Hamilton DW, Moor RM (1973) The origin of equine endometrial cups. II. Invasion of the endometrium by trophoblast. Anat Rec 177:485–501
Allen WR, Kydd J, Miller J, Antczak DF (1984) Immunological studies on feto maternal relationships in equine pregnancy. Buttersworths, London
Allen WR, Brown L, Wright M, Wilsher S (2007) Reproductive efficiency of Flatrace and National Hunt Thoroughbred mares and stallions in England. Equine Vet J 39:438–445
Allen WR, Gower S, Wilsher S (2017) Localisation of epidermal growth factor (EGF), its specific receptor (EGF-R) and aromatase at the materno-fetal interface during placentation in the pregnant mare. Placenta 50:53–59
Amoroso EC (1958) Placentation. In: Parkes AS (ed) Marshall's physiology of reproduction. Longmans, Green, and Co, London
Antczak DF (1989) Maternal antibody responses in pregnancy. Curr Opin Immunol 1:1135–1140
Antczak DF, Bright SM, Remick LH, Bauman BE (1982) Lymphocyte alloantigens of the horse. I. Serologic and genetic studies. Tissue Antigens 20:172–187
Antczak DF, Miller JM, Remick LH (1984) Lymphocyte alloantigens of the horse. II Antibodies to ELA antigens produced during equine pregnancy. J Reprod Immunol 6:283–297
Apps R, Murphy SP, Fernando R, Gardner L, Ahad T, Moffett A (2009) Human leucocyte antigen (HLA) expression of primary trophoblast cells and placental cell lines, determined using single antigen beads to characterize allotype specificities of anti-HLA antibodies. Immunology 127:26–39
Armstrong DL, Mcgowen MR, Weckle A, Pantham P, Caravas J, Agnew D, Benirschke K, Savage-Rumbaugh S, Nevo E, Kim CJ, Wagner GP, Romero R, Wildman DE (2017) The core transcriptome of mammalian placentas and the divergence of expression with placental shape. Placenta 57:71–78
Aurich C, Weber J, Nagel C, Merkl M, Jude R, Wostmann S, Ollech D, Baron U, Olek S, Jansen T (2014) Low levels of naturally occurring regulatory T lymphocytes in blood of mares with early pregnancy loss. Reprod Fertil Dev 26:827–833
Baker JM, Bamford AI, Antczak DF (1999) Modulation of allospecific CTL responses during pregnancy in equids: an immunological barrier to interspecies matings? J Immunol 162:4496–4501
Ball BA (1988) Embryonic loss in mares. Incidence, possible causes, and diagnostic considerations. Vet Clin North Am Equine Pract 4:263–290
Barton SC, Surani MA, Norris ML (1984) Role of paternal and maternal genomes in mouse development. Nature 311:374–376
Battut I, Colchen S, Fieni F, Tainturier D, Bruyas JF (1997) Success rates when attempting to nonsurgically collect equine embryos at 144, 156 or 168 hours after ovulation. Equine Vet J Suppl 29:60–62
Bhavnani BR, Short RV, Solomon S (1969) Formation of estrogens by the pregnant mare. I. Metabolism of 7-3H-dehydroisoandrosterone and 4-14C-androstenedione injected into the umbilical vein. Endocrinology 85:1172–1179
Bhavnani BR, Short RV, Solomon S (1971) Formation of estrogens by the pregnant mare. II. Metabolism of 14C-acetate and 3H-cholesterol injected into the fetal circulation. Endocrinology 89:1152–1157
Blois SM, Ilarregui JM, Tometten M, Garcia M, Orsal AS, Cordo-Russo R, Toscano MA, Bianco GA, Kobelt P, Handjiski B, Tirado I, Markert UR, Klapp BF, Poirier F, Szekeres-Bartho J, Rabinovich GA, Arck PC (2007) A pivotal role for galectin-1 in fetomaternal tolerance. Nat Med 13:1450–1457

Bousfield GR, Sugino H, Ward DN (1985) Demonstration of a COOH-terminal extension on equine lutropin by means of a common acid-labile bond in equine lutropin and equine chorionic gonadotropin. J Biol Chem 260:9531–9533
Bousfield GR, Butnev VY, Gotschall RR, Baker VL, Moore WT (1996) Structural features of mammalian gonadotropins. Mol Cell Endocrinol 125:3–19
Bright S, Antczak DF, Ricketts S (1978) Studies on equine leukocyte antigens. J Equine Med Surg Suppl 1:229–236
Brosnahan MM, Miller DC, Adams M, Antczak DF (2012) IL-22 is expressed by the invasive trophoblast of the equine (Equus caballus) chorionic girdle. J Immunol 188:4181–4187
Brosnahan MM, Silvela EJ, Crumb J, Miller DC, Erb HN, Antczak DF (2016) Ectopic trophoblast allografts in the horse resist destruction by secondary immune responses. Biol Reprod 95:135
Burton GJ, Watson AL, Hempstock J, Skepper JN, Jauniaux E (2002) Uterine glands provide histiotrophic nutrition for the human fetus during the first trimester of pregnancy. J Clin Endocrinol Metab 87:2954–2959
Cabrera-Sharp V, Read JE, Richardson S, Kowalski AA, Antczak DF, Cartwright JE, Mukherjee A, De Mestre AM (2014) SMAD1/5 signaling in the early equine placenta regulates trophoblast differentiation and chorionic gonadotropin secretion. Endocrinology 155:3054–3064
Carnevale EM, Maclellan LJ, Stokes JAE (2019) In vitro culture of embryos from horses. Methods Mol Biol 2006:219–227
Carter AM, Enders AC (2013) The evolution of epitheliochorial placentation. Annu Rev Anim Biosci 1:443–467
Carter AM, Mess A (2007) Evolution of the placenta in eutherian mammals. Placenta 28:259–262
Clegg MT, Boda JM, Cole HH (1954) The endometrial cups and allantochorionic pouches in the mare with emphasis on the source of equine gonadotrophin. Endocrinology 54:448–463
Clutton-Brock J (1992) Horse power: a history of the horse and the donkey in human societies. Harvard University Press, Cambridge, MA
Cole LA (2010) Biological functions of hCG and hCG-related molecules. Reprod Biol Endocrinol 8:102
Cole HH, Goss H (1943) The source of equine gonadotropin. Essays in Biology in Honor of Herbert M Evans. University of California Press, Berkeley
Cole HH, Hart GH (1930) The potency of blood serum of mares in progressive stages of pregnancy in effecting the sexual maturity of the immature rat. Am J Physiol 93:57–68
Croy BA, Chantakru S, Esadeg S, Ashkar AA, Wei Q (2002) Decidual natural killer cells: key regulators of placental development (a review). J Reprod Immunol 57:151–168
Daels PF, Albrecht BA, Mohammed HO (1998) Equine chorionic gonadotropin regulates luteal steroidogenesis in pregnant mares. Biol Reprod 59:1062–1068
Davies CJ, Antczak DF, Allen WR (1985) Reproduction in mules: embryo transfer using sterile recipients. Equine Vet J Suppl 3:63–67.
De Mestre AM, Miller D, Roberson MS, Liford J, Chizmar LC, Mclaughlin KE, Antczak DF (2009) Glial cells missing homologue 1 is induced in differentiating equine chorionic girdle trophoblast cells. Biol Reprod 80:227–234
De Mestre A, Noronha L, Wagner B, Antczak DF (2010) Split immunological tolerance to trophoblast. Int J Dev Biol 54:445–455
De Mestre AM, Hanlon D, Adams AP, Runcan E, Leadbeater JC, Erb HN, Costa CC, Miller D, Allen WR, Antczak DF (2011) Functions of ectopically transplanted invasive horse trophoblast. Reproduction 141:849–856
De Mestre AM, Rose BV, Chang YM, Wathes DC, Verheyen KLP (2019) Multivariable analysis to determine risk factors associated with early pregnancy loss in thoroughbred broodmares. Theriogenology 124:18–23
Devroey P, Bourgain C, Macklon NS, Fauser BC (2004) Reproductive biology and IVF: ovarian stimulation and endometrial receptivity. Trends Endocrinol Metab 15:84–90
Dewolf S, Sykes M (2017) Alloimmune T cells in transplantation. J Clin Invest 127:2473–2481

Dini P, Carossino M, Balasuriya UBR, Ali HE, Loux SC, Esteller-Vico A, Scoggin KE, Loynachan AT, Kalbfleisch T, De Spiegelaere W, Daels P, Ball BA (2021) Paternally expressed retrotransposon gag like 1 gene, RTL1, is one of the crucial elements for placental angiogenesis in horses. Biol Reprod:ioab039

Donaldson WL, Zhang CH, Oriol JG, Antczak DF (1990) Invasive equine trophoblast expresses conventional class I major histocompatibility complex antigens. Development 110:63–71

Donaldson WL, Oriol JG, Plavin A, Antczak DF (1992) Developmental regulation of class I major histocompatibility complex antigen expression by equine trophoblastic cells. Differentiation 52:69–78

Enders AC, Liu IKM (1991) Lodgement of the equine blastocyst in the uterus from fixation through endometrial cup formation. J Reprod Fertil Suppl 44:427–438

Evans MJ, Irvine CH (1975) Serum concentrations of FSH, LH and progesterone during the oestrous cycle and early pregnancy in the mare. J Reprod Fertil Suppl 23:193–200

Ewart J (1897) A critical period in the development of the horse. Adam and Charles Black, London

Fedorka CE, Loux SL, Scoggin KE, Adams AA, Troedsson MHT, Ball BA (2019) Alterations in T cell-related transcripts at the feto-maternal interface throughout equine gestation. Placenta 89:78–87

Filant J, Spencer TE (2014) Uterine glands: biological roles in conceptus implantation, uterine receptivity and decidualization. Int J Dev Biol 58:107–116

Flaminio MJ, Antczak DF (2005) Inhibition of lymphocyte proliferation and activation: a mechanism used by equine invasive trophoblast to escape the maternal immune response. Placenta 26:148–159

Fonseca JF, Souza-Fabjan JM, Oliveira ME, Leite CR, Nascimento-Penido PM, Brandão FZ, Lehloenya KC (2016) Nonsurgical embryo recovery and transfer in sheep and goats. Theriogenology 86:144–151

Frost JM, Moore GE (2010) The importance of imprinting in the human placenta. PLoS Genet 6: e1001015

Futas J, Oppelt J, Janova E, Musilova P, Horin P (2020) Complex variation in the KLRA (LY49) immunity-related genomic region in horses. HLA 96:257–267

Gambini A, Maserati M (2017) A journey through horse cloning. Reprod Fertil Dev 30:8–17

Geisert RD, Whyte JJ, Meyer AE, Mathew DJ, Juarez MR, Lucy MC, Prather RS, Spencer TE (2017) Rapid conceptus elongation in the pig: an interleukin 1 beta 2 and estrogen-regulated phenomenon. Mol Reprod Dev 84:760–774

Gerri C, Mccarthy A, Alanis-Lobato G, Demtschenko A, Bruneau A, Loubersac S, Fogarty NME, Hampshire D, Elder K, Snell P, Christie L, David L, Van De Velde H, Fouladi-Nashta AA, Niakan KK (2020) Initiation of a conserved trophectoderm program in human, cow and mouse embryos. Nature 587:443–447

Ginther OJ (2014) How ultrasound technologies have expanded and revolutionized research in reproduction in large animals. Theriogenology 81:112–125

Gray AP (1972) Mammalian hybrids: a check-list with bibliography. Commonwealth Agricultural Bureaux, Slough

Gridelet V, Perrier D'hauterive S, Polese B, Foidart JM, Nisolle M, Geenen V (2020) Human chorionic gonadotrophin: new pleiotropic functions for an "old" hormone during pregnancy. Front Immunol 11:343

Grunig G, Triplett L, Canady LK, Allen WR, Antczak DF (1995) The maternal leucocyte response to the endometrial cups in horses is correlated with the developmental stages of the invasive trophoblast cells. Placenta 16:539–559

Hamilton DW, Allen WR, Moor RM (1973) The origin of equine endometrial cups. 3. Light and electron microscopic study of fully developed equine endometrial cups. Anat Rec 177:503–517

Hanlon DW, Stevenson M, Evans MJ, Firth EC (2012) Reproductive performance of thoroughbred mares in the Waikato region of New Zealand: 1. Descriptive analyses. N Z Vet J 60:329–334

Hansen PJ (2020) Implications of assisted reproductive technologies for pregnancy outcomes in mammals. Annu Rev Anim Biosci 8:395–413

Hendriks WK, Colleoni S, Galli C, Paris D, Colenbrander B, Stout TAE (2019) Mitochondrial DNA replication is initiated at blastocyst formation in equine embryos. Reprod Fertil Dev 31:570–578
Hinrichs K (2018) Assisted reproductive techniques in mares. Reprod Domest Anim 53:4–13
Huynh KD, Lee JT (2001) Imprinted X inactivation in eutherians: a model of gametic execution and zygotic relaxation. Curr Opin Cell Biol 13:690–697
Iqbal K, Chitwood JL, Meyers-Brown GA, Roser JF, Ross PJ (2014) RNA-seq transcriptome profiling of equine inner cell mass and trophectoderm. Biol Reprod 90:61
Jones CJ, Choudhury RH, Aplin JD (2015) Tracking nutrient transfer at the human maternofetal interface from 4 weeks to term. Placenta 36:372–380
Kammerer R, BallesterosA BD, Warren J, Williams JM, Moore T, Dveksler G (2020) Equine pregnancy-specific glycoprotein CEACAM49 secreted by endometrial cup cells activates TGFB. Reproduction 160:685–694
Kinder JM, Turner LH, Stelzer IA, Miller-Handley H, Burg A, Shao TY, Pham G, Way SS (2020) CD8(+) T cell functional exhaustion overrides pregnancy-induced fetal antigen Alloimmunization. Cell Rep 31:107784
Klein C (2016) Early pregnancy in the mare: old concepts revisited. Domest Anim Endocrinol 56 (Suppl):S212–S217
Klohonatz KM, Hess AM, Hansen TR, Squires EL, Bouma GJ, Bruemmer JE (2015) Equine endometrial gene expression changes during and after maternal recognition of pregnancy. J Anim Sci 93:3364–3376
Kydd J, Miller J, Antczak DF, Allen WR (1982) Maternal anti-fetal cytotoxic antibody responses of equids during pregnancy. J Reprod Fertil Suppl 32:361–369
Lea RG, Bolton AE (1991) The effect of horse placental tissue extracts and equine chorionic gonadotrophin on the proliferation of horse lymphocytes stimulated in vitro. J Reprod Immunol 19:13–23
Leemans B, Gadella BM, Stout TA, De Schauwer C, Nelis H, Hoogewijs M, Van Soom A (2016) Why doesn't conventional IVF work in the horse? The equine oviduct as a microenvironment for capacitation/fertilization. Reproduction 152:R233–R245
Matsui T, Sugino H, Miura M, Bousfield GR, Ward DN, Titani K, Mizuochi T (1991) Beta-subunits of equine chorionic gonadotropin and lutenizing hormone with an identical amino acid sequence have different asparagine-linked oligosaccharide chains. Biochem Biophys Res Commun 174:940–945
Mcgrath J, Solter D (1984) Completion of mouse embryogenesis requires both the maternal and paternal genomes. Cell 37:179–183
Medicine PCOASFR (2008) Gonadotropin preparations: past, present, and future perspectives. Fertil Steril 90:S13–S20
Meeusen EN, Bischof RJ, Lee CS (2001) Comparative T-cell responses during pregnancy in large animals and humans. Am J Reprod Immunol 46:169–179
Mikkola M, Hasler JF, Taponen J (2019) Factors affecting embryo production in superovulated Bos taurus cattle. Reprod Fertil Dev 32:104–124
Moffett A, Colucci F (2014) Uterine NK cells: active regulators at the maternal-fetal interface. J Clin Invest 124:1872–1879
Morris LHA (2018) The development of in vitro embryo production in the horse. Equine Vet J 50:712–720
Nakamura T, Shirouzu T, Nakata K, Yoshimura N, Ushigome H (2019) The role of major histocompatibility complex in organ transplantation-donor specific anti-major histocompatibility complex antibodies analysis goes to the next stage. Int J Mol Sci 20:4544
Naranjo Chacón F, Montiel Palacios F, Canseco Sedano R, Ahuja-Aguirre C (2020) Embryo production after superovulation of bovine donors with a reduced number of FSH applications and an increased eCG dose. Theriogenology 141:168–172
Noronha LE, Antczak DF (2010) Maternal immune responses to trophoblast: the contribution of the horse to pregnancy immunology. Am J Reprod Immunol 64:231–244

Noronha LE, Antczak DF (2012) Modulation of T-cell reactivity during equine pregnancy is antigen independent. Am J Reprod Immunol 68:107–115
Noronha LE, Harman RM, Wagner B, Antczak DF (2012a) Generation and characterization of monoclonal antibodies to equine NKp46. Vet Immunol Immunopathol 147:60–68
Noronha LE, Huggler KE, De Mestre AM, Miller DC, Antczak DF (2012b) Molecular evidence for natural killer-like cells in equine endometrial cups. Placenta 33:379–386
Olivera R, Moro LN, Jordan R, Pallarols N, Guglielminetti A, Luzzani C, Miriuka SG, Vichera G (2018) Bone marrow mesenchymal stem cells as nuclear donors improve viability and health of cloned horses. Stem Cells Cloning 11:13–22
Oriol JG (1994) The equine embryonic capsule: practical implications of recent research. Equine Vet J 26:184–186
Oriol JG, Betteridge KJ, Clarke AJ, Sharom FJ (1993) Mucin-like glycoproteins in the equine embryonic capsule. Mol Reprod Dev 34:255–265
Ortiz-Rodriguez JM, Ortega-Ferrusola C, Gil MC, Martin-Cano FE, Gaitskell-Phillips G, Rodriguez-Martinez H, Hinrichs K, Alvarez-Barrientos A, Roman A, Pena FJ (2019) Transcriptome analysis reveals that fertilization with cryopreserved sperm downregulates genes relevant for early embryo development in the horse. PLoS One 14:e0213420
Palmer E, Bézard J, Magistrini M, Duchamp G (1991) In vitro fertilization in the horse. A retrospective study. J Reprod Fertil Suppl 44:375–384
Papkoff H (1981) Variations in the properties of equine chorionic gonadotropin. Theriogenology 15:1–11
Pashen RL, Allen WR (1979) The role of the fetal gonads and placenta in steroid production, maintenance of pregnancy and parturition in the mare. J Reprod Fertil Suppl 27:499–509
Pashen RL, Sheldrick EL, Allen WR, Flint AP (1982) Dehydroepiandrosterone synthesis by the fetal foal and its importance as an oestrogen precursor. J Reprod Fertil Suppl 32:389–397
Piliszek A, Madeja ZE (2018) Pre-implantation development of domestic animals. Curr Top Dev Biol 128:267–294
Policastro P, Ovitt CE, Hoshina M, Fukuoka H, Boothby MR, Boime I (1983) The beta subunit of human chorionic gonadotropin is encoded by multiple genes. J Biol Chem 258:11492–11499
Posfai E, Rovic I, Jurisicova A (2019) The mammalian embryo's first agenda: making trophectoderm. Int J Dev Biol 63:157–170
Rambags BP, Krijtenburg PJ, Drie HF, Lazzari G, Galli C, Pearson PL, Colenbrander B, Stout TA (2005) Numerical chromosomal abnormalities in equine embryos produced in vivo and in vitro. Mol Reprod Dev 72:77–87
Rawn SM, Cross JC (2008) The evolution, regulation, and function of placenta-specific genes. Annu Rev Cell Dev Biol 24:159–181
Read JE, Cabrera-Sharp V, Kitscha P, Cartwright JE, King PJ, Fowkes RC, De Mestre AM (2018a) Glial cells missing 1 regulates equine chorionic gonadotrophin Beta subunit via binding to the proximal promoter. Front Endocrinol (Lausanne) 9:195
Read JE, Cabrera-Sharp V, Offord V, Mirczuk SM, Allen SP, Fowkes RC, De Mestre AM (2018b) Dynamic changes in gene expression and signalling during trophoblast development in the horse. Reproduction 156:313–330
Renfree MB, Suzuki S, Kaneko-Ishino T (2013) The origin and evolution of genomic imprinting and viviparity in mammals. Philos Trans R Soc Lond B Biol Sci 368:20120151
Rizzo M, Ducheyne KD, Deelen C, Beitsma M, Cristarella S, Quartuccio M, Stout TAE, De Ruijter-Villani M (2019) Advanced mare age impairs the ability of in vitro-matured oocytes to correctly align chromosomes on the metaphase plate. Equine Vet J 51:252–257
Robbin MG, Wagner B, Noronha LE, Antczak DF, De Mestre AM (2011) Subpopulations of equine blood lymphocytes expressing regulatory T cell markers. Vet Immunol Immunopathol 140:90–101
Rock KL, Reits E, Neefjes J (2016) Present yourself! By MHC class I and MHC class II molecules. Trends Immunol 37:724–737

Roser JF, Meyers-Brown G (2019) Enhancing fertility in mares: recombinant equine gonadotropins. J Equine Vet Sci 76:6–13
Rowe JH, Ertelt JM, Xin L, Way SS (2013) Regulatory T cells and the immune pathogenesis of prenatal infection. Reproduction 146:R191–R203
Rutz S, Eidenschenk C, Ouyang W (2013) IL-22, not simply a Th17 cytokine. Immunol Rev 252:116–132
Salvany-Celades M, Van Der Zwan A, Benner M, Setrajcic-Dragos V, Bougleux Gomes HA, Iyer V, Norwitz ER, Strominger JL, Tilburgs T (2019) Three types of functional regulatory T cells control T cell responses at the human maternal-fetal interface. Cell Rep 27:2537–2547 e5
Samuel CA, Allen WR, Steven DH (1975) Ultrastructural development of the equine placenta. J Reprod Fertil Suppl 23:575–578
Samuel CA, Allen WR, Steven DH (1976) Studies on the equine placenta II. Ultrastructure of the placental barrier. J Reprod Fertil 48:257–264
Schauder W (1912) Untersuchungen uber die eithaute und Embryotrophe des pferdes. Arch Anat Physiol 192:259–302
Scholtz EL, Krishnan S, Ball BA, Corbin CJ, Moeller BC, Stanley SD, Mcdowell KJ, Hughes AL, Mcdonnell DP, Conley AJ (2014) Pregnancy without progesterone in horses defines a second endogenous biopotent progesterone receptor agonist, 5alpha-dihydroprogesterone. Proc Natl Acad Sci U S A 111:3365–3370
Schumacher A (2017) Human chorionic gonadotropin as a pivotal endocrine immune regulator initiating and preserving fetal tolerance. Int J Mol Sci 18:2166
Schumacher A, Zenclussen AC (2019) Human chorionic gonadotropin-mediated immune responses that facilitate embryo implantation and placentation. Front Immunol 10:2896
Sherman GB, Wolfe MW, Farmerie TA, Clay CM, Threadgill DS, Sharp DC, Nilson JH (1992) A single gene encodes the beta-subunits of equine luteinizing hormone and chorionic gonadotropin. Mol Endocrinol 6:951–959
Shilton CA, Kahler A, Davis BW, Crabtree JR, Crowhurst J, Mcgladdery AJ, Wathes DC, Raudsepp T, De Mestre AM (2020) Whole genome analysis reveals aneuploidies in early pregnancy loss in the horse. Sci Rep 10:13314
Short RV (1975) The contribution of the mule to scientific thought. J Reprod Fertil Suppl 23:359–364
Smits K, De Coninck DI, Van Nieuwerburgh F, Govaere J, Van Poucke M, Peelman L, Deforce D, Van Soom A (2016) The equine embryo influences immune-related gene expression in the oviduct. Biol Reprod 94:36
Smits K, Willems S, Van Steendam K, Van De Velde M, De Lange V, Ververs C, Roels K, Govaere J, Van Nieuwerburgh F, Peelman L, Deforce D, Van Soom A (2018) Proteins involved in embryo-maternal interaction around the signalling of maternal recognition of pregnancy in the horse. Sci Rep 8:5249
Sommer JR, Collins EB, Estrada JL, Petters RM (2007) Synchronization and superovulation of mature cycling gilts for the collection of pronuclear stage embryos. Anim Reprod Sci 100:402–410
Squires E (2020) Current reproductive technologies impacting equine embryo production. J Equine Vet Sci 89:102981
Squires EL, Ginther OJ (1975) Follicular and luteal development in pregnant mares. J Reprod Fertil Suppl 23:429–433
Stabenfeldt GH, Hughes JP, Evans JW, Neely DP (1974) Spontaneous prolongation of luteal activity in the mare. Equine Vet J 6:158–163
Stout TAE (2020) Clinical application of in vitro embryo production in the horse. J Equine Vet Sci 89:103011
Stout TA, Lamming GE, Allen WR (2000) The uterus as a source of oxytocin in cyclic mares. J Reprod Fertil Suppl 56:281–287

Tachibana Y, Nakano Y, Nagaoka K, Kikuchi M, Nambo Y, Haneda S, Matsui M, Miyake Y, Imakawa K (2013) Expression of endometrial immune-related genes possibly functioning during early pregnancy in the mare. J Reprod Dev 59:85–91
Takagi N, Sasaki M (1975) Preferential inactivation of the paternally derived X chromosome in the extraembryonic membranes of the mouse. Nature 256:640–642
Takeo T, Nakagata N (2015) Superovulation using the combined administration of inhibin antiserum and equine chorionic gonadotropin increases the number of ovulated oocytes in C57BL/6 female mice. PLoS One 10:e0128330
Than NG, Romero R, Goodman M, Weckle A, Xing J, Dong Z, Xu Y, Tarquini F, Szilagyi A, Gal P, Hou Z, Tarca AL, Kim CJ, Kim JS, Haidarian S, Uddin M, Bohn H, Benirschke K, Santolaya-Forgas J, Grossman LI, Erez O, Hassan SS, Zavodszky P, Papp Z, Wildman DE (2009) A primate subfamily of galectins expressed at the maternal-fetal interface that promote immune cell death. Proc Natl Acad Sci U S A 106:9731–9736
Toyooka Y (2020) Trophoblast lineage specification in the mammalian preimplantation embryo. Reprod Med Biol 19:209–221
Tucci V, Isles AR, Kelsey G, Ferguson-Smith AC, Erice Imprinting G (2019) Genomic imprinting and physiological processes in mammals. Cell 176:952–965
Urwin VE, Allen WR (1982) Pituitary and chorionic gonadotrophic control of ovarian function during early pregnancy in equids. J Reprod Fertil Suppl 32:371–381
Valenzuela OA, Couturier-Tarrade A, Choi YH, Aubriere MC, Ritthaler J, Chavatte-Palmer P, Hinrichs K (2018) Impact of equine assisted reproductive technologies (standard embryo transfer or intracytoplasmic sperm injection (ICSI) with in vitro culture and embryo transfer) on placenta and foal morphometry and placental gene expression. Reprod Fertil Dev 30:371–379
Van Niekerk CH, Gerneke WH (1966) Persistence and parthenogentic cleavage of tubal ova in the mare. Onderstepoort J Vet Res 33:195–232
Vitez SF, Forman EJ, Williams Z (2019) Preimplantation genetic diagnosis in early pregnancy loss. Semin Perinatol 43:116–120
Wade CM, Giulotto E, Sigurdsson S, Zoli M, Gnerre S, Imsland F, Lear TL, Adelson DL, Bailey E, Bellone RR, Blocker H, Distl O, Edgar RC, Garber M, Leeb T, Mauceli E, Macleod JN, Penedo MC, Raison JM, Sharpe T, Vogel J, Andersson L, Antczak DF, Biagi T, Binns MM, Chowdhary BP, Coleman SJ, Della Valle G, Fryc S, Guerin G, Hasegawa T, Hill EW, Jurka J, Kiialainen A, Lindgren G, Liu J, Magnani E, Mickelson JR, Murray J, Nergadze SG, Onofrio R, Pedroni S, Piras MF, Raudsepp T, Rocchi M, Roed KH, Ryder OA, Searle S, Skow L, Swinburne JE, Syvanen AC, Tozaki T, Valberg SJ, Vaudin M, White JR, Zody MC, Broad Institute Genome Sequencing P, Broad Institute Whole Genome Assembly T, Lander ES, Lindblad-Toh K (2009) Genome sequence, comparative analysis, and population genetics of the domestic horse. Science 326:865–867
Wang X, Miller DC, Clark AG, Antczak DF (2012) Random X inactivation in the mule and horse placenta. Genome Res 22:1855–1863
Wang X, Miller DC, Harman R, Antczak DF, Clark AG (2013) Paternally expressed genes predominate in the placenta. Proc Natl Acad Sci U S A 110:10705–10710
Weber JA, Freeman DA, Vanderwall DK, Woods GL (1991) Prostaglandin E2 secretion by oviductal transport-stage equine embryos. Biol Reprod 45:540–543
Wildman DE (2011) Review: toward an integrated evolutionary understanding of the mammalian placenta. Placenta 32:S142–S145
Wildman DE, Chen C, Erez O, Grossman LI, Goodman M, Romero R (2006) Evolution of the mammalian placenta revealed by phylogenetic analysis. Proc Natl Acad Sci U S A 103:3203–3208
Wilsher S, Allen WR (2003) The effects of maternal age and parity on placental and fetal development in the mare. Equine Vet J 35:476–483
Wilsher S, Allen WR (2011) Intrauterine administration of plant oils inhibits luteolysis in the mare. Equine Vet J 43:99–105

Wilsher S, Allen WR (2012) Factors influencing placental development and function in the mare. Equine Vet J Suppl 41:113–119

Wilsher S, Newcombe JR, Allen WR (2019) The immunolocalization of Galectin-1 and progesterone-induced blocking factor (PIBF) in equine trophoblast: possible roles in trophoblast invasion and the immunological protection of pregnancy. Placenta 85:32–39

Wooding FB, Morgan G, Fowden AL, Allen WR (2001) A structural and immunological study of chorionic gonadotrophin production by equine trophoblast girdle and cup cells. Placenta 22:749–767

Woods GL, Baker CB, Baldwin JL, Ball BA, Bilinski J, Cooper WL, Ley WB, Mank EC, Erb HN (1987) Early pregnancy loss in brood mares. J Reprod Fertil Suppl 35:455–459

Implantation and Placentation in Ruminants

Jonathan A. Green, Rodney D. Geisert, Greg A. Johnson, and Thomas E. Spencer

Contents

Abstract In comparison to many other mammalian species, ruminant ungulates have a unique form of placentation. Ruminants initially display an epitheliochorial type of placentation; however, during the period of placental attachment, trophoblast giant binucleate cells (BNC) develop within the chorion to migrate and fuse with the uterine surface epithelium to form syncytial plaques. Binucleate cell migration and fusion continues throughout pregnancy but never appears to breach the basal lamina,

J. A. Green (✉)
Division of Animal Sciences, College of Agriculture, Food, Natural Resources, S138 Animal Science Research Center, University of Missouri, Columbia, MO, USA
e-mail: greenjo@missouri.edu

R. D. Geisert
Division of Animal Sciences, College of Agriculture, Food, Natural Resources, S121 Animal Science Research Center, University of Missouri, Columbia, MO, USA
e-mail: geisertr@missouri.edu

G. A. Johnson
Department of Veterinary Integrative Biosciences, College of Veterinary Medicine and Biomedical Sciences, Texas A&M University, College Station, TX, USA
e-mail: gjohnson@cvm.tamu.edu

T. E. Spencer
Division of Animal Sciences, College of Agriculture, Food, Natural Resources, S135 Animal Science Research Center, University of Missouri, Columbia, MO, USA
e-mail: spencerte@missouri.edu

R. D. Geisert, T. E. Spencer (eds.), *Placentation in Mammals*, Advances in Anatomy, Embryology and Cell Biology 234, https://doi.org/10.1007/978-3-030-77360-1_7

beneath the uterine surface or luminal epithelium. Therefore, the semi-invasive type of placentation in ruminants is classified as synepitheliochorial. The endometrium of ruminant species also contains unique specialized aglandular structures termed "caruncles" in which the chorioallantois (cotyledons) interdigitates and forms highly vascularized fetal–maternal "placentomes." This chapter will discuss the current knowledge of early conceptus development during the peri-attachment period, establishment of pregnancy, conceptus attachment, and placentation in ruminant ungulates. The features of placentomes, BNCs, fetomaternal hybrid cells, and multinucleated syncytial plaques of the cotyledonary placenta of ruminant species will be reviewed to highlight the unique form of placentation compared to the placentae of other artiodactyls.

Keywords Ruminant · Bovine · Ovine · Pregnancy · Conceptus · Uterus · Implantation · Placentation

1 Introduction

Attachment of the embryo or conceptus (embryo and extraembryonic membranes) to the maternal uterine endometrium followed subsequently by development of the placenta is the means by which eutherians provide the environment needed to support the prolonged period of in utero fetal development until parturition. It is notable that the early stages of embryo development between fertilization of the oocyte and blastocyst formation are conserved among mammalian viviparous species. It is only following hatching of the blastocyst from its zona pellucida coat that distinct deviations in the steps of mammalian implantation and placental development become apparent. Indeed, there is great variation in the area and depth of placental interactions with the maternal endometrium. Examples of such differences range from the noninvasive, epitheliochorial attachment in the pig and horse to the invasive, endotheliochorial, and hemochorial types of implantation in canines, rodents, and primates (Grosser 1909; Mossman 1937; Amoroso 1952; Wimsatt 1974). The acquisition of the epitheliochorial placenta in some larger species provides an interesting contrast relative to any advantages that an invasive placental form might offer (Mossman 1974; Carter and Enders 2013). Ruminants have a unique semi-invasive type of attachment in which formation of trophoblast giant binucleate cells (BNCs) migrate and fuse with the luminal epithelium (LE). The "invasiveness" of the BNC appears to be limited to fusion with the uterine epithelium to form syncytial plaques that do not penetrate the basal lamina present beneath the LE. The "synepitheliochorial" type of attachment is functionally intermediate between the noninvasive placentation in the pig and the invasive endotheliochorial placenta in canines (Chavatte-Palmer and Guillomot 2007). In addition to the unique form of conceptus attachment, the endometrium of the bicornuate uterus of ruminant

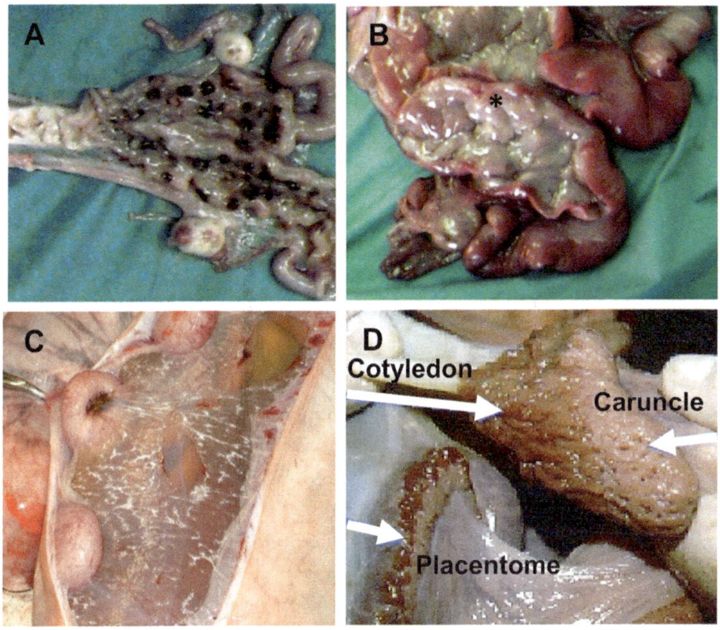

Fig. 1 Bicornuate uterus of the ewe (**a**) and heifer (**b**) opened to show the endometrial caruncles (asterisk) within the uterine lining. The presence of melanocytes in caruncles of blackface ewes results in the dark-pigmented color of the caruncles in panel **a**. (**c**) Sheep "concave" placentomes on day 120 of pregnancy. (**d**) Fetal cotyledons and maternal caruncles which form the "convex" placentome of the bovine placenta on day 240 of pregnancy

species is lined with specialized aglandular structures termed "caruncles" (Fig. 1a, b). Interdigitation of chorioallantois (cotyledons) and formation of crypts within the caruncles ultimately leads to the formation of highly vascularized fetal-maternal "placentomes" (Fig. 1c, d). The caruncles of ruminants develop in the endometrium after birth, and the total number varies between individual animals and species (Wiley et al. 1987; Hradecky et al. 1988). During pregnancy, the bovine, ovine, caprine, and *Antilopinae* (gazelles, blackbucks) placenta can have 75 to 125 placentomes, while *Cervidae* (deer) and other ruminant species may contain only 4–6 placentomes (Hradecky 1983). The number of cotyledons present in the giraffe can range from 123 to 191. The overall shape of the placentomes ranges from primarily flat (deer, antelope), convex (cattle) to concave (sheep, goat) (Mossman 1987). The camel is considered a ruminant. However, it is quite distinct evolutionarily from the *Ruminantia* and its placenta resembles the diffuse form of pigs and horses (Skidmore et al. 1996).

In this chapter, the current knowledge related to early embryonic development, implantation, and placentation in the cattle and sheep is discussed. The synepitheliochorial type of uterine attachment and the cotyledonary placenta are unique adaptations in nearly all ruminant ungulate species. The ruminant placenta represents a form of placental interaction with the maternal uterus that is distinct in comparison to other mammals.

2 Early Conceptus Development

After fertilization, the embryo undergoes cell divisions to reach the morula stage, which occurs around day 5 of gestation in cattle (Brackett et al. 1980). Initially both outer and inner cells of the morula do not have well-developed intercellular junctions. Instead, they are in contact only through cellular focal and gap junctions. Subsequently, the outer cells establish tight junctions, with a concomitant increase in the extent of cell-cell contacts. Meanwhile the inner cells retain the focal and gap junction. This process represents compaction of the morula. A blastocoelic cavity is formed through fluid movement into the intercellular spaces of the compact morula. The osmotic gradient driving blastocoel formation is established by Na^+/K^+-ATPase (a "sodium pump") present within the basement membrane of the outer (trophectoderm) cells. Formation of outer polarized and inner nonpolarized cells of the blastocyst triggers differential cellular activity of the Ras and Hippo signaling pathways to induce lineage segregation (see Telugu, Chapter "Development of Pre-Implantation Mammalian Blastocyst").

In mice, cellular expression of NANOG and SOX2 prevents induction of a trophectoderm lineage and leads to the formation of the inner cell mass (ICM), while activation of the caudal-related homeobox gene *Cdx2* by Tead4/Yap stimulates trophectoderm lineage (Cockburn and Rossant 2010; see Telugu, Chapter "Development of Pre-Implantation Mammalian Blastocyst"). However, care must be taken in extending the mouse model to other mammals as there is variability among species (Spencer et al. 2006; Berg et al. 2011; Negrón-Pérez et al. 2018; Llobat 2020). In mice, expression of the octamer-binding transcription factor Pou5f1, also known as Oct4, is restricted to the ICM, which is not the case during early bovine blastocyst development (Berg et al. 2011). Pou5f1 is co-expressed with Cdx2 in the trophectoderm, even after blastocyst development. The possible co-expression of Pou5f1 in the trophectoderm may be related to the extended period of peri-attachment and growth that is observed in the conceptuses of ungulates.

The bovine blastocyst hatches from its zona pellucida (Fig. 2a) on day 9–10 of gestation by weakening the zona through enzymes produced by both the blastocyst and endometrium as well as expanding and contracting to create a fracture in the zona glycoprotein coating. After hatching, the conceptus takes on a spherical and

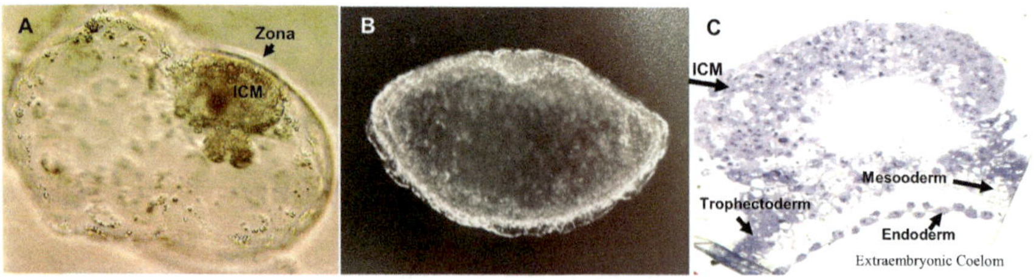

Fig. 2 Photomicrographs of (**a**) hatching day-10 bovine blastocyst, (**b**) ovoid 1 mm day-14 bovine conceptus, and (**c**) histological section of ICM and trophoblast of a day-15 bovine conceptus

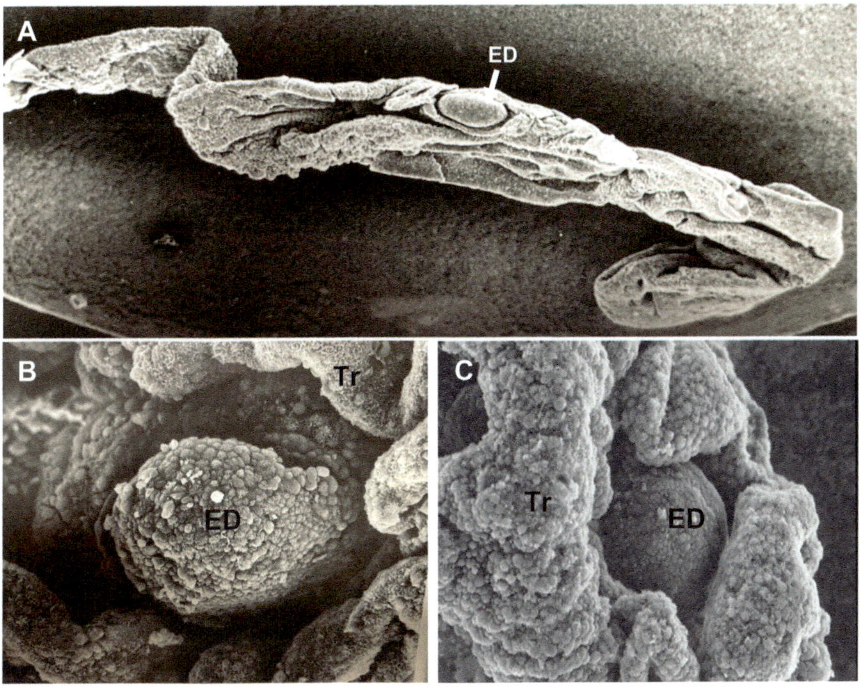

Fig. 3 Scanning electron microscopy photograph of (**a**) elongated 15 mm day-15 bovine conceptus, (**b**) embryonic disc of a day-15 bovine conceptus, and (**c**) embryonic disc descending into the trophoblast folds to form the amnionic sac of a bovine conceptus. *ED* embryonic disc, *Tr* trophoblast. (SEM photomicrographs courtesy Michael F. Smith University Missouri)

then an ovoid morphology (Fig. 2b). Differentiation of germ layers (Fig. 2c) begins in the ICM soon afterwards (Perry 1981). The establishment of the germ layers during gastrulation is the prelude to formation of the chorion, amnion, yolk sac, and allantoic membranes typical of the chorioallantoic type of placenta in ungulates (Perry 1981). During this period, the spherical conceptus slowly grows from 1 mm to 8 mm between day 12 and 14 of gestation (Betteridge et al. 1980). By day 15, endoderm completely lines the basal membrane of the trophectoderm of the ovoid conceptus (Fig. 2b). Mesoderm from the ICM migrates between endoderm and trophectoderm layers to form a trilaminar trophoblast (Fig. 2c). The mesoderm separates to associate with the endoderm (forming the vascular splanchnopleure yolk sac) and the outer trophectoderm (to form the avascular somatopleure) of the developing chorion. Thus, the origin of the extraembryonic membranes begins to form in ruminants prior to firm attachment to the uterine luminal surface epithelium.

The trophectoderm covering the ICM (Rauber's layer) is lost during the period of early conceptus expansion (Betteridge and Flechon 1988). The growing embryonic disc appears as a rounded sphere on the surface of the day 16 elongated conceptus (Fig. 3a). On days 16 to 18, the embryonic disc appears to descend (Fig. 3b) into the conceptus as the developing trophoblast layer (somatopleure) begins to fold around and engulf it (Fig. 3c). Closure of the trophoblast fold over the embryonic disc establishes the amnion and forms the cavity for fluid accumulation within the amniotic sac. Development of the vascular yolk sac, which is in close apposition to the mesenchyme of the chorion (Fig. 2c), serves an essential function to transfer nutrients from the chorion surface (choriovitelline) to the embryos during early development.

3 Conceptus Elongation, Apposition, and Establishment of Pregnancy

The changes in early embryo development, attachment, and placental expansion in cattle and sheep have been documented and reviewed (Leiser 1975; King et al. 1982; Betteridge and Flechon 1988; Assis Neto et al. 2010). Growth and elongation of the bovine conceptus within the uterine horn is accelerated from day 15 to 18 of gestation. During this period, the 2 mm diameter filamentous conceptus reaches lengths over 200 mm (Fig. 4) to extend into the contralateral horn by day 18 to 19 of gestation (Greenstein et al. 1958; Betteridge et al. 1980). Conceptus elongation in sheep is initiated on day 12, and the conceptus membranes extend into the contra-lateral horn by day 17 (Rowson and Moore 1966). Unlike the rapid morphological

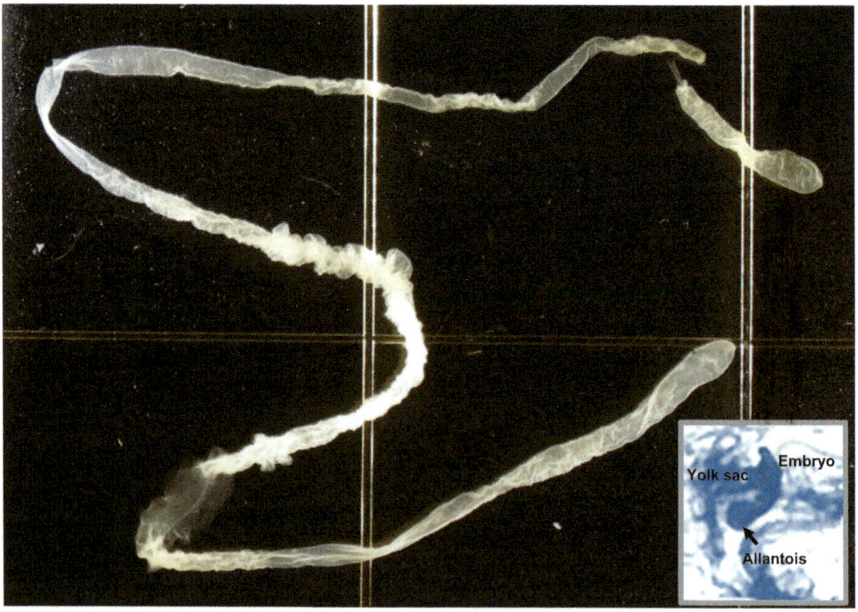

Fig. 4 Photomicrograph of an elongated 220 mm day-19 bovine conceptus. Insert: early embryo with yolk sac and outgrowth of allantois from the hindgut of a bovine conceptus. (Photo courtesy of Sofia Ortega, University of Missouri)

elongation of pig conceptuses (Geisert et al. 2015), the increase in the length of cattle and sheep conceptuses occurs through cell proliferation, migration, and morphological change in shape over 2–3 days (Wang et al. 2009; Bazer et al. 2010).

The bovine conceptus is rather weakly attached to the uterine surface epithelium during the period of elongation until approximately day 23 of gestation. During this timeframe, connections with the endometrium begin near the embryonic disk and spread from there (King et al. 1980). For the most part, apposition and attachment is similar among ruminants. A few differences exist, however. In sheep, the apposition steps begin with a decrease in trophoblast microvilli between days 13 and 15 (Guillomot et al. 1981, 1982, 1993). The transient loss of microvilli may promote apposition of trophoblasts with the LE. Besides cellular contacts between trophoblast and endometrial surface epithelium, the portion of the trophoblast overlying the uterine gland openings extend papillae into the superficial mouths of the glands between days 15 and 17 of pregnancy (Fig. 5a). Development of the trophoblastic papillae is proposed to anchor the peri-attachment conceptus to the uterine surface (Fig. 5b) during elongation throughout the uterine horns in the cow and sheep (Guillomot et al. 1981; Guillomot and Guay 1982). The trophoblastic papillae may also serve to absorb gland secretions until the formation of the areolae in the developing chorion.

The increase in the rate of conceptus elongation during this peri-attachment period is regulated by changes in the transcriptome and secretions of the endometrium which are programmed by progesterone following ovulation (Forde et al. 2011; Spencer and Hansen 2015; Spencer et al. 2016). In cattle, changes in endometrial secretions are sensitive to the rise of progesterone following ovulation and timing of the downregulation of progesterone receptor (PGR) in the endometrial luminal (LE) and glandular (GE) epithelium (see Lonergan et al. 2016). An earlier increase in circulating progesterone following ovulation is associated with advancing conceptus development and elongation. Exogenous treatment of bred cows with progesterone immediately after ovulation advances conceptus elongation on day 15 of gestation (Garrett et al. 1988) which is similar to the timing of elongation in sheep (Satterfield et al. 2006). Interactions between the conceptus and the endometrium during the peri-implantation period of conceptus elongation are sensitive to endometrial changes because dysregulation of these changes leads to downstream effects on conceptus survival and later embryonic loss in subfertile cows (Moraes et al. 2018).

During the peri-attachment period, sheep and bovine endometrium and conceptuses secrete prostaglandins as expression of prostaglandin-endoperoxide synthase 2 (PTGS2) is enhanced (Charpigny et al. 1997a, b). Prostaglandins modify the endometrial transcriptome and influence endometrial secretions to stimulate conceptus growth and survival (Simmons et al. 2009). Brooks et al. (2015) demonstrated that peroxisome proliferator activator receptor gamma (PPARG), a nuclear factor that binds to prostaglandins, is an essential regulator of conceptus proliferation and survival. In addition, conceptus development and elongation in sheep is severely comprised when endometrial and conceptus prostaglandin production is inhibited

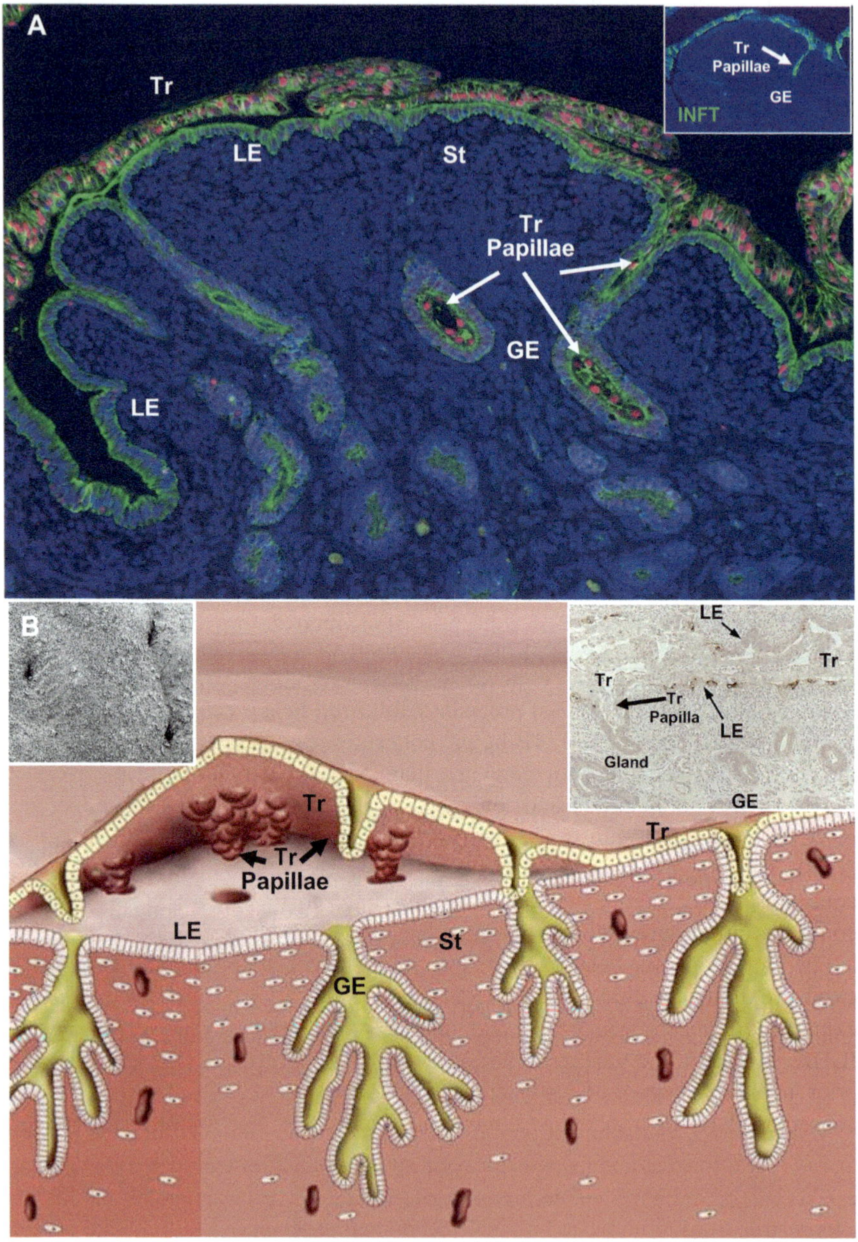

Fig. 5 During the peri-attachment period of conceptus elongation and expansion within the uterine lumen, apposition of the conceptus trophoblast (Tr) to the uterine luminal epithelium (LE) is aided by the development of trophoblastic papillae into the endometrial glands of sheep and cattle. (**a**) Immunofluorescence staining E-cadherin (green) and PCNA (red) of a serial section from a day-18 sheep pregnant uterus containing an expanded conceptus (10×). The intensive PCNA staining of the growing conceptus shows the close of apposition of the trophoblast to the LE and the extension of a trophoblastic papilla into the mouth of an endometrial gland. The insert image displays the trophoblast expression of IFNT (green fluorescence) to illustrate the depth of the papilla penetration

with a PTGS2-specific inhibitor during the peri-attachment period (Dorniak et al. 2011). The progesterone-driven uterine transcriptome, proteome, lipidome, and metabolome play a critical role in conceptus growth and elongation essential to downstream placental development and survival.

Elongation of the bovine and ovine conceptus through the ipsilateral uterine horn on days 15 to 18 of pregnancy is critical to not only provide the essential uterine surface area for adequate nutrient and oxygen exchange but to inhibit luteolysis and extend progesterone production beyond the length of the estrous cycle. Maternal recognition of pregnancy in ruminants requires not only rapid growth, elongation, and apposition across the uterine lumen surface (Fig. 6a), but conceptus expression and secretion of the type 1 interferon tau (IFNT) (Fig. 6d) to prevent regression of the CL (see Spencer and Hansen 2015).

During the estrous cycle and early pregnancy, progesterone receptor (PGR) is expressed in the endometrial stroma cells, but no PGR expression is observed in the LE or GE (Fig. 6c). Loss of epithelial PGR in cyclic females allows upregulation of estrogen receptor, and follicular estrogen production stimulates an increase in endometrial oxytocin receptor expression. The release of prostaglandin F2α (PGF2α) initiates release of oxytocin from the CL which forms a positive feedback loop for increased pulsatile release of endometrial PGF2α leading to luteolysis (Spencer and Hansen 2015). Briefly, during establishment of pregnancy (days 12–15 in sheep and 15–21 in cattle), conceptus growth (Fig. 5b) drives expansion and apposition to the endometrial LE (Fig. 6a) and positions the trophoblast so that secreted IFNT (Fig. 6a, d) can bind to type 1 IFN receptors present on the endometrial LE inhibiting the transcription of estrogen receptor alpha (ESR1) (Fig. 6e). Inhibition of ESR1 prevents establishment of the positive feedback loop necessary for the pulsatile release of PGF2α from the endometrium preventing luteolysis (Spencer and Hansen 2015). Conceptus secretion of IFNT can also help regulate the endometrial transcriptome and possibly regulate immune function to prevent rejection of the semi-allograft fetus (Bazer et al. 2015, 2020).

Fig. 5 (continued) down into the endometrial gland (10×). (**b**) Cartoon illustration of apposition of the ruminant conceptus before attachment of the trophoblast to the uterine LE on day 20 of gestation. The development of the cone-shaped trophoblastic papilla into the mouth of the uterine glands is proposed to anchor the conceptus during elongation and expansion and provides absorption of glandular secretions. Left insert: scanning ECM of the openings of uterine glands on the endometrial surface of a heifer. Right insert: immunohistochemical section of a papilla extending into uterine gland during early pregnancy. *LE* luminal epithelium, *GE* glandular epithelium, *St* stroma

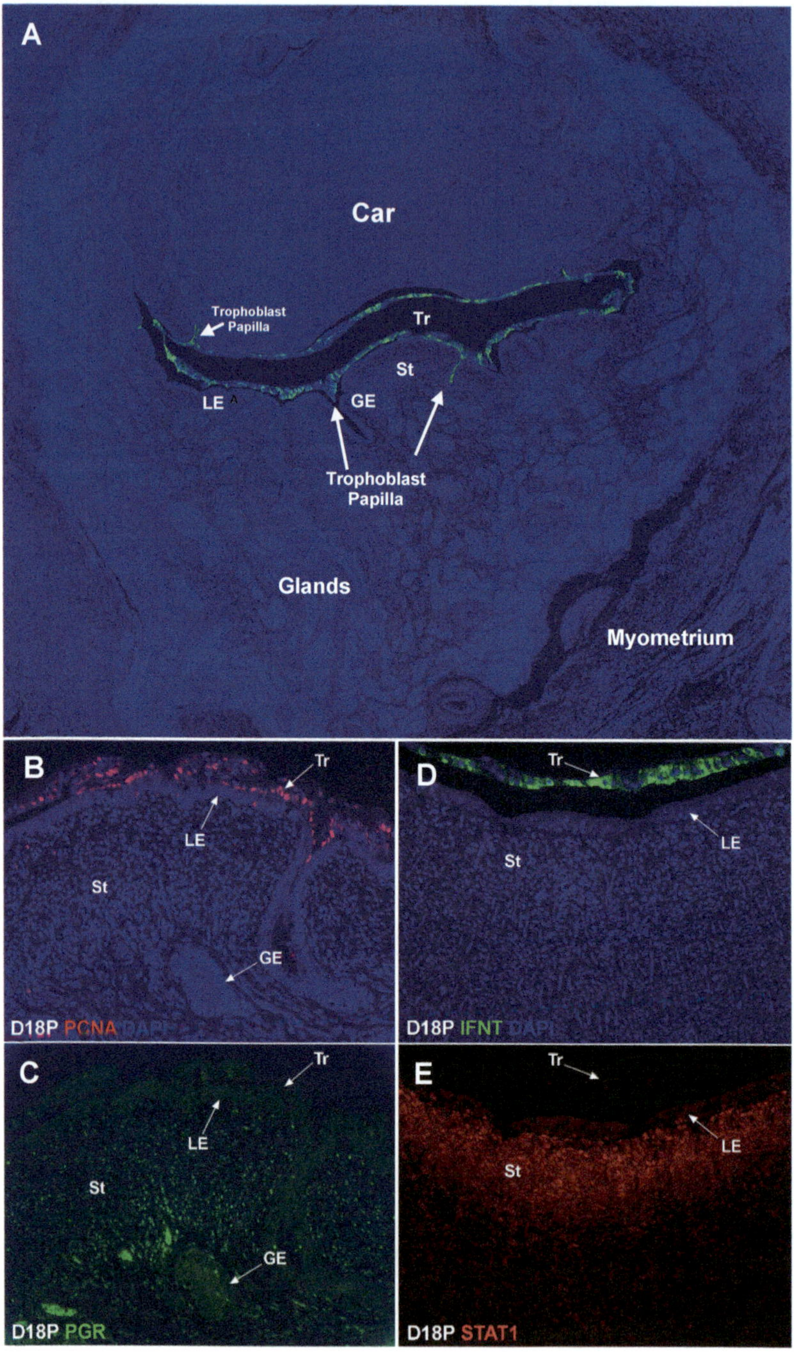

Fig. 6 Top panel: During early pregnancy in sheep, the mononuclear cells of the placental trophectoderm synthesize and secrete interferon tau (IFNT), the signal for maternal recognition of pregnancy. (**a**) Paraffin-embedded thin section of an elongated, attaching conceptus within the uterine lumen of a day-18 pregnant sheep that has been immunofluorescence stained for IFNT

4 Conceptus Attachment and Placental Formation

The initial attachment between bovine trophoblast and uterine surface epithelium is evident on days 18 to 21 (King et al. 1980). With continued development and differentiation of the chorion, firm attachment occurs through interdigitation of the microvilli with the epithelium of the caruncular and intercaruncular areas of the uterine lining. At this point in development, the tips of the chorion extend throughout both uterine horns (King et al. 1982).

During implantation in the mouse, attachment to the uterine luminal epithelial surface involves progesterone induced loss of the antiadhesive MUC1 which serves as a physical interference at the LE for attachment (Surveyor et al. 1995). Expression of MUC1 is not downregulated by progesterone in the LE of sheep during the midluteal stage of the estrous cycle but is reduced at the conceptus–endometrial surface interface in association with conceptus IFNT production (Raheem et al. 2016). Loss of MUC1 allows cell to cell interactions with adhesion factors such as integrins, which are needed to facilitate attachment (Kim et al. 2010; Johnson et al. 2001, 2014). Multiple molecules such as GlyCAM (Spencer et al. 1999a, b), cadherins, galectin-15 (LGALS15), and secreted phosphoprotein 1 (SPP1) are proposed to form complexes with integrins and glycoconjugates expressed on the aligned apical surfaces of the trophoblast and LE (Johnson et al. 2018; Spencer et al. 2004a, b).

The apical surfaces of the interface between the endometrial LE and placenta can express integrin receptors $\alpha v\beta 3$, $\alpha v\beta 1$, avb5, $\alpha 4\beta 1$, and $\alpha 5\beta 1$ during conceptus attachment and throughout gestation in sheep (Johnson et al. 2001; Burghardt et al. 2009). Endometrial expression of SPP1, also known as osteopontin, serves as a crosslinker to bind the cellular integrin receptors together (Fig. 7a). Seo et al. (2020) found that $\alpha v\beta 3$, $\alpha v\beta 5$, and $\alpha 4\beta 1$ (Fig. 7b, c) serve as the integrin receptors for conceptus attachment during the peri-implantation period of pregnancy in sheep. Only $\beta 3$ is exclusively limited to the apical surface of endometrial LE during the period of peri-attachment suggesting a specific role of $\beta 3$ during attachment of conceptus. Interestingly, $\beta 5$ expression is present in the LE and extends down into the openings of the GE. It is possible that $\beta 5$ integrin associates with trophectoderm

Fig. 6 (continued) (green), and stained with DAPI (blue) for histological reference. The conceptus is attached to the LE of a glandular intercaruncular area of the endometrium with the papillae extending in the mouth of uterine glands and luminal epithelium (LE) of the nonglandular caruncle (Car). Higher magnification of the same implantation site illustrating (**b**) multiple proliferating conceptus trophectoderm cells [proliferating cell nuclear antigen (PCNA)], but few proliferating endometrial LE cells, (**c**) progesterone receptors (PGRs) are expressed by the endometrial stromal cells, but no PGR expression is observed in the LE or glandular epithelium (GE), (**d**) IFNT expression by trophectoderm cells, and (**e**) The classical interferon stimulated gene [signal transducer and stimulator of transcription 1 (STAT1)] expression by cells in the stroma (St), but no expression of STAT1 by LE cells. (Immunofluorescence staining was performed by Dr. Heewon Seo, Department of Veterinary Integrative Biosciences, Texas A&M University)

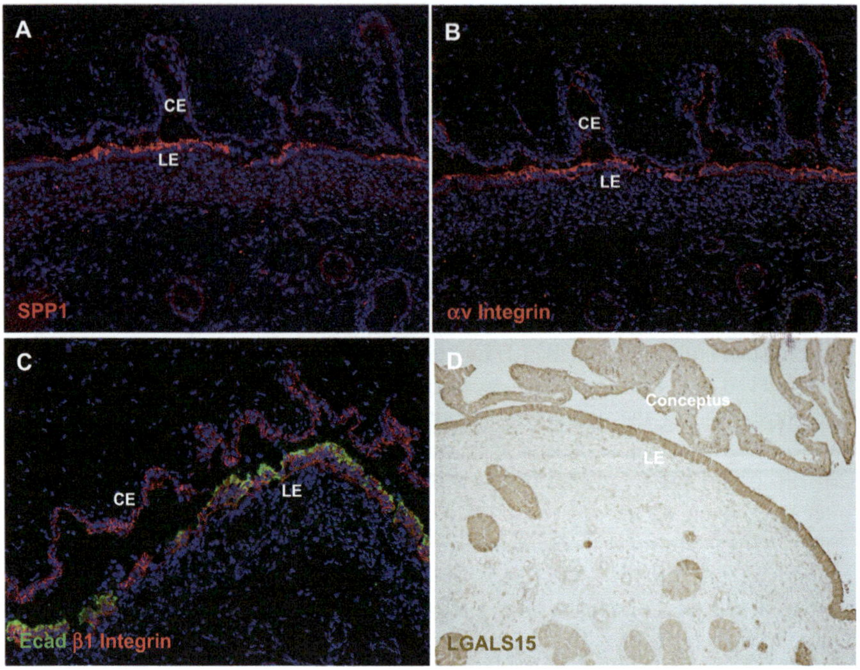

Fig. 7 Immunofluorescence microscopy illustrating the distribution of (**a**) SPPI, (**b**) αv integrin, and (**c**) β1 integrin (red) and E-cad (green, immunostained endometrial luminal epithelial (LE) cells) at the endometrial–placental interface on day 60 of pregnancy in the sheep (10×). (**d**) Immunohistochemical localization of galactin-15 (LGALS15) in the sheep endometrium and conceptus on day 16 of pregnancy. *CE* chorionic epithelium. (The images in this figure adapted from Seo et al. 2020 and Gray et al. 2004)

papillae that anchor the conceptus during elongation and help provide absorption of glandular secretions during early development.

Galectin-15 has a carbohydrate recognition domain that binds to beta-galactosides (Cooper 2002) and contains sequences that can interact with integrins (Spencer et al. 2004a, b). In sheep, the conceptus and placenta do not produce LGALS15 which is specifically expressed by endometrial LE and superficial GE (Gray et al. 2004). The endometrial LGALS15 is secreted into the uterine lumen throughout pregnancy where it is absorbed by trophoblasts and utilized in cell adhesion or internalized to regulate cell growth, differentiation, and apoptosis (Fig. 7d). LGALS15 is absent in placentomes when the caruncular surface is devoid of LE during syncytial formation (Gray et al. 2004). However, LGALS15 is present in the intercaruncular area of the endometrium when the LE are reestablished.

During the peri-attachment period, the yolk sac develops from days 18 to 23 of gestation (Wrobel and Suess 1998) to provide the initial vascular system to transfer nutrients to the developing embryo (Assis Neto et al. 2010). The yolk sac provides the vascular transport of blood and nutrients to the developing embryo and it is also the source of primordial germ cells (Kritzenberger and Wrobel 2004). It also contributes specific proteins and metabolites to the developing embryo (Mançanares et al. 2013; Galdos-Riveros et al. 2015). Attachment of the chorion to the uterine surface is closely followed by a projection of extraembryonic splachnopleure from the embryo hindgut (Fig. 4a) which initiates allantois formation on day 21 of gestation (Maddox-Hyttel et al. 2003). The fluid-filled vascular allantois rapidly expands throughout the extraembryonic coelom (Fig. 8), while the yolk sac concurrently regresses becoming vestigial by day 55 although it is still present at term (Assis Neto et al. 2010). The chorion and allantois mesenchyme fuse to form the vascular chorioallantoic placenta from day 26 to 60 of gestation which will be present throughout the remainder of gestation (Fig. 8). The accumulation of fluid in the allantois serves to expand the chorioallantoic membranes forcing apposition into the intercaruncular and caruncular surface of the maternal endometrium. Fluid within the amnion is a different composition compared to the allantois (Fig. 8) where it functions to provide a protective environment for embryo/fetal development. Ligation of the urachus from the fetal kidney within the umbilical cord results in the loss of allantoic fluid during late pregnancy in sheep. The passage of fluid from the embryonic kidney through the urachus in the umbilical cord has led to the misconception that the allantois functions as a reservoir for fetal waste. However, the kidney only redistributes fluid as the maternal vasculature provides the water for the fetus and placenta (Bazer 1989). The allantois contains ions, carbohydrates (glucose, fructose), and amino acids as well as selective proteins and hormones (Bazer et al. 1981; Bazer and Johnson 2018). Therefore, the allantois provides a reserve of nutrients essential for the rapid demands of fetal growth throughout gestation in ungulates. Fluid in the amnion and allantois may also be transported directly through aquaporin water channels expressed in the chorioallantois (Zhu et al. 2015). Expression of transporters for ions, glucose, and amino acids in the chorioallantoic membranes contributes to maternal transfer of electrolytes, glucose, amnio acids, and nutrients (Bazer et al. 2012).

As the placental membranes continue to grow, fluid volumes of the amnion and allantois increase and change throughout gestation in cattle and sheep. In cattle, allantoic fluid volume increases from 50 mL on day 30 to 600 mL by day 100 while amnion fluid volume reaches 450 mL (Eley et al. 1978). In cattle, fluid volume is over 5 L within the allantois and 2 L in the amnion during late gestation (day 260) with total volumes reaching 9 L near term (Rasby et al. 1990). In sheep, there is a similar increase in allantoic fluid volume from day 25 to 40 (21 to 91 mL) as the allantois expands within the chorion, but the fluid volume declines by day 70 (32 mL) (Gresham et al. 1972). A second increase in allantoic fluid volume continues to near term on day 140 (438 mL). These dynamic changes in allantoic fluid volume likely impact development and growth of the placenta and fetus.

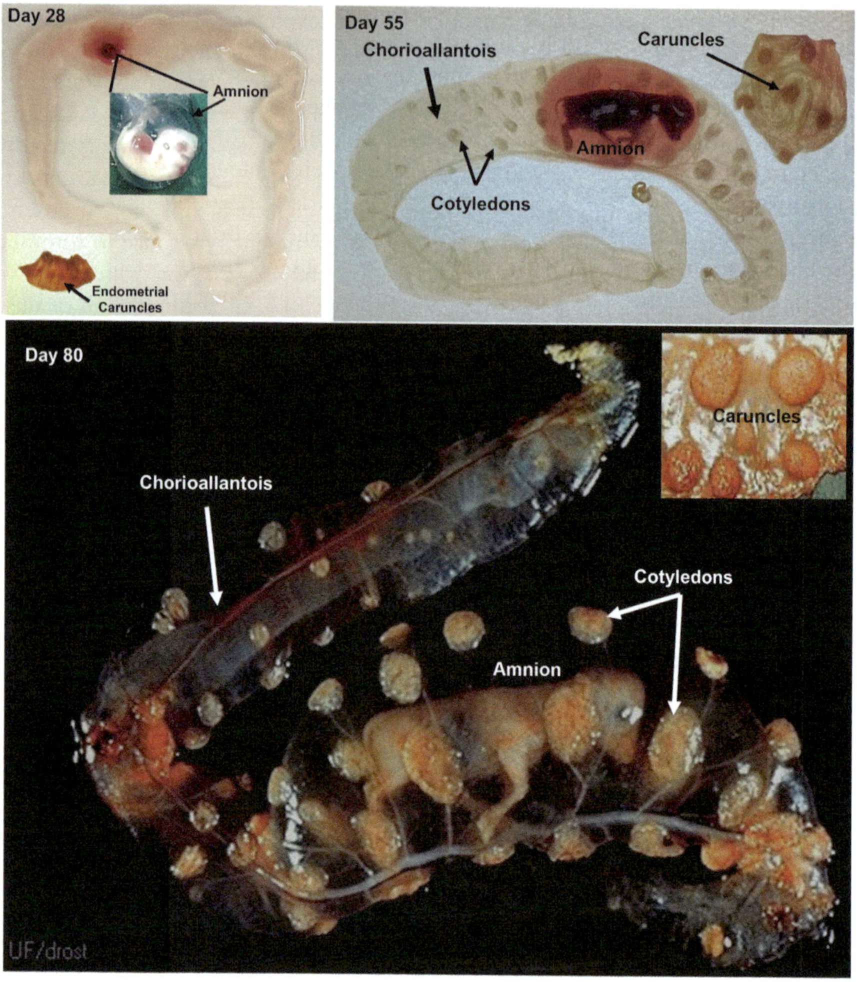

Fig. 8 Placental development during pregnancy in cattle. The fluid-filled allantois expands throughout the chorion which is attached to the entire uterine lumen by day 28 (Insert demonstrates the embryo enclosed in amnion). Although there is some growth of the maternal caruncles, chorionic cotyledons are not visually obvious at this time. The developing embryo is contained within the fluid-filled amnion (the red color which helps highlight the amnion is cause by hemolysis during handling). Cotyledons, which are largest near the amnion, are present on the chorion by day 55. Growth of the chorionic cotyledons occurs from the amnion toward the tips of the chorioallantois. The cotyledons are well developed in the contralateral uterine horn at this time. By day 80, the growing cotyledons and uterine caruncles forming the placentomes are developed throughout both uterine horns. (Photo of day 28 bovine embryo (insert) and day 80 bovine placenta were provided by The Drost Project The Visual Guides of Animal Reproduction (visgar, vetmed.ufl.edu))

5 Placentome Development

The expansion of the chorioallantois brings the chorion into close contact with glandular intercaruncular regions and aglandular caruncular (consisting of connective tissue formations covered by columnar epithelium) areas of the ruminant endometrium (Fig. 9a). The firm attachment of the chorion involves interdigitation at the microvillar junctions between the LE and chorionic epithelium (Fig. 9b). The chorion overlaying the openings to the uterine glands differentiate into areolae (Fig. 9c), which are involved with uptake of secretions into the capillaries of the chorioallantosis (Johnson et al. 2018). In cattle, the chorion does not significantly infiltrate into the endometrial caruncles until after day 30 of gestation (King et al. 1979), and the cotyledonary attachments are not visible on the chorioallantois at this time (Fig. 8). From day 33 to 50 of gestation, the initial attachments between trophoblasts and caruncles are reinforced, and the surface area increases as the chorionic villi (cotyledons) penetrate into septa within the caruncular endometrium. The villi increase in length and undergo extensive branching (Figs. 8 and 9d–f). Simultaneously, development of an extensive blood supply in the villi produces mature "placentomes," which serve as the main site of exchange for easily diffusible nutrients, waste disposal, and for meeting the oxygen demands of the developing fetus (King 1993; Leiser et al. 1998). The size of the placental cotyledons is largest near the amnion where cotyledon formation is first initiated and then spreads towards the tips of the chorion (Fig. 8). Since cotyledons represent the most visually apparent aspect of the mature ruminant placenta, it is also described as a "cotyledonary placenta." The only animals in *Ruminantia* that do not have a cotyledonary placenta are in the *Tragulidae* family (e.g., the mouse deer); however, these animals do have giant binucleate trophoblast cells or BNC (described below) (Kimura et al. 2004). Potentially, the tragulid placenta represents an intermediate form between the epitheliochorial and the cotyledonary placentas of swine and most ruminants, respectively.

6 Trophobast BNC Differentiation and Syncytial Formation

Grosser initially described the ruminant placenta as "syndesmochorial" (Grosser 1909, 1927). He had concluded that the uterine epithelium was lost during ruminant placental establishment and that there was direct apposition of trophoblast with the maternal connective tissue. Subsequently, Wooding and others concluded that the uterine epithelium was altered, but appeared to have remained, albeit sometimes in altered form (Wooding 1982a, b, 1984). Instead of an erosion of uterine epithelium, these histological scientists suggested that there was a fetomaternal syncytial cell layer. Consequently, the term "syndesmochorial placenta" was revised to "synepitheliochorial placentation"; where "syn" stands for the syncytium and

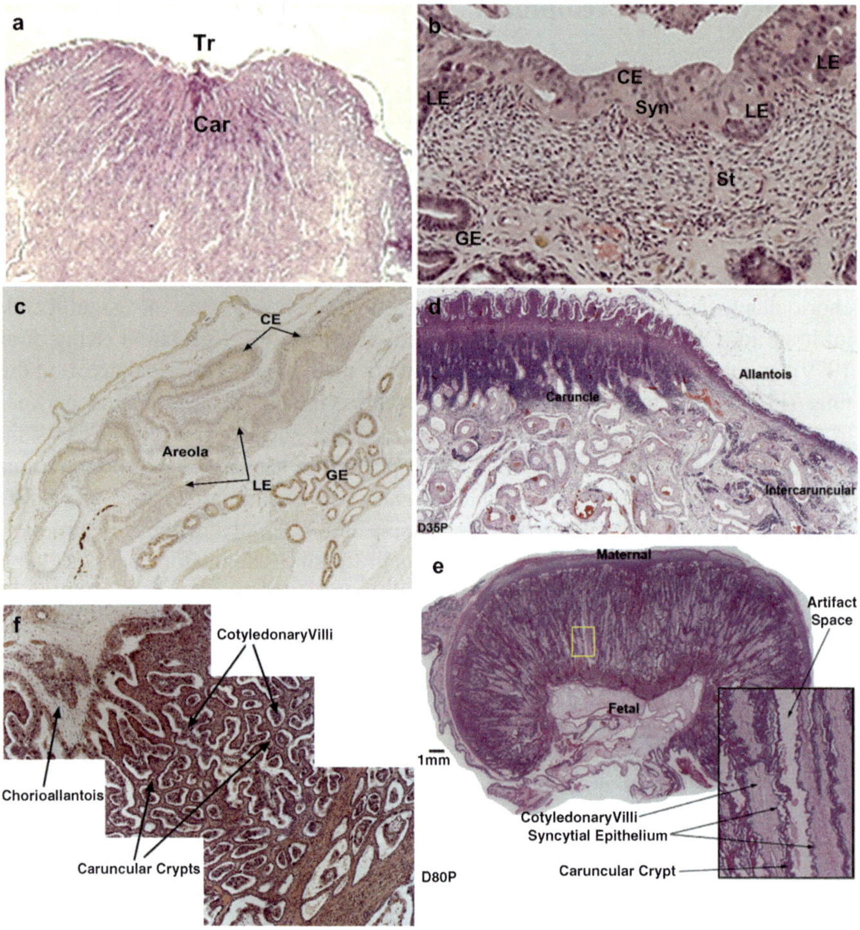

Fig. 9 Attachment and development of the chorioallantois to the intercaruncular and caruncular regions of the uterus. (**a**) Attachment of the trophoblast to the aglandular carunclar region of the uterus on day 21 of pregnancy in cattle (Reprinted/adapted with permission from Journal of Reproduction and Fertility, King et al. 1980), (**b**) bovine chorionic epithelium attachment to the uterine intercaruncular LE and syncytial plaque formation on day 20 of gestation, (**c**) day 60 chorionic areola of a sheep showing absorption of SPP1 secreted by the uterine glands (see Frank et al. 2020 for a detailed description of this pattern of expression), (**d**) day 35, (**e**) 80 placentome development in sheep, and (**f**) Penetration of the cotyledonary villi into the caruncular crypts of a bovine placentome. *Tr* trophoblast, *LE* luminal epithelium, *GE* glandular epithelium, *St* stroma, *CE* chorionic epithelium, *Syn* syncytial plaque

"epitheliochorial" signifies the persistence of recognizable fetal and maternal epithelia at the fetal–maternal interface (Wooding 1992).

During the peri-attachment period of conceptus development, a number of mononuclear cells within the trophoectoderm begin to differentiate into BNC (Fig. 10a) (Spencer and Hansen 2015). The BNCs appear around day 16 pc in sheep (Boshier

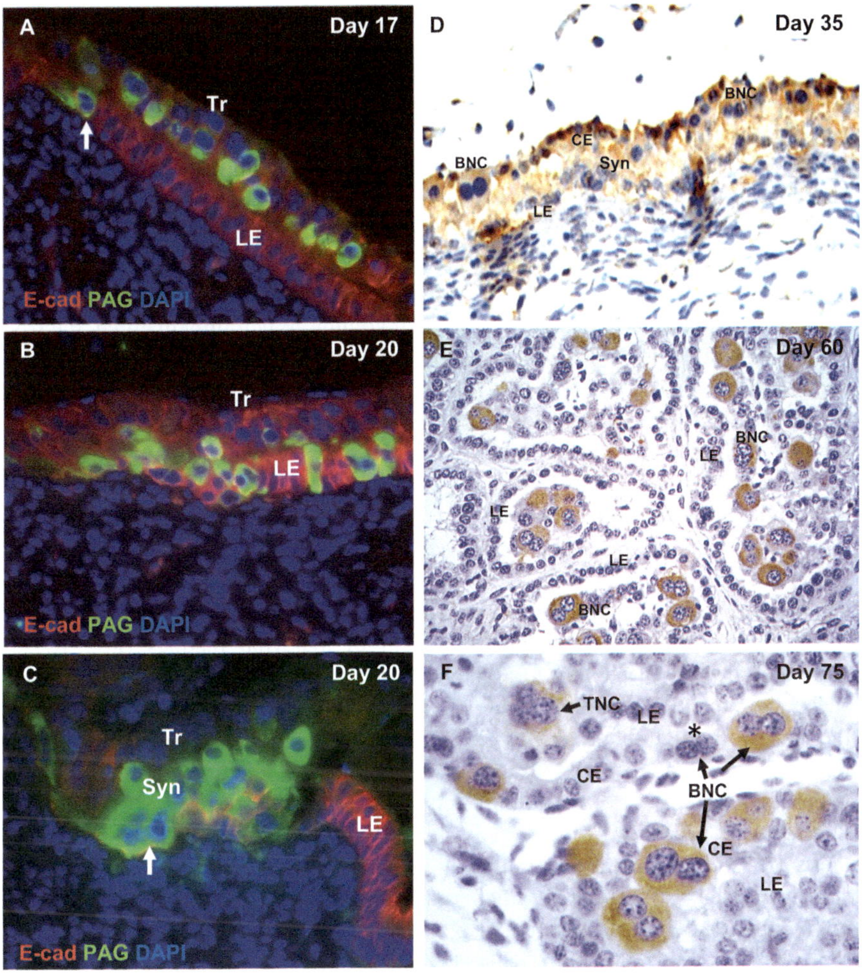

Fig. 10 Interactions between trophoblast giant "binucleated" cells (BNC) and uterine LE cells at the interface between fetal trophoblasts and maternal endometrium. Panels **a**, **b**, and **c** are interface images from sheep that were harvested during early pregnancy (after attachment) which were double immunofluorescence stained for PAG (green) and E-cadherin (red) at attachment sites between day 17 and day 20 of pregnancy. Panels **d**, **e**, and **f** are interface images from cattle that were harvested at various times during the first trimester of pregnancy which were immunostained for PAG with eosin counterstaining to reveal the cells present at the interface. (**a**) Within the trophoblast layer, multiple PAG-stained BNCs are present on day 17 of pregnancy. A BNC has migrated through the microvillar junction between the trophoblast and LE into the uterine LE cell layer (white arrow). (**b**) On day 20, there is extensive migration of BNCs into the uterine LE. (**c**) Uterine attachment site on day 20 where the uterine LE cells are being absorbed and replaced by migrating BNC forming a syncytial plaque. (**d**) Immunostaining of PAG at the bovine caruncular interface of the chorion and LE on day 35 of pregnancy. At this stage, cotyledons are just beginning to project into caruncles to begin placentome formation. A syncytium is present at the interface but syncytia are rarely observed after approximately day 40 of gestation in cattle. However, BNC fusion resulting in the formation of short-lived trinucleated cells continues throughout the remainder of the pregnancy. (**e**) Immunostaining of BNC PAG within a day-60 placentome. The section illustrates the cotyledonary villi containing BNC that have projected into the crypts of the uterine caruncles.

1969), day 18 pc in goat (Wango et al. 1990), and day 19 to 20 pc in cattle (Greenstein et al. 1958). The BNC arise from the mononuclear trophectoderm cells through karyokinesis without cytokinesis (endoreduplication) or from mitotic polyploidy (Wooding and Flint 1994; Wooding et al. 1997; Spencer et al. 2010).

After formation of a vascularized chorioallantois, the BNCs represent approximately 15–20% of the chorionic epithelium, and these numbers remain relatively constant throughout gestation (Wooding et al. 1986; Wooding 1992; Wooding and Flint 1994). The BNC is a terminally differentiated cell, but it does undergo additional maturation. Along with two nuclei, it has an extensive rough endoplasmic reticulum and large golgi bodies that give rise to large numbers of cytoplasmic granules that make up almost 50% of the cell volume. When fully matured, the BNC migrates through the microvillar junctions between the trophoblast and luminal epithelia (Fig. 10a, b and f). They fuse with an uterine epithelial cell (Fig. 10b, f) to create a fetomaternal trinucleate cell (TNC) (Wooding 1987) (Fig. 10c, f). After the fusion, the secretory granules of the BNC transit towards the maternal side of the nascent TNC (Wooding 1987) to be expelled toward the maternal stroma. Following fusion, the TNC reforms part of the apposition with the adjacent trophoblast cells, which maintains an intact barrier between the two interfacing epithelia (Wooding 1987).

After the formation of trinucleate cells, binucleate cell migration and fusion can continue, which leads to the establishment of syncytial plaques (Fig. 10c, d) (Wooding 1984). The extent of syncytium formation differs between species. In sheep, the syncytial plaques are quite extensive; these contain up to 20–24 nuclei and they persist throughout pregnancy (Morgan and Wooding 1983). In cattle, syncytial formation is actually limited (Fig. 10d). No extensive syncytium is present beyond approximately day 40 of gestation (King et al. 1979; Wathes and Wooding 1980; King et al. 1982). After that time, only short-lived TNCs are formed (Fig. 10e, f) and these are soon replaced by epithelial cells (Wooding and Wathes 1980). The situation differs somewhat in small ruminants, such as sheep and goats. In the intercotyledonary regions, syncytial plaques are found initially, but they are soon replaced by uterine epithelium (King et al. 1981; King and Atkinson 1987; Wooding and Flint 1994). Afterwards, migrations and fusion of BNCs with uterine epithelia result in the formation of TNCs, but not a syncytium. In the placentomal regions of sheep, the syncytia persist at the interface throughout the pregnancy (Fig. 10c). The origins of the syncytia continue to be studied, and recent work suggests that further

---

Fig. 10 (continued) Note the LE of the crypts is intact and there are no syncytial plaques as occurs throughout pregnancy in sheep. **(f)** Immunostaining of BNC PAG within a day-75 placentome. Note a newly formed BNC (*) that has initiated PAG synthesis. A trinucleated cell arising from recent migration of a BNC that has integrated within the LE of the caruncular crypt is indicated by black arrow. *Tr* mononuclear trophoblast cells, *BNC* binucleated trophoblast giant cell, *TNC* trinucleated cell, *LE* luminal epithelium, *CE* cotyledonary epithelium (trophoblast), *Syn* syncytium, *E-cad* E-cadherin (red fluorescence, stains mononuclear trophoblasts, and LE), *PAG* pregnancy-associated glycoproteins

refinements of our understanding of ruminant placentation may be warranted. There is some evidence that uterine epithelia in the caruncles may not be part of the syncytial plaques at the interface of sheep (Seo et al. 2019). Of note, colocalization studies provided some evidence that the LE are eliminated and the syncytium itself is comprised of multinucleated syncytiotrophoblast entirely of placental origin (Seo et al. 2019).

As was mentioned previously, the giant trophoblast BNCs represent a cell type that contributes to the unique aspects of the placental form of ruminants. Numerous studies have started to reveal some of the mechanisms underlying the origins and functions of the ruminant-specific BNCs. For example, endometrial and conceptus expression of endogenous retroviruses (ERVs) are proposed to serve a role in placenta morphogenesis (Black et al. 2010) and the migration and fusion of BNC to the surface epithelium in the carunculuar and intercaruncular areas of the uterus (Dunlap et al. 2006a; Spencer et al. 2010). Syncytins, which are expressed by envelope genes of retroviruses and have fusogenic activity, may have a role in placentation. *Syncytin-Rum1, for* which expression is specific to BNC, is associated with the synepitheliochorial placenta of ruminants and is not detected in other *Cetartiodactyla* or primates and rodents (Cornelis et al. 2013; Mi et al. 2000). Endogenous Jaagsiekte sheep retrovirus (enJSRVs) is expressed by both uterine luminal and glandular epithelia (Palmarini et al. 2001; Spencer et al. 2010) as well as in the conceptus trophectoderm (Dunlap et al. 2005). enJSRVs are abundantly expressed in the trophoblast giant BNC and the synytial plaques that arise from them. The receptor for both JSRV and enJSRVs Env, hyaluronoglucosaminidase 2 (HYAL2), is detectable in the BNC and the multinucleated syncytial plaques (Dunlap et al. 2005). Spencer et al. (2010) proposed that the formation of trinucleated fetomaternal hybrid cells arose when co-expressed enJSRVs env and HYAL2 initially fuse with endometrial luminal epithelial (LE) cells that are expressing enJSRVs env. It is worth noting that loss of enJSRVs env expression compromises conceptus growth and implantation in sheep (Dunlap et al. 2006b).

In addition to their contribution to the syncytia and formation of placentomes, it is important to note that BNCs produce steroid (progesterone), prostaglandin (PGI2, PGE2), and protein hormones [placental lactogens or chorionic somatomammotropin hormone one (CSH1)], as well as proteins without a clearly understood function (e.g., pregnancy-associated glycoproteins; PAGs) (Duello et al. 1986; Green et al. 2000; Reimers et al. 1985; Wallace et al. 2015; Wooding et al. 2005; Xie et al. 1991) (Fig. 10d–f). The latter are packaged in the BNC secretory granules. After fusion, the granules are released toward the uterine connective tissue. The PAGs can accumulate in the uterine stroma; they can also enter the maternal circulation (Wooding et al. 2005; Green et al. 2005). Their accumulation in the maternal blood within a few days after BNCs begin to appear in the chorion has made these proteins useful markers of pregnancy in ruminants (Green et al. 2005; Sasser et al. 1986; Zoli et al. 1992). Indeed, the circulating abundance of the PAGs appears to be correlated with placental function or viability of the pregnancy. The amount of certain PAGs at key stages in early pregnancy of cattle can serve to predict which cows are likely to spontaneously abort the pregnancy (Pohler et al. 2013,

2016a, b). Recent evidence supports a dual role for placental lactogen (CSH1) for paracrine effects to stimulate uterine gland secretions for embryonic and placental development (Spencer et al. 1999a, b, 2004a, b; Kelleher et al. 2019).

Thus, trophoblast BNCs have two principal functions: (1) primary contributors to the fetomaternal syncytium required for successful implantation and placentome formation and (2) production and delivery of protein and steroid hormones and numerous other proteins of unknown function to the maternal and fetal systems.

7 Summary

It is hoped that this chapter has provided the reader with an appreciation of the similarities and differences in embryo/conceptus development, attachment to the uterine endometrium, and subsequent placental development, between ruminant ungulates and the other mammals described in this book. Placentomes, BNCs, fetomaternal hybrid cells, and multinucleated syncytial plaques are distinguishing features of the cotyledonary placenta of *Ruminantia*. These modifications relative to other epitheliochorial forms within the *Artiodactyla* order likely reflect the rapid evolution of placental forms, even within the same phylogenetic order. The placenta is considered one of the most rapidly evolving mammalian organs (Roberts et al. 2016; Haig 1996). Among other things, these differences are probably driven by the selection of large, rapidly evolving gene families, the acquisition of endogenous retrovirus-derived genes, and the rapid evolution of placenta-specific enhancers (Roberts et al. 2016; Telugu and Green 2007). There are no structures comparable to placentomes in the placentas of other artiodactyls. Likewise, giant trophoblasts and the syncytial plaques they form are unique aspects of this placental form.

The fusion of BNCs with uterine epithelial cells or the establishment of syncytial plaques in place of an intact uterine epithelium is the extent of invasive implantation in ruminant ungulates. Even so, these alterations would change how the ruminant conceptus interacts with the maternal system. The nature of these interactions (immunological, endocrinological, etc.) is very different in many ways to that observed in pigs, whales, and camels, where erosion of the uterine epithelium does not occur (Roberts et al. 2016; Telugu and Green 2007).

References

Amoroso EC (1952) Placentation. In: Parkes AS (ed) Marshall's physiology of reproduction, vol 2, 3rd edn. Longmans, London, pp 127–311

Assis Neto AC, Pereira FTV, Santos TC, Ambrosio CE, Leiser R, Miglino MA (2010) Morphophysical recording of bovine conceptus (Bos indicus) and placenta from days 20 to 70 of pregnancy. Reprod Domest Anim 45:760–772

Bazer FW (1989) Allantoic fluid: regulation of volume and composition. In: Brace RA (ed) Fetal and neonatal body fluids. Perinatology Press, Cornell, NY, pp 135–157

Bazer FW, Johnson GA (2018) Allantois. In: Skinner MK (ed) Encyclopedia of reproduction, vol 2. Academic Press, Elsevier, New York, NY, pp 559–561
Bazer FW, Goldstein MH, Barron DH (1981) Water and electrolyte transport by pig chorioallantois. In: Mastroianni L, Biggers JD (eds) Fertilization and embryonic development in vitro. Plenum, New York, pp 299–321
Bazer FW, Wu G, Spencer TE, Johnson GA, Burghardt RC, Bayless K (2010) Novel pathways for implantation and establishment and maintenance of pregnancy in mammals. Mol Hum Reprod 16:135–152
Bazer FW, Kim J, Ka H, Johnson GA, Wu G, Song G (2012) Select nutrients in the uterine lumen of sheep and pigs affect conceptus development. J Reprod Dev 58:180–188
Bazer FW, Ying W, Wang X, Dunlap KA, Zhou B, Johnson JA, Wu G (2015) The many faces of interferon tau. Amino Acids 47:449–460
Bazer FW, Seo H, Wu G, Johnson GA (2020) Interferon tau: influences on growth and development of the conceptus. Theriogenology 150:75–83
Berg DK, Smith CS, Pearton DJ, Wells DN, Broadhurst R, Donnison M, Pfeffer PL (2011) Trophectoderm lineage determination in cattle. Dev Cell 20:244–255
Betteridge KJ, Flechon JE (1988) The anatomy and physiology of the pre-attachment bovine embryos. Theriogenology 29:155–187
Betteridge KJ, Eaglesome MD, Randall GC, Mitchell DJ (1980) Collection, description and transfer of embryos from cattle 10-16 days after oestrus. Reprod Fertil 59:205–216
Black SG, Arnaud F, Palmarini M, Spencer TE (2010) Endogenous retroviruses in trophoblast differentiation and placental development. Am J Reprod Immunol 64:255–264
Boshier DP (1969) A histological and histochemical examination of implantation and early placentome formation in sheep. J Reprod Fertil 19:51–61
Brackett BG, Oh YK, Evans JF, Donawick WJ (1980) Fertilization and early development of cow ova. Biol Reprod 23:189–205
Brooks KE, Burns GW, Spencer TE (2015) Peroxisome proliferator activator receptor gamma (PPARG) regulates conceptus elongation in sheep. Biol Reprod 42:1–13
Burghardt RC, Burghardt JR, Taylor JD II, Reeder AT, Nguyen BT, Spencer TE, Johnson GA (2009) Enhanced focal adhesion assembly reflects increased mechanosensation and mechanotransduction along the maternal/conceptus interface during pregnancy in sheep. Reproduction 137:567–582
Carter AM, Enders AC (2013) The evolution of epitheliochorial placentation. Annu Rev Anim Biosci 1:443–467
Charpigny G, Reinaud P, Tamby JP, Creminon C, Guillomot M (1997a) Cyclooxygenase-2 unlike cyclooxygenase-1 is highly expressed in ovine embryos during the implantation period. Biol Reprod 57:1032–1040
Charpigny G, Reinaud P, Tamby JP, Creminon C, Martal J, Maclouf J, Guillomot M (1997b) Expression of cyclooxygenase-1 and -2 in ovine endometrium during the estrous cycle and early pregnancy. Endocrinology 138:2163–2217
Chavatte-Palmer P, Guillomot M (2007) Comparative implantation and placentation. Gynecol Obstet Invest 64:166–174
Cockburn K, Rossant J (2010) Making the blastocyst: lessons from the mouse. J Clin Invest 120:995–1003
Cooper DN (2002) Galectinomics: finding themes in complexity. Biochim Biophys Acta 1572:209–231
Cornelis G, Heidmann O, Degrelle SA, Vernochet C, Lavialle C, Letzelter C, Bernard-Stoecklin S, Hassanin A, Mulot B, Guillomot M, Hue I, Heidmann T, Dupressoir A (2013) Captured retroviral envelope syncytin gene associated with the unique placental structure of higher ruminants. PNAS 110:E828–E837
Dorniak P, Bazer FW, Spencer TE (2011) Prostaglandins regulate conceptus elongation and mediate effects of interferon tau on the ovine uterine endometrium. Biol Reprod 84:1119–1127

Duello TM, Byatt JC, Bremel RD (1986) Immunohistochemical localization of placental lactogen in binucleate cells of bovine placentomes. Endocrinology 119:1351–1355

Dunlap KA, Palmarini M, Adelson DL, Spencer TE (2005) Sheep endogenous betaretroviruses (enJSRVs) and the hyaluronidase 2 (HYAL2) receptor in the ovine uterus and conceptus. Biol Reprod 73:271–279

Dunlap KA, Palmarini M, Spencer TE (2006a) Ovine endogenous betaretroviruses (enJSRVs) and placental morphogenesis. Placenta 27(Suppl a):S135–S140

Dunlap KA, Palmarini M, Varela M, Burghardt RC, Hayashi K, Farmer JL, Spencer TE (2006b) Endogenous retroviruses regulate periimplantation placental growth and differentiation. Proc Natl Acad Sci 103:14390–14395

Eley RM, Thatcher WW, Bazer FW, Wilcox CJ, Becker RB, Head HH, Adkinson RW (1978) Development of the conceptus in the bovine. J Dairy Sci 61:467–473

Forde N, Carter F, Spencer TE, Bazer FW, Sandra O, Mansouri-Attia N, Okumu LA, McGettigan PA, Mehta JP, McBride R, O'Gaora P, Roche JF, Lonergan P (2011) Conceptus-induced changes in the endometrial transcriptome: how soon does the cow know she is pregnant? Biol Reprod 85:144–156

Frank JW, Steinhauser CB, Wang X, Bughardt RC, Bazer FW, Johnson GA (2020) Loss of ITGB3 in ovine conceptuses decreases conceptus expression of NOS3 and SPP1: implications for the developing placental vasculature. Biol Reprod:ioaa212

Galdos-Riveros AC, Favaron PO, Seal W, Miglino MA, Maria DA (2015) Bovine yolk sac: from morphology to metabolomic and proteomic profiles. Genet Mol Res 14:6223–6238

Garrett JE, Geisert RD, Zavy MT, Morgan GL (1988) Evidence for maternal regulation of early conceptus growth and development in beef cattle. J Reprod Fertil 84:437–446

Geisert RD, Johnson GA, Burghardt RC (2015) Implantation and establishment of pregnancy in the pig. Adv Anat Embryol Cell Biol 216:137–163

Gray CA, Adelson DL, Bazer FW, Burghardt RC, Meeusen EN, Spencer TE (2004) Discovery and characterization of an epithelial-specific galectin in the endometrium that forms crystals in the trophectoderm. PNAS 101:7982–7987

Green JA, Xie S, Quan X, Bao B, Gan X, Mathialagan N, Beckers J-F, Roberts RM (2000) Pregnancy-associated bovine and ovine glycoproteins exhibit spatially and temporally distinct expression patterns during pregnancy. Biol Reprod 62:1624–1631

Green JA, Parks TE, Avalle MP, Telugu BP, McLain AL, Peterson AJ, McMillan W, Mathialagan N, Xie S, Hook RR, Roberts RM (2005) The establishment of an ELISA for the detection of pregnancy-associated glycoproteins (PAGs) in the serum of pregnant cows and heifers. Theriogenology 63:1481–1503

Greenstein JS, Murray RW, Foley RC (1958) Observations on the morphogenesis and histochemistry of the bovine pre-attachment placenta between 16 and 33 days of gestation. Anat Rec 132:321–341

Gresham EL, Rankin JHG, Makowski EL, Meschia G, Battaglia FC (1972) An evaluation of fetal renal function in a chronic sheep preparation. J Clin Invest 51:149–156

Grosser O (1909) Vergleichende Anatomie und Entwicklungsgeschichte der Eihaute und der Placenta. W. Braumuller, Vienna and Leipzig

Grosser O (1927) Fruhentwicklung, Eihautbidung und Placentation des Menschen und der Saugetiere. J. F. Bergmann, Munchen

Guillomot M, Guay P (1982) Ultrastructural features of the cell surfaces of uterine and trophoblastic epithelia during embryo attachment in the cow. Anat Rec 204:315–322

Guillomot M, Fechon JE, Wintenberger-Torres S (1981) Conceptus attachment in the ewe: an ultrastructural study. Placenta 2:169–182

Guillomot M, Fléchon JE, Wintenberger-Torres S (1982) Cytochemical studies of uterine and trophoblastic surface coats during blastocyst attachment in the ewe. J Reprod Fertil 65:1–8

Guillomot M, Flechon JE, Leroy F (1993) Blastocyst development and implantation. In: Thibault C, Levasseur MC, Hunter RHF (eds) Reproduction in mammals and man. Ellipses, Paris, pp 387–411

Haig D (1996) Altercation of generations: genetic conflicts of pregnancy. Am J Reprod Immunol 35:226–232
Hradecky P (1983) Placental morphology in African antelopes and giraffes. Theriogenology 20:725–734
Hradecky P, Mossman HW, Stott GG (1988) Comparative development of the ruminant placentomes. Theriogenology 29:715–7129
Johnson GA, Bazer FW, Jaeger LA, Ka H, Garlow JE, Pfarrer C, Spencer TE, Burghardt RC (2001) Muc-1, integrin and osteopontin expression during the implantation cascade in sheep. Biol Reprod 65:820–828
Johnson GA, Burghardt RC, Bazer FW (2014) Osteopontin: a leading candidate adhesion molecule for implantation in pigs and sheep. J Anim Sci Biotech 5:56
Johnson GA, Bazer FW, Burghardt RC, Wu G, Seo H, Kramer AC, McLendon BA (2018) Cellular events during ovine implantation and impact for gestation. Anim Reprod 15:843–855
Kelleher AK, Demayo FJ, Spencer TE (2019) Uterine glands: developmental biology and functional roles in pregnancy. Endocr Rev 40:1424–1445
Kim J, Erikson DW, Burghardt RC, Spencer TE, Wu G, Bayless KJ, Johnson GA, Bazer FW (2010) Secreted phosphoprotein 1 binds integrins to initiate multiple cell signaling pathways, including FRAP1/mTOR, to support attachment and force-generated migration of trophectoderm cells. Matrix Biol 29:369–382
Kimura J, Sasaki M, Endo H, Fukuta K (2004) Anatomical and histological characterization of the female reproductive organs of mouse deer (Tragulidae). Placenta 25:705–711
King GJ (1993) Comparative placentation in ungulates. J Exp Zool 266:588–602
King GJ, Atkinson BA (1987) The bovine intercaruncular placenta throughout gestation. Anim Reprod Sci 12:241–254
King GJ, Atkinson BA, Robertson HA (1979) Development of the bovine placentome during the second month of gestation. J Reprod Fertil 55:173–180
King GJ, Atkinson BA, Robertson HA (1980) Development of the bovine placentome from days 20 to 29 of gestation. J. Reprod Fertil 59:95–100
King GJ, Atkinson BA, Robertson HA (1981) Development of the intercaruncular area during early gestation and the establishment of the bovine placenta. J Reprod Fertil 61:469–474
King GJ, Atkinson BA, Robertson HA (1982) Implantation and early placentation in domestic ungulates. J Reprod Fertil Suppl 31:17–30
Kritzenberger M, Wrobel K-H (2004) Histochemical in situ identification of bovine embryonic blood cells reveals differences to the adult hematopoietic system and suggests a close relationships between hematopoietic stem cells and primordial germ cells. Histochem Cell Biol 121:273–289
Leiser R (1975) Development of contact between trophoblast and uterine epithelium during the early stages on implantation in the cow. Zentralbl Veterinarmed C 4:63–86
Leiser R, Pfarrer C, Abd-Elnaeim M, Dantzer V (1998) Feto-maternal anchorage in epitheliochorial and endotheliochorial placental types studied by histology and microvascular corrosion casts. Placenta 9:21–39
Llobat L (2020) Embryo gene expression in pig pregnancy. Reprod Domest Anim 55:523–529
Lonergan P, Forde N, Spencer T (2016) Role of progesterone in embryo development in cattle. Reprod Fertil Dev 28:66–74
Maddox-Hyttel P, Alexopoulos NI, Vajta G, Lewis I, Rogers P, Cann L, Callesen A, Tveden-Nyborg P, Trounson A (2003) Immunohistochemical and ultrastructural characterization of the initial post-hatching development of bovine embryos. Reproduction 125:607–623
Mançanares CAF, Leiser R, Favaron PO, Carvalho AF, Oliveira VC, De Santos JM et al (2013) A morphological analysis of the transition between the embryonic primitive intestine and yolk sac in bovine embryos and fetuses. Microsc Res Tech 76:756–766
Mi S, Lee X, X-p L, Veldman GM, Finnerty H, Racie L, LaVallie E, Tang X-Y, Edouard P, Howes S, Keith JC, McCoy JM (2000) Syncytin is a captive retroviral envelope protein involved in human placental morphogenesis. Nature 403:785–789

Moraes JGN, Behura SK, Geary TW, Hansen PJ, Neibergs HL, Spencer TE (2018) Uterine influences on conceptus development in fertility-classified animals. Proc Natl Acad Sci 115: E1749–E1758

Morgan G, Wooding FBP (1983) Cell migration in the ruminant placenta, a freeze fracture study. J Ultrastruct Res 83:148–160

Mossman HW (1937) Comparative morphogenesis of the fetal membranes and accessory uterine structures. Contr Embryol Carnegie Instn 158:129–246

Mossman HW (1974) Structural changes in vertebrate fetal membranes associated with the adoption of viviparity. Obstet Gynecol Annu 3:7–32

Mossman HW (1987) Vertebrate fetal membranes. Rutgers University Press, München, p 383

Negrón-Pérez VM, Rodrigues LT, Mingoti GZ, Hansen PJ (2018) Role of ROCK signaling in formation of the trophectoderm of the bovine preimplantation embryo. Mol Reprod Dev 85:374–375

Palmarini M, Gray CA, Carpenter K, Fan H, Bazer FW, Spencer TE (2001) Expression of endogenous betaretroviruses in the ovine uterus: effects of neonatal age, estrous cycle, pregnancy, and progesterone. J Virol 75:11319–11327

Perry JS (1981) The mammalian fetal membranes. J Reprod Fertil 62:321–335

Pohler KG, Geary TW, Johnson CL, Atkins JA, Jinks EM, Busch DC, Green JA, MacNeil MD, Smith MF (2013) Circulating bovine pregnancy associated glycoproteins are associated with late embryonic/fetal survival but not ovulatory follicle size in suckled beef cows. J Anim Sci 91:4158–4167

Pohler KG, Pereira MHC, Lopes FR, Lawrence JC, Keisler DH, Smith MF, Vasconcelos JLM, Green JA (2016a) Circulating concentrations of bovine pregnancy-associated glycoproteins and late embryonic mortality in lactating dairy herds. J Dairy Sci 99:1584–1594

Pohler KG, Peres RFG, Green JA, Graff H, Martins T, Vasconcelos JLM, Smith MF (2016b) Use of bovine pregnancy-associated glycoproteins to predict late embryonic mortality in postpartum Nelore beef cows. Theriogenology 85:1652–1659

Raheem KA, Marei WFA, Campbell BK, Fouladi-Nashta AA (2016) *In vivo* and *in vitro* studies of MUC1 regulation in sheep endometrium. Theriogenology 85:1635–1643

Rasby RJ, Wettemann RP, Geisert RD, Rice LE, Wallace CR (1990) Nutrition, body condition and reproduction in beef cows: fetal and placental development, and estrogens and progesterone in plasma. J Anim Sci 68:4267–4276

Reimers T, Ullmann M, Hansel W (1985) Progesterone and prostanoid production by bovine binucleate trophoblastic cells. Biol Reprod 33:1227–1236

Roberts RM, Green JA, Schulz LC (2016) The evolution of the placenta. Reproduction 152:R179–R189

Rowson LEA, Moore RM (1966) Embryo transfer in sheep; the significance of synchronizing oestrus in the donor and recipient animal. J Reprod Fertil 11:207–212

Sasser RG, Ruder CA, Ivani KA, Butler JE, Hamilton WC (1986) Detection of pregnancy by radioimmunoassay of a novel pregnancy-specific protein in serum of cows and a profile of serum concentrations during gestation. Biol Reprod 35:936–942

Satterfield MC, Bazer FW, Spencer TE (2006) Progesterone regulation of preimplantation conceptus growth and galectin 15 (LGALS15) in the ovine uterus. Biol Reprod 75:289–296

Seo H, Bazer FW, Burghardt RC, Johnson GA (2019) Immunohistochemical examination of trophoblast syncytialization during early placentation in sheep. Int J Mol Sci 20:4530–4543

Seo H, James W, Frank JW, Robert C, Burghardt RC, Fuller W, Bazer FW, Johnson GA (2020) Integrins and OPN localize to adhesion complexes during placentation in sheep. Reproduction 160:521–532

Simmons RM, Erikson DW, Kim J, Burghardt RC, Bazer FW, Johnson GA, Spencer TE (2009) Insulin-like growth factor binding protein-1 in the ruminant uterus: potential endometrial marker and regulator of conceptus elongation. Endocrinology 150:4295–4305

Skidmore JA, Wooding FBP, Allen WR (1996) Implantation and early placentation in the one-humped camel (Camelus dromedarius). Placenta 17:253–262

Spencer TE, Hansen TR (2015) Implantation and establishment of pregnancy in ruminants. Adv Anat Embryol Cell Biol 216:105–135

Spencer TE, Bartol FF, Bazer FW, Johnson GA, Joyce MM (1999a) Identification and characterization of glycosylation-dependent cell adhesion molecule 1-like protein expression in the ovine uterus. Biol Reprod 60:241–250

Spencer TE, Stagg AG, Taylor KM, Johnson GA, Gertler A, Gootwine E, Bazer FW (1999b) Effects of recombinant ovine interferon tau, placental lactogen and growth hormone on ovine endometrial function. Biol Reprod 61:1409–1418

Spencer TE, Johnson GA, Bazer FW, Burghardt RC (2004a) Implantation mechanisms: insights from the sheep. Reproduction 128:657–668

Spencer TE, Johnson GA, Burghardt RC, Bazer FW (2004b) Progesterone and placental hormone actions on the uterus: insights from domestic animals. Biol Reprod 71:2–10

Spencer DS, Ross JW, Ashworth MD, Geisert RD, Rickords LF (2006) Porcine conceptus Oct-4 mRNA expression during peri-implantation development. Reprod Domest Anim 41:571–572

Spencer TE, Black SG, Arnaud F, Palmarini M (2010) Endogenous retroviruses of sheep: a model system for understanding physiological adaptation to an evolving ruminant genome. Soc Reprod Fertil Suppl 67:95–104

Spencer TE, Forde N, Lonergan P (2016) The role of progesterone and conceptus-derived factors in uterine biology during early pregnancy in ruminants. J Dairy Sci 99:5941–5950

Surveyor GA, Gendler SJ, Pemberton L, Das SK, Chakraborty I, Julian J, Pimental RA, Wegner CC, Dey SK, Carson DD (1995) Expression and steroid hormonal control of Muc-1 in the mouse uterus. Endocrinology 136:3639–3647

Telugu BP, Green JA (2007) Comparative placentation. In: Constantinescu G, Schatten H (eds) Veterinary reproductive biology. Chapter 12. Blackwell, Ames, IA

Wallace RM, Pohler KG, Smith MF, Green JA (2015) Placental PAGs: gene origins, expression patterns and use as markers of pregnancy. Reproduction 149:R115–R126

Wang J, Guillomot M, Hue I (2009) Cellular organization of the trophoblastic epithelium in elongating conceptuses of ruminants. C R Biol 332:986–997. https://doi.org/10.1016/j.crvi.2009.09.004

Wango EO, Wooding FBP, Heap RB (1990) The role of trophoblastic cells in implantation in the goat: a quantitative study. Placenta 11:381–394

Wathes DC, Wooding FBP (1980) An electron microscopic study of implantation in the cow. Am J Anat 159:285–306

Wiley AA, Bartol FF, Barron DH (1987) Histogenesis of the ovine uterus. J Anim Sci 64:1262–1269

Wimsatt WA (1974) Some comparative aspects of implantation. Biol Reprod 12:1–40

Wooding FBP (1982a) Structure and function of placental binucleate (giant) cells. Bibl Anat 22:134–139

Wooding FBP (1982b) The role of binucleate cell in ruminant placental structure. Reprod Fertil Suppl 31:31–39

Wooding FBP (1984) Role of binucleate cells in fetomaternal cell fusion at implantation ın the sheep. Am J Anat 170:233–250

Wooding FBP (1987) Ultrastructural evidence for placental lactogen transport and secretion in ruminants. J Physiol 386:26

Wooding FBP (1992) Current topic: the synepitheliochorial placenta of ruminants: binucleate cell fusions and hormone production. Placenta 13:101–113

Wooding FBP, Flint APF (1994) Placentation. In: Lamming GE (ed) Marshall's physiology of reproduction. Chapman and Hall, London, pp 233–460

Wooding FBP, Wathes DC (1980) Binucleate cell migration in the bovine placentome. J Reprod Fertil 59:425–430

Wooding FBP, Flint APF, Heap RB, Morgan G, Buttle HL, Young IR (1986) Control of binucleate cell migration in the placenta of ruminants. J Reprod Fertil 76:499–512

Wooding FB (1992) Current topic: the synepitheliochorial placenta of ruminants: binucleate cell fusions and hormone production. 13:101–113

Wooding FBP, Morgan G, Adam CL (1997) Structure and function in the ruminant synepitheliochorial placenta: central role of the trophoblast binucleate cell in deer. Microsc Res Tech 38:88–99

Wooding FBP, Roberts RM, Green JA (2005) Light and electron microscope immunocytochemical studies of the distribution of pregnancy-associated glycoproteins (PAGs) throughout pregnancy in the cow: possible functional implications. Placenta 26:807–827

Wrobel K-H, Suess F (1998) Identification and temporospatial distribution of bovine primordial germ cells prior to gonadal sexual differentiation. Anat Embryol 197:451–467

Xie SC, Low BG, Nagel RJ, Kramer KK, Anthony RV, Zoli AP, Beckers JF, Roberts RM (1991) Identification of the major pregnancy-specific antigens of cattle and sheep as inactive members of the aspartic proteinase family. Proc Natl Acad Sci 88:10247–10251

Zhu C, Jiang Z, Bazer FW, Johnson GA, Burghardt RC, Wu G (2015) Aquaporins in the female reproductive system of mammals. Front Biosci 20:838–871

Zoli AP, Guilbault LA, Delahaut P, Ortiz WB, Beckers JF (1992) Radioimmunoassay of a bovine pregnancy-associated glycoprotein in serum: its application for pregnancy diagnosis. Biol Reprod 46:83–92

Canine Endotheliochorial Placenta: Morpho-Functional Aspects

Mariusz P. Kowalewski, Ali Kazemian, Karl Klisch, Tina Gysin, Miguel Tavares Pereira, and Aykut Gram

Contents

Abstract In the domestic dog, placentation arises from central implantation, passing through a transitional, yet important stage of choriovitelline placenta (yolk sac placenta), on the way to the formation of the definite, deciduate, zonary (girdle) allantochorionic endotheliochorial placenta.

Sharing some similarities with other invasive types of placentation, e.g., by revealing decidualization, it is characterized by restricted (shallow) invasion of trophoblast not affecting maternal capillaries and maternal decidual cells. Thus,

The original version of this chapter was revised with some text corrections. The corrections to this chapter can be found at https://doi.org/10.1007/978-3-030-77360-1_12

M. P. Kowalewski (✉) · A. Kazemian · K. Klisch · T. Gysin · M. Tavares Pereira · A. Gram
Institute of Veterinary Anatomy, Vetsuisse-Faculty, University of Zurich, Zurich, Switzerland
e-mail: kowalewski@vetanat.uzh.ch

R. D. Geisert, T. E. Spencer (eds.), *Placentation in Mammals*, Advances in Anatomy, Embryology and Cell Biology 234, https://doi.org/10.1007/978-3-030-77360-1_8

being structurally and functionally placed between noninvasive epitheliochorial placentation and the more invasive hemochorial type, it presents an interesting and important model for understanding the evolutionarily determined aspects of mammalian placentation. More profound insights into the biological mechanisms underlying the restricted invasion of the fetal trophoblast into maternal uterine structures and the role of decidual cells in that process could provide better understanding of some adverse conditions occurring in humans, like *preeclampsia* or *placenta accreta*. As an important endocrine organ actively responding to ovarian steroids and producing its own hormones, e.g., serving as the source of gestational relaxin or prepartum prostaglandins, the canine placenta has become an attractive research target, both in basic and clinical research. In particular, the placental feto-maternal communication between maternal stroma-derived decidual cells and fetal trophoblast cells (i.e., an interplay between *placenta materna* and *placenta fetalis*) during the maintenance and termination of canine pregnancy serves as an interesting model for induction of parturition in mammals and is an attractive subject for translational and comparative research. Here, an updated view on morpho-functional aspects associated with canine placentation is presented.

Keywords Domestic dog · *Canis familiaris* · Placenta · Morphology · Function

1 Introduction

Despite its distinctive structure among domestic animal species, rendering it a unique model for comparative and evolutionary studies, it is only recently that the canine endotheliochorial placenta has become an attractive target for researchers, in particular those dealing with canine reproduction. The clinical interest and research efforts in understanding canine reproduction were initially mostly directed towards endocrine mechanisms governing ovarian function, as a measure for monitoring and/or controlling reproduction (Hoffmann et al. 2004; Papa and Hoffmann 2011; Klein et al. 2003). This was predominantly due to the fact that ovaries are the only source of circulating steroids in pregnant and nonpregnant canine cycles (Hoffmann et al. 1994). Currently, deeper knowledge has been acquired regarding the functional relationship between ovarian steroids and placenta in the establishment and maintenance of canine pregnancy (Kowalewski 2014; Kowalewski et al. 2020). Moreover, starting with early placentation, the fetal part of the placenta becomes a source of relaxin (RLN), the only currently known endocrine marker of canine pregnancy, originating in the fetal trophoblast (Nowak et al. 2017). The luteolytic parturition cascade is initiated from the placenta responding to decreasing circulating progesterone levels and leading to prepartum output of prostaglandins (Kowalewski et al. 2010; Gram et al. 2013, 2014a). Thus, expectedly, although still less explored than in other species, the canine placental feto-maternal exchange goes way beyond being a communication interface between the mother and the fetus. Despite not producing its own steroids (in fact dog is the only domestic animal species devoid of placental steroidogenic activity), the canine placenta responds to circulating steroids and produces its own hormones, and thus, clearly serves as an important, previously underestimated, endocrine organ. The responsiveness to steroids is an important part of the placental feto-maternal communication and is associated with initialization of the prepartum luteolytic cascade. The species-specific decidualization and the maternal decidual cell-bound expression of nuclear progesterone and estrogen receptors play an important role in this cascade by regulating fetal prostaglandin synthesis (Nowak et al. 2019).

The two perspectives, evolutionarily determined morphology positioning the canine placenta between the less invasive epitheliochorial and more invasive hemochorial placentation types, together with species-specific endocrine function, make the canine placenta an interesting model indeed for comparative studies. Some clinical conditions, like the human *placenta accreta* or canine subinvolution of placental sites (SIPS), bear similarities, while a deeper understanding of the shallow/restricted invasion of the fetal trophoblast into maternal endometrial structures and the role of decidual cells in that process could provide a better understanding of human *preeclampsia* (Kutzler et al. 2012).

Taken together, understanding the morphology and function of the canine placenta may play a key role in managing pregnancy outcomes, thereby regulating reproduction in dogs. Additionally, a canine model of decidualization and shallow invasion could serve important roles in translational research. Here, an updated view on the morpho-functional aspects of canine placentation is presented, highlighting the underlying feto-maternal interrelationship in response to endocrine insult. Accordingly, some related comprehensive reviews have been published recently, dealing with canine conceptus-maternal communication (see e.g., (Kowalewski et al. 2015, 2020)) and highlighting the endocrine events associated with canine pregnant and nonpregnant cycles (Kowalewski 2014, 2017, 2018; Papa and Kowalewski 2020).

2 Endocrinological Milieu During Implantation and Placentation

In order to better understand the morpho-functional aspects of canine placentation, it is important to consider the species-specific endocrinological context of the establishment and maintenance of canine pregnancy.

Quite peculiar compared with other domestic animal species, the hormonal environment associated with the canine reproductive cycle is similar in pregnant and pseudopregnant bitches, particularly in the early diestrus when embryonic development, implantation, and placentation take place. Similar circulating progesterone levels are observed in both situations, and luteolysis does not occur in nonpregnant dogs (Concannon et al. 1975). Due to similar progesterone profiles throughout most of the canine pregnancy, there is no classical recognition of pregnancy in this species, understood as a biological mechanism avoiding luteolysis and extending the luteal lifespan beyond the *corpus cyclicum* stage. Instead, the morphological and functional interplay between the three entities, i.e., the uterus, embryo, and the corpus luteum, preventing embryonic rejection, has recently been proposed as a more suitable definition of maternal recognition of pregnancy in the dog (Kowalewski et al. 2015). Progesterone levels observed in the peripheral blood are high compared with other domestic animal species. They increase prior to ovulation as a sign of exceptionally strong preovulatory luteinization (compared to other species) and reach peripheral levels of at least 5 ng/mL at the time of ovulation (reviewed in (Kowalewski 2018)). Ovulation takes place 2–3 days after the initial LH surge, and unlike other domesticated species, primary oocytes are released from the ovaries. As in other *Canidae,* the first meiotic division is completed within 2–3 days in the oviduct and secondary oocytes can be fertilized (Kowalewski et al. 2015). Following ovulation, progesterone concentrations continue to rise rapidly, reaching levels of above 30 ng/mL (Kowalewski 2018) when corpora lutea are fully developed. The pregnancy lasts on average 60 days and is entirely dependent on the luteal progesterone supply, as the placenta is devoid of steroidogenic activity (mentioned elsewhere) (reviewed in (Kowalewski 2014, 2017, 2018; Kowalewski et al. 2015)). Following their highest steroidogenic output, during the second half of diestrus, corpora lutea (same for the *corpus cyclicum* and *corpus graviditatis*) undergo gradual regression, mirrored in decreasing peripheral progesterone levels (Kowalewski 2014, 2017). The progesterone secretion profiles begin to differ between pregnant and nonpregnant dogs during prepartum luteolysis, when the steep progesterone decline, 12–24 h prior to term, signals the onset of parturition. Conversely, there is no active luteolytic process in nonpregnant bitches, and therefore, pseudopregnancy can last as long as pregnancy or longer. Then, during the extended luteal regression, progesterone levels fade slowly, until circulating concentrations of <1 ng/mL, *per definitionem*, signal the onset of anestrus (an obligatory quiescence period).

Considering the strongly increasing progesterone levels before, and continuing after ovulation, the canine uterus is exposed to high concentrations of this hormone prior to implantation and placentation. However, the spontaneous, progesterone-driven decidualization observed in humans does not exist in the dog. Instead, similar to rodents, decidualization is induced by embryo presence (Graubner et al. 2017a, b; Kautz et al. 2014). Similarly, in vitro, progesterone does not seem to be a strong inducer of decidualization in the dog (Graubner et al. 2020), although it does appear to modulate the expression of some decidualization markers, e.g., prolactin receptor (PRLR), as well as the expression of its own nuclear receptor, PGR (Graubner et al. 2020), thereby apparently regulating its own biological availability. Furthermore, there is no pregnancy-associated increase in estrogens, nor is parturition associated with elevated estrogen levels (Hoffmann et al. 1994). Interestingly, in naturally estrogenized bitches, further estrogen supply is not required for implantation and placentation to occur (Concannon et al. 2001). Conversely, luteal progesterone is continuously required for the oviductal development of early embryos and throughout pregnancy. Interfering with its functionality leads to retardation of embryo development or, at later stages, to prepartum luteolysis/abortion (Kowalewski et al. 2009, 2010; Baan et al. 2005; Reynaud et al. 2015). The similar secretion profiles of circulating steroids in pregnant and pseudopregnant bitches preclude them as markers of pregnancy. Similarly, increasing levels of prolactin (PRL), typically higher in pregnant bitches than during pseudopregnancy, cannot be used as a marker of pregnancy, because higher concentrations of PRL can be observed in overt pseudopregnancy. This renders RLN as the only usable marker of pregnancy so far (Steinetz et al. 1989).

3 Embryo–Maternal Communication Before Implantation and Placentation

Despite the lack of classical recognition of pregnancy, as in other species, the canine embryo needs to communicate its presence in order to not be rejected.

Canine embryos cross the uterotubal junction relatively late, around 7–8 (up to 10) days after fertilization. At this time, they can be at a wide range of embryonic stages, e.g., 16 cells, morula or blastocyst (reviewed in (Kowalewski et al. 2015)). Their uterine migration can last for a further 9–10 days, during which their development becomes more synchronized, so that by day 14, at the latest, all blastocysts are hatched. The apposition starts within 12–14 days of embryonic life. Subsequently, about 17–18 days after fertilization, implantation takes place and embryos invade the uterine wall, leading to placentation (reviewed in (Kowalewski et al. 2015)).

A deeper insight into early embryo–maternal communication has been provided by genomic analysis of global transcriptome changes in the uterus, associated with the presence of early free-floating (i.e., preimplantation) embryos (days 10–12) (Graubner et al. 2017a). The two functional terms most strongly enriched in the preimplantation uterus in response to embryo presence are related to extracellular matrix remodeling, followed by immune and inflammatory responses (for details see in (Graubner et al. 2017a)). Based on these observations, it seems that, in contrast with humans, for example, early free-floating embryos induce biochemical and functional, rather than morphological changes in the uterus (Graubner et al. 2017a; Kautz et al. 2014). It is only following implantation that the more intimate embryo–maternal contact is initiated, as reflected in the first morphological signs of decidualization (Graubner et al. 2017b) (elaborated more in detail below). As a sign of ongoing matrix remodeling prior to implantation, the expression of some ECM-related factors, e.g., ECM1, TIMP2/4, or LAMA2, was positively affected by the presence of free-floating embryos, whereas FN1 expression was decreased, cumulatively apparently contributing to the uterine tissue rearrangement (Graubner et al. 2017a, 2018). On the other hand, expression of structural collagens (COL) such as COL1, −3, and −4 and factors involved in cell-to cell communication (e.g., connexin (CX) 26 and −43) was not affected by the presence of embryos during early (pre-attachment) pregnancy (Graubner et al. 2018). Apart from the altered expression of some immune-related factors in response to free-floating embryos (e.g., IL4, IFNγ, AIF1, CXCL16, CXCR7, or IDO; (Graubner et al. 2017a; Schafer-Somi et al. 2008, 2020)), most recent data from our laboratory (Tavares Pereira et al. 2021a) suggest the presence of embryo-driven, controlled pro-inflammatory signaling during the pre-attachment period, with T helper cells prevailing over cytotoxic CD8 cells. The immunosuppressive regulatory T and natural killer (NK) cells appear to play an important role at the time of implantation (Tavares Pereira et al. 2021a).

4 Structural Remodeling of the Uterus and Development of an Intimate Embryo–Maternal Interface During Implantation and Placentation

4.1 Implantation, Initialization of Decidualization, and Formation of Maternal Decidual Cells

Implantation is central in the dog; blastocysts grow, remain in the uterine cavity, and come into circumferential contact with the uterine surface (Amoroso 1952). As a sign of the direct embryo–maternal communication, the subepithelial morphological transformation of the stromal compartment begins at the time of implantation (days 17–18 after fertilization) (Graubner et al. 2017b). The mesenchymal-epithelial transition of the uterine stromal compartment, described previously for humans (Yu et al. 2016), is an apparent sign of decidualization and is characterized by increased expression of COL4, ECM1, and CX43 in decidualizing cells and suppression of NEP/CD10 expression (Graubner et al. 2020; Payan-Carreira et al. 2014). During this process, the uterine stromal cells undergo morphologic changes in their transition from fibroblast-shaped cells to large, swollen, and round-shaped cells with enriched secretory activity (Graubner et al. 2017b; Kautz et al. 2015). This process, ultimately, leads to the development of highly specialized maternally derived decidual cells, which together with the maternal vascular endothelial cells escape the trophoblast invasion on the way to the formation of canine endotheliochorial placenta. Despite the mesenchymal-epithelial transition, the cells continue to manifest their mesenchymal character and express respective markers, like vimentin (Vim) or alpha-smooth muscle actin (alphaSMA) (Graubner et al. 2017b; Kautz et al. 2015). Decidual cells are also the only cells of the canine placenta expressing the nuclear progesterone (PGR) and estrogen (ERα) receptors, as well as the oxytocin receptor (OXTR) (Kowalewski et al. 2010; Vermeirsch et al. 2000a, b; Gram et al. 2014b), characteristics that determine their importance for the maintenance of canine pregnancy (elaborated in more detail elsewhere).

4.2 Development of Fetal Membranes, Trophoblast Invasion, and Placentation

In its initial response to the implanting embryo, the so-far proliferative uterus becomes hypertrophied, with abundant mucosal folds and corresponding crypt-like spaces. The lumina of the uterine glands develop into chambers with narrow necks (Fig. 1). Apart from the limitations of inter-glandular connective tissue, this enlargement leads to the development of the spongy appearance of the deeper layer, in contrast to the superficial compact zone (Amoroso 1952).

Following implantation, the feto-maternal contact becomes more intimate during invasion and placentation (Fig. 1). As in other mammalian species, the four extra-embryonic membranes are formed in the dog: chorion, amnion, vitelline-, and allantoic sac (Fig. 2).

During embryonic development, the secondary and tertiary chorion are formed by the addition of a mesoderm and vasculature to the outer monolayer epithelium of the blastocyst (trophoblastic layer). The amnion is formed by the ectodermal vesicle, which is lined with mesoderm, keeping the embryo in a liquid environment (Miglino et al. 2006). The amniotic folds fuse at around day 15 after fertilization. The chorionic cavity is formed at day 20 after the LH peak (i.e., approximately 16–17 days after fertilization), heartbeat starts at days 23–25, and skeleton is visible 33–39 days after LH peak (Hermanson et al. 2020). Extraembryonic endoderm expands and lines the trophoblast during the early blastocyst (preimplantation) stage, and the yolk sac becomes completely formed at the time of implantation (Hermanson et al. 2020; McGeady et al. 2017). Together with the trophoblast layer it initiates the physiological interaction with the endometrium, resulting in primary placentation. This leads to the formation of the choriovitelline placenta (yolk sac placenta) (Fig. 2) (Amoroso 1952) in three developmental stages, starting with the development of a nonvascular bilaminar omphalopleure, when the trophoblast and endoderm-derived yolk sac are in apposition to the uterine wall (i.e., around day 14). This is followed by a transitional nonvascular trilaminar omphalopleure, formed when mesoderm extends between the two layers, that then converts to a vascular trilaminar omphalopleure when vitelline vessels are developed (Amoroso 1952). The vitelline vessels in the dog embryo play an important role in feto-maternal exchange (Miglino et al. 2006; Leiser and Kaufmann 1994). Even though implantation is quite fast, and the yolk sac is relatively small, it still provides the necessary nutrition for the embryo in the very early stages of its development, without which the embryo would be resorbed during the third week of pregnancy (Amoroso 1952). The nutritive function of the yolk sac (the so-called yolk sac placenta) ends on about day 25–28 after the LH peak (21–24 days after fertilization) (Amoroso 1952; Hermanson et al. 2020). The allantois then starts to sprout from the endoderm and expands into the chorionic cavity (McGeady et al. 2017). It is completely formed by days 27–31 after LH peak (23–27 days after fertilization).

The canine chorioallantoic placenta starts to form immediately after implantation, on day 17–18 after fertilization (Amoroso 1952; Kehrer 1973). At that time, the

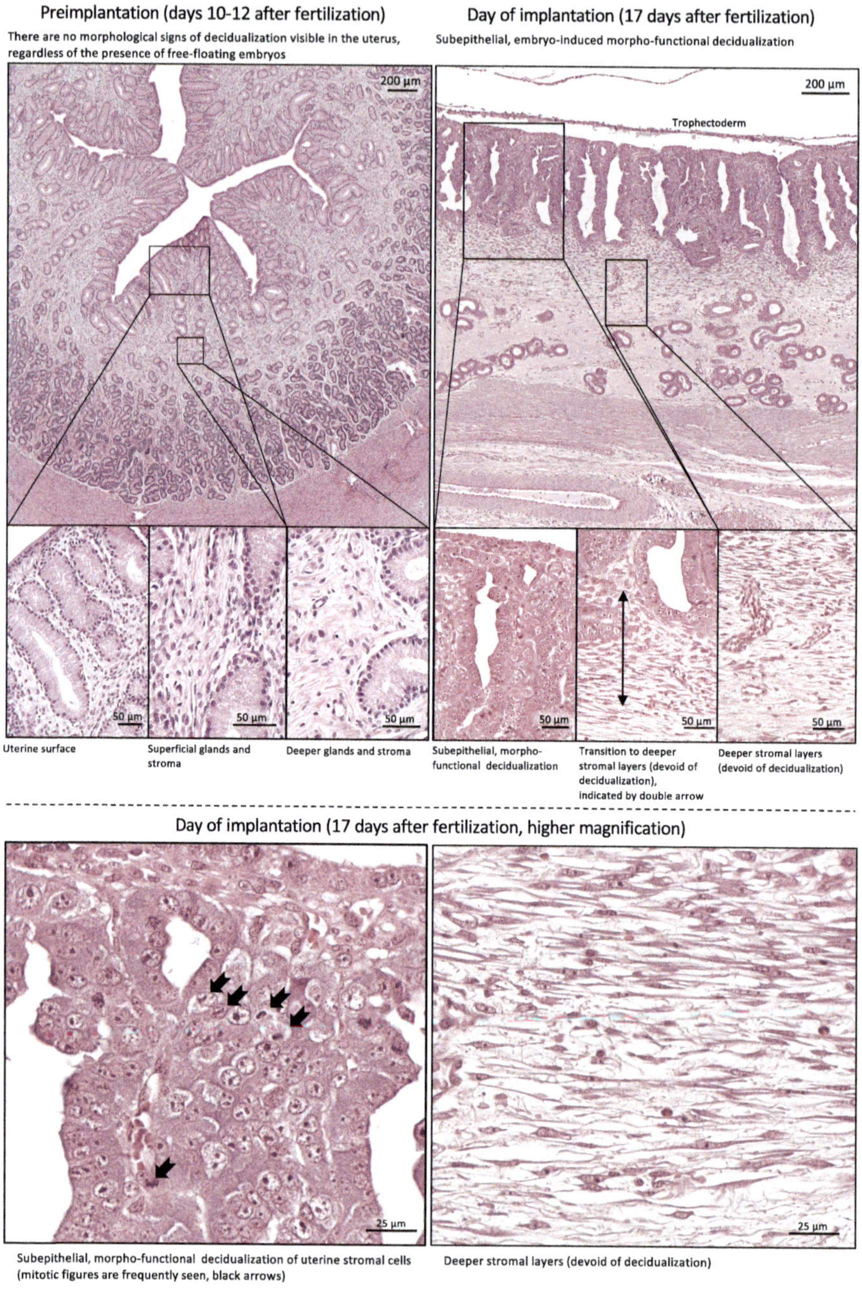

Fig. 1 Morphological appearance of the canine uterus during early pregnancy. Preimplantation stage (days 10–12 after fertilization), and at the time of implantation (day 17 of embryonal life). Inserts show different fragments of uterine compartments. Morphological signs of decidualization are seen in subepithelial stromal compartments during implantation (enlarged cells, embryo-induced morpho-functional decidualization), but not at the preimplantation stage

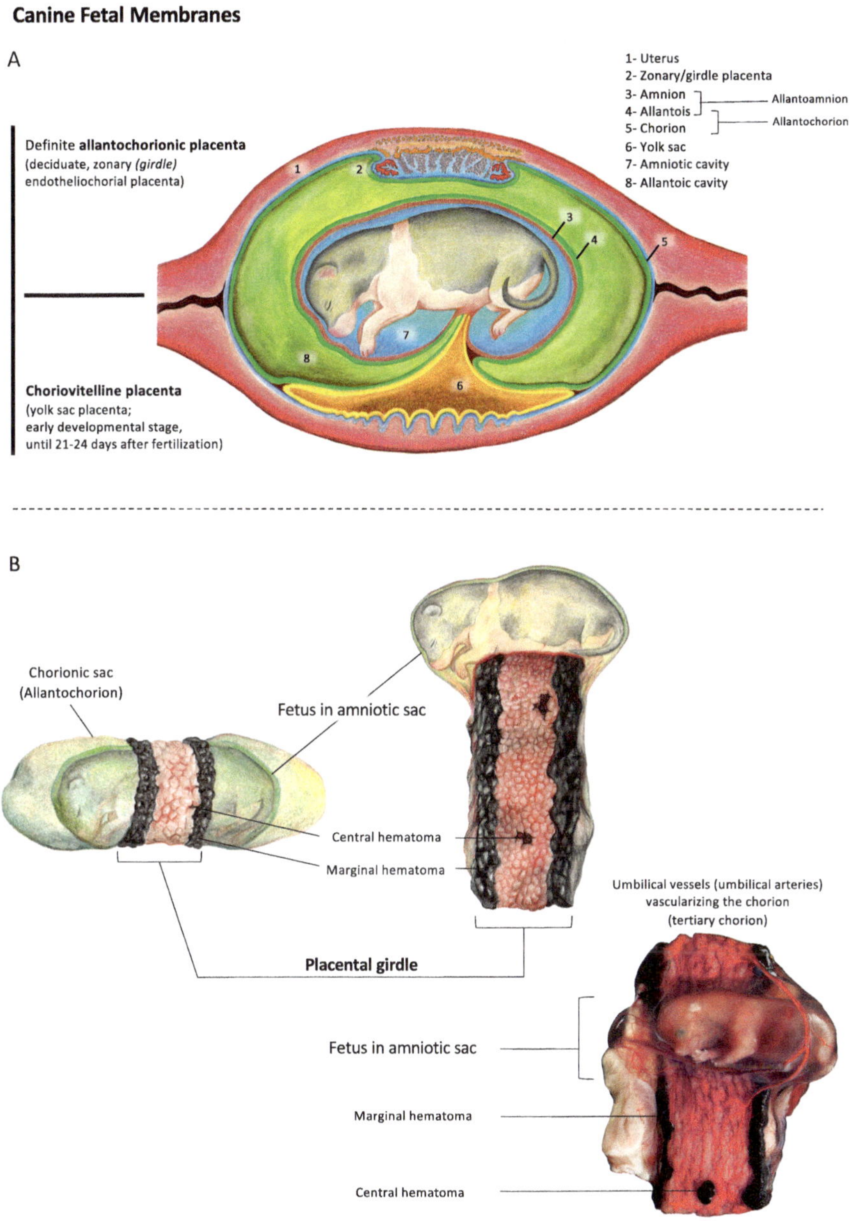

Fig. 2 Canine fetal membranes: (**A**) combined schematic representation of canine fetal membranes at an early stage of placentation (yolk sac placenta, lower half of the drawing) and after the formation of definite zonary placenta (upper half of the drawing); (**B**) drawing and photograph of canine fetal membranes including zonary (girdle) placenta (the latter, i.e., photograph, shared by courtesy of Prof. Dr. Iris Reichler, Small Animal Reproduction and Michelle Oesch, Dept. of Scientific Communication, Vetsuisse Faculty, University of Zurich)

developing chorionic layer produces abundant invading trophoblasts, resulting in an increased area of feto-maternal contact (Amoroso 1952). Following the vascularization of the chorion (tertiary chorion) with umbilical vessels (umbilical arteries) (Fig. 2), the allantois attaches to the chorion and contributes to the formation of the placental labyrinth (chorioallantoic placentation) (Amoroso 1952; Miglino et al. 2006). At the implantation site, the trophoblast invades the uterus (deciduate placenta) and contacts the superficial layer of the stroma, which proliferates and its cells increase (decidualization, see above), as the uterine epithelium has already been destroyed. Invasion continues in the form of chorioallantoic villi capped by cellular trophoblast (cytotrophoblast), until it reaches the endothelial wall of the maternal capillaries, forming the definitive endotheliochorial placenta (Amoroso 1952; Barrau et al. 1975). Part of the decidualizing tissue is also removed so that the maternal capillaries become completely surrounded by syncytial trophoblast, which surrounds both the maternal vessels and embedding decidual cells (below and Fig. 3). Decidual cells are also detectable between undifferentiated stromal cells (Amoroso 1952).

Macro- and Microscopical Morphology of the Canine Placenta

Macroscopically, the canine placenta presents a central type of placentation in which the hatched blastocyst makes a broad surface contact with the luminal epithelium of the uterus, resulting in a zonary (girdle) placenta (Leiser and Kaufmann 1994; Enders and Carter 2004; Wynn and Corbett 1969).

Histologically, utero-placental compartments consist of the following different areas: placental labyrinth, endometrium (including junctional zone, glandular zone, connective tissue layer separating superficial from deep glands, and a layer of deep uterine glands), and myometrium (Fig. 3).

Interdigitations between opposing fetal and maternal tissues are formed by multiple branched folds, creating a lamellar pattern (Leiser and Koob 1993). The villi are extensively branched and give a lobular appearance to the labyrinth zone (Amoroso 1952). In the labyrinth, there are two trophoblast layers: cytotrophoblast, which is closer to the fetal stroma, and syncytiotrophoblast (syncytial trophoblast), which is closer to the maternal tissues (surrounding maternal vessels) (Fig. 3). The syncytiotrophoblast is formed by the fusion of cytotrophoblasts and does not undergo cell division.

The canine placenta is a hemophagus organ with marginal hematoma areas (at the border of the girdle) and small central hematomas (inside the placenta) in which maternal blood is extravasated from capillaries. Thus, in hematomas, developed (vascularized) chorioallantoic villi come into contact with the maternal blood (Amoroso 1952). The phagocytic (cyto)trophoblast encloses hematomas and absorbs blood; hematomas are green in appearance (uteroverdin, biliverdin). It appears plausible that the hemophagous areas (hematomas together with the phagocytic trophoblasts) in the canine placenta may be involved in maternal-fetal iron transport. Due to the close contact between the maternal and fetal structures, the exchange of immunoglobulins between the mother and fetus is possible in the canine endotheliochorial placenta (see below for more details). Surrounded by

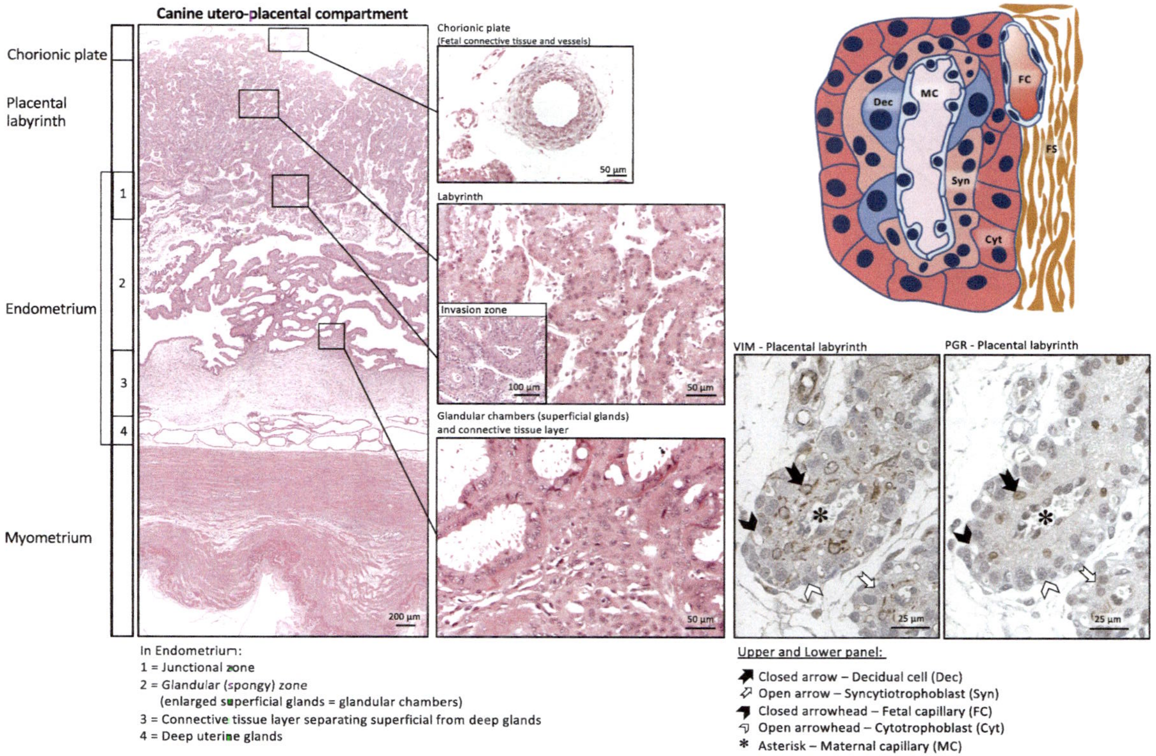

Fig. 3 Histological appearance of the canine fully developed placenta. Cross-section throughout the canine utero-placental compartment (full thickness, left-side panels, hematoxylin, and eosin staining). Schematic representation of the canine placental labyrinth (upper right panel). Localization of maternal stroma-derived decidual cells within the placental labyrinth is shown in right-side lower panels: vimentin staining (VIM, mesenchymal cell marker) and immunohistochemical detection of the nuclear localization of the progesterone receptor (PGR). Decidual cells are the only cells of the canine placenta expressing PGR. During differentiation these cells retain their mesenchymal character and express VIM

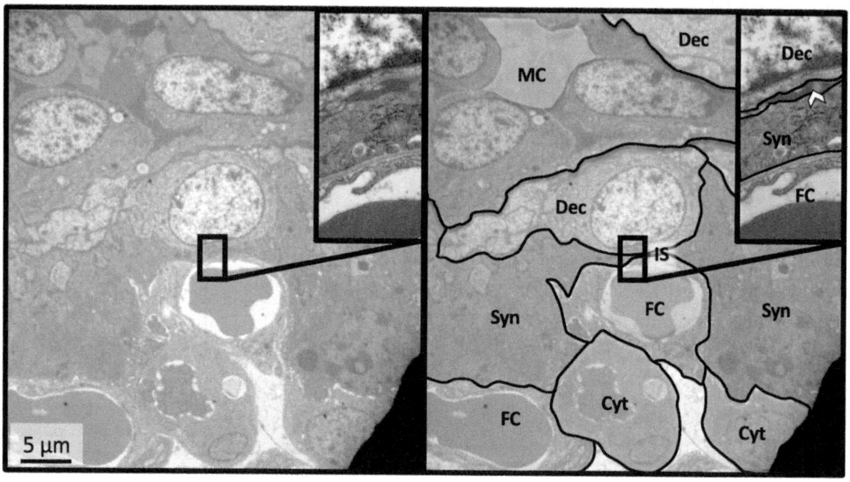

Fig. 4 Transmission electron microscopic (TEM) appearance of the canine placental labyrinth (day 35 of pregnancy). *MC* maternal capillary, *FC* fetal capillary, *Dec* decidual cell, *Syn* syncytiotrophoblast, *Cyt* cytotrophoblast, *IS* interstitial space (basal lamina, open arrowhead)

syncytiotrophoblast and with decidual cells in close proximity, the maternal capillaries in the placental labyrinth are quite peculiar and easily recognized by their relatively large diameter (20–45 µm) (Kehrer 1973). The nuclei of endothelial cells are enlarged and seem to protrude into the lumen (Figs. 3 and 4).

The junctional zone is a transitional area between the labyrinth and glandular zone, in which the columnar phagocytic cytotrophoblast from the terminal part of the lamellae invades the endometrial uterine glands (spongy zone) (Amoroso 1952; Furukawa et al. 2014). It consists of invading villous tips, maternal vessels, and maternal cellular debris (Amoroso 1952, 1961). The spongy zone is located under the junctional zone and consists of the hypertrophied, greatly enlarged, uterine glands (Amoroso 1952; Furukawa et al. 2014) (Fig. 3). They form a 1–2 mm thick layer and are separated from the deep uterine glands by a dense connective tissue layer (Fig. 3) (Grether et al. 1998). They are lined with tall columnar epithelium and reach their maximal size around day 40 of pregnancy (Grether et al. 1998; Grether and Friess 1993), after which their size decreases until parturition (Grether et al. 1998). Chorioallantoic villi project into the lumen of the glandular chambers (Grether et al. 1998; Grether and Friess 1993).

Ultrastructural Morphology

The ultrastructural morphology of a mid-term placenta is presented in Fig. 4. At the ultrastructural level, maternal endothelial cells are filled with abundant irregular vacuoles as well as bloated mitochondria and the golgi apparatus. The decidual cells have a large, roundish nucleus, pale cytoplasm, prominent nucleoli, wide endoplasmic reticulum, and quite enlarged golgi vesicles (Anderson 1969) and are surrounded by a shapeless layer of interstitial matrix, apparently containing COL4 and similar substances existing in the basement membrane of maternal capillaries (Graubner et al. 2018, 2020; Amoroso 1961). The cytotrophoblast cells display a rather large and round nucleus and stain lightly in electron microscopic preparations (Amoroso 1961 and Fig. 4). Their cytoplasm is sporadically filled with the mitochondria and their cell profile shows small endoplasmic reticulum as well as golgi apparatus (Anderson 1969). The number of the cytotrophoblast cells decreases during the progression of pregnancy (Amoroso 1961). Fetal capillaries in the connective tissue are separated from the trophoblastic branches by a well-developed basal lamina (Amoroso 1961; Anderson 1969). Syncytial trophoblast has an intensively stained cytoplasm and has an irregular cell membrane that is in contact with the perivascular basement membrane (Wynn and Corbett 1969; Amoroso 1961). The cellular composition is more or less similar to that of cytotrophoblasts, except that they have more than one nucleus. By their extended cellular branches, syncytial trophoblasts are in contact with both fetal stroma and the basal lamina of fetal capillaries (Amoroso 1961). Although fetal and maternal endothelial cells present similar cellular components, they are hugely different in size. As indicated above, the maternal endothelial cells are much larger and their nuclei bulge/protrude into the lumen (Wynn and Corbett 1969; Anderson 1969; Stoffel et al. 1998, Figs. 3 and 4). In some areas of the syncytium, fetal capillaries can be found in close proximity to maternal ones, while a relatively thick layer of basal lamina (extracellular matrix) exists between them (Barrau et al. 1975).

4.3 Utero-Placental Extracellular Matrix During Canine Placentation

An overview of the current knowledge regarding the distribution and gestational expression patterns of some of the extracellular matrix related factors in the dog can be found under (Kutzler et al. 2012; Graubner et al. 2018; Beceriklisoy et al. 2007; Fellows et al. 2012).

The canine trophoblast has an invasive character. Thus, in the absence of adequate extracellular matrix (ECM) remodeling, invasion and proliferation of the trophoblast cells would not be possible, thereby preventing pregnancy from being maintained.

Following the functional induction of the ECM components during preimplantation (Graubner et al. 2018), the uterus undergoes strong tissue remodeling during trophoblast invasion, as expected, in association with modulation of the ECM components. Thus, placentation is associated with upregulation of ECM1, COL1, -3, -4, and FN1 expression in the uterine wall at interplacental sites. At the implantation site, however, there is decreased expression of two major collagens, COL1 and -3, and of LAMA2. COL1 and -3 are more strongly represented in stromal compartments compared with their epithelial expression, as would be expected, and this pattern is also true for LAMA2 (Graubner et al. 2018). Considering its similar expression patterns in other species with invasive placentation, e.g., rats (Clark et al. 1993), it seems that COL1 suppression is essential for tissue remodeling and "softening" during placentation, presumably facilitating the proliferation and further invasion of the trophoblast into the softened uterus. Gap junctions are important components of intercellular communication during implantation, decidualization, and placentation, with some of them, like CX43 or CX26, playing superior roles (Yu et al. 2014, 2016; Grummer et al. 1994, 1996, 2004; Laws et al. 2008). Although not affected by embryo presence in the preimplantation uterus, the expression and localization of CX43 in decidualizing cells and in other cellular components of the placenta suggests its involvement during decidualization and placental angiogenesis in the dog (Graubner et al. 2018, 2020).

As in other species with invasive placentation, and as mentioned elsewhere, the dog trophoblast exhibits strong invasive properties (Beceriklisoy et al. 2007), although not into the maternal vascular endothelial cells and decidual cells, which escape the proteolytic activity of the trophoblast. The invasive properties of trophoblast appear to be controlled by as yet unknown mechanisms. As a matter of fact, the trophoblast migration (i.e., invasion) into maternal structures does not stop, and strong invasion of the trophoblast at the bottom of the placental labyrinth is observed until term. Interestingly, whereas major matrix metalloproteinases (MMP2 and -9) are expressed in fetal trophoblast cells (Beceriklisoy et al. 2007; Fellows et al. 2012), their inhibitors, in particular TIMP2, are expressed in both counterparts of the placenta (i.e., in the *placenta materna* and *placenta fetalis)* (Graubner et al. 2018), including the fetal trophoblast, implying a possible self-regulatory feedback loop protecting the maternal compartments from excessive trophoblast invasion (Graubner et al. 2018). Accordingly, even more peculiar is the strong presence of TIMP2 and -4, together with the ECM1, in the endometrial connective tissue layer separating deeper tissues from the superficial endometrial layers. Recently, a role of this layer as an active, biochemical and not only physical protection barrier preventing unrestrained invasion of fetal cells into uterine layers was suggested from our studies (Graubner et al. 2018). Noteworthy among its many roles, ECM1 has already been shown to be involved in the reduction of MMP9 proteolytic activity in humans (Fujimoto et al. 2006). It would therefore be interesting to consider the expression and function of the MMP-ECM1-TIMP balancing system during conditions characterized by disturbed or exaggerated trophoblast expansion (like, for example, in SIPS).

At the time of prepartum luteolysis, COL1 and -3 expression is strongly suppressed (Graubner et al. 2018), which has been interpreted as an indicator of preparation for parturition (placentolysis) and release of fetal membranes, a phenomenon that has been previously implied for cattle (Attupuram et al. 2016).

Finally, although not yet fully understood, it is clear that modification of the extracellular matrix reflects the dynamic feto-maternal interactions that occur during the establishment of the canine endotheliochorial placenta with strong, yet restricted, invasion.

5 Vascularization and Cell-to-Cell Communication

While there is de novo formation of a vascular network (vasculogenesis) in the *placenta fetalis*, growth of the new blood vessels from preexisting ones (angiogenesis) occurs in the *placenta materna*. Branches of blood vessels originating from the umbilical cord are seen in both the yolk sac and the placental girdle at 25–30 days of pregnancy in the dog (Miglino et al. 2006). Although the amnion remains poorly vascularized, the allantois becomes prominent and highly vascularized around day 45 of canine pregnancy (Miglino et al. 2006). At day 53 of gestation, the vessels of the highly vascularized allantois cover the central zone of the placental girdle (Miglino et al. 2006). The formation of this dense vascular network in the canine placenta is essential to facilitate the metabolic exchange between the mother and fetus, and many locally produced angiogenic and vasculogenic factors are involved in its regulation. In particular, vascular endothelial growth factors, like VEGFA and its receptors (VEGFR1/Flt-1 and VEGFR2/Flk-1/KDR), have been investigated extensively for their potential involvement in regulating placental angiogenesis and vasculogenesis. Thus, the expression and localization of the VEGFA system was investigated in the canine uterus and utero-placental compartments throughout pregnancy (Gram et al. 2015). VEGF-165, -182, and -188 and their respective receptors, VEGFR1/-2, were found to be induced in response to the presence of preimplantation embryos (day 10 of gestation), indicating a high vascularization rate and supporting the proliferative state of the preimplantation uterus (Bukowska et al. 2011). Besides apparently facilitating increased vascularization, localization of VEGFA and its two receptors at this stage of pregnancy in the endometrial luminal surface epithelial cells and vascular endothelial cells, as well as in superficial and deep uterine glands implies increased vascular permeability and uterine edema (Gram et al. 2015). The latter are essential for angiogenesis, uterine remodeling, embryo attachment, and implantation (Gram et al. 2015). These findings of increased vasculogenesis and angiogenesis were corroborated by microarray analysis, as among the highly represented functional groups detected in the preimplantation canine uterus were those associated with the regulation of vasculature development, angiogenesis, and blood vessel morphogenesis (Graubner et al. 2017a). By regulating vascular functionality, endothelins (ETs) also play important roles during implantation, placental angiogenesis, blood supply, as well as during trophoblast proliferation and invasion (Cervar-Zivkovic et al. 2011). Accordingly, expression and localization of ET1 and -2, ET receptors (ETA and ETB), and endothelin converting enzyme 1 (ECE1) were detected in the canine uterus and utero-placental compartments during pregnancy (Gram et al. 2017). While ET1 was constantly expressed in the canine uterus and utero-placental compartments throughout pregnancy, expression of ET2 appeared to be linked to early pregnancy (Gram et al. 2017). Like the VEGF system, there was increased availability of ET2 in the preimplantation uterus (localized predominantly in the endometrial luminal surface epithelial cells, and in superficial and deep uterine glands), which appears to underline the importance of ETs as well in the regulation of vascular permeability, endothelial cell proliferation, and uterine edema prior to implantation. Following implantation and at mid-gestation, VEGFA and its receptors (VEGFR1/Flt-1 and VEGFR2/Flk-1/KDR) were strongly expressed in fetal trophoblast cells (cytotrophoblast and syncytiotrophoblast) and in placental vascular endothelial cells, clearly indicating the importance of the VEGFA system in the autocrine and/or paracrine regulation of placental angiogenesis and vasculogenesis, and thus in fetal survival and growth (Gram et al. 2015). In particular, localization of strong VEGFR2/Flk-1/KDR signals in endothelial cells in both the *placenta materna* and *fetalis* implies involvement of this receptor in VEGFA-mediated angiogenesis and vasculogenesis. In addition, there was an interesting observation related to the expression of ETB, which gradually increased with the progression of canine gestation (Gram et al. 2017). Its expression was prominent within the syncytiotrophoblast together with ET1 and the ET activator, ECE1 (Gram et al. 2017). This distribution pattern of ETs indicates their possible involvement in the regulation of invasion and proliferation of trophoblast cells within the canine placenta. In particular, binding of ETs to ETB activates the nitric oxide (NO) release that is responsible for vasodilation (Verhaar et al. 1998). Furthermore, the vasodilatory properties of ETB, along with its constantly increasing expression during pregnancy and co-localization with other ET-system members in the *placenta fetalis* (trophoblast cells), strongly suggest an increased ET-mediated blood supply, which is necessary for fetal survival and growth. This also relates to the strong presence of CX26 and -43 in the intima and media of blood vessels in the canine placenta, indicating their involvement in the placental blood supply and angiogenesis (Graubner et al. 2018).

During prepartum luteolysis, downregulation of VEGFA and VEGFR1/Flt-1 might initiate regression of placental blood vessels. Accordingly, as implied from a transcriptomic approach (Nowak et al. 2019), strongly upregulated ANGPT2 together with decreased VEGFA expression in the prepartum canine placenta suggests loosening of peri-endothelial cell contact, thereby increasing endothelial cell apoptosis and vascular degeneration. The contribution of ANGPTs to the disruption of canine placental vascular function is an interesting question that requires further clarification. Similarly, the expression and function of members of the endothelin system (ET1, ECE1, ETA, and ETB) are strongly upregulated in canine utero-placental compartments during the onset of parturition (Gram et al. 2017). They are co-localized in trophoblast together with members of the prostaglandin (PG)-system, e.g., COX2/PTGS2, PTGFS/AKR1C3, or PTGES, that are responsible for the provision of prepartum luteolytic prostaglandins (Kowalewski et al. 2010; Gram et al. 2013, 2014a). This clearly indicates a close relationship between the onset of canine labor and disruption of placental vascular function. Accordingly, several endothelial-associated immune and pro-inflammatory pathways appear to be activated during prepartum luteolysis in the canine placenta, e.g., ICAM1, CXCL8 (IL8), IL3-, eNOS-, RhoA-, and NO-signaling (Nowak et al. 2019). Furthermore, the uterine immune milieu appears to be associated with an increased presence of immune cells during parturition, involving macrophages associated with tissue remodeling, as well as Th and Treg (Tavares Pereira et al. 2021b). Also, these findings indicate the disruption of placental vascular function at the onset of parturition as an important factor leading to parturition and placenta release.

Furthermore, blocking of progesterone signaling by the application of nuclear progesterone-receptor (PGR) blocker (aglepristone) initiates production of PGs, similar to that observed during natural prepartum luteolysis (Kowalewski et al. 2010, 2015; Kowalewski 2014, 2018). Pursuing this idea, we investigated the effects of antigestagen-induced luteolysis on the VEGFA system and ET system in the placenta of pregnant dogs on days 40–45 of gestation. Although VEGFA and VEGFR2/Flk-1/KDR were not affected, VEGFR1/Flt-1 was downregulated by aglepristone treatment (Gram et al. 2015), while ET1 and its vasoconstrictive receptor (ETA) were upregulated, and ET2 and the vasodilator receptor (ETB) were downregulated (Gram et al. 2017). These important and novel observations clearly show a contribution of progesterone-mediated pathways to the dynamic utero-placental vascular activity and functionality during the maintenance and termination of canine pregnancy. Therefore, the involvement of progesterone in the regulation of canine utero-placental vascular function needs more attention.

6 Transplacental Transportation of Maternal Immunoglobulins in the Dog

Very little information is available regarding transplacental transportation of immunoglobulins (IgGs) during pregnancy in the dog. Due to the high contact between maternal and fetal structures, they are transported to the fetus in the last third of pregnancy (Krakowka et al. 1978; Stoffel et al. 2000). During this time, IgGs are predominantly localized in the placental labyrinth, in all layers of the feto-maternal barrier (i.e., in the *placenta materna* and *fetalis*), and particularly in the maternal basement membrane surrounding maternal vessels and in the syncytiotrophoblast. The absence of IgGs in intercellular spaces between maternal endothelial cells, together with its presence within the cells, implies transcellular transport (Stoffel et al. 2000). In the hemophagus zone (marginal hematoma), IgGs are localized within phagolysosomes in the cytotrophoblast, but not in fetal vessels, therefore, apparently they cannot be transferred to the fetus via marginal hematoma (Stoffel et al. 2000). Consequently, maternal IgGs seem to pass from mother to fetus primarily though the placental labyrinth (Stoffel et al. 2000). Nevertheless, the feto-placental contact allows a passage of no more than 2%–18% of the maternal immunoglobulins to the offspring in the dog (Krakowka et al. 1978; Stoffel et al. 2000). Therefore, most of the IgGs can be transferred to the puppies after birth via the colostrum.

7 Canine Placenta: An Important Endocrine Organ During Maintenance and Termination of Pregnancy

In addition to its lack of steroidogenesis (mentioned elsewhere), the canine placenta also lacks PRL synthesis. However, it does express the respective receptors for steroids (PGR and ERα), as well as PRLR (Kowalewski et al. 2010, 2011; Vermeirsch et al. 2000a, b), RLN (Steinetz et al. 1987, 1989), and prostaglandins (Kowalewski et al. 2010; Gram et al. 2013, 2014a), and so remains also an important endocrine organ. The increasing levels of PRL during pregnancy appear to be of pituitary origin, and its placental receptor (PRLR) is primarily localized in the cytotrophoblast, including that localized in the invasive junctional zone (invading cytotrophoblast). This has suggested a possible role for PRL in placental development, through influencing the invasion of fetal trophoblast cells (Kowalewski et al. 2011). Also, during the second half of diestrus (both in pregnant and nonpregnant dogs), PRL plays a role as the main luteotropic factor (reviewed in (Kowalewski 2017)), and its inhibition by provision of dopamine-agonists induces abortion (Concannon et al. 1987). Due to its clinical relevance as a marker of pregnancy, RLN appears to be one of the most important placenta-derived hormones in the dog. In pregnant bitches, an increase in plasma RLN concentrations is observed around days 20–25 of pregnancy, i.e., following placentation (Steinetz et al. 1987, 1989). Its level continues to rise during pregnancy reaching the highest levels 2–3 weeks prior to parturition (Steinetz et al. 1987, 1989). After parturition, RLN vanishes from peripheral blood, becoming undetectable within a few days (Steinetz et al. 1987, 1989). Thus, in the dog as in some other domestic animals (e.g., cats), the placenta is the major source of RLN. As for its localization within the placenta, strong immunohistochemical signals for RLN expression were found in the cytotrophoblast cells (Nowak et al. 2017). Its receptors (RXFP1 and RXFP2) show a clear compartmentalization: whereas RXFP1 is localized in placental endothelial cells and trophoblast cells, RXFP2 was primarily found in decidual cells (Nowak et al. 2017). This latter finding in particular, in addition to possible auto/paracrine effects of RLN in regulating utero-placental function, implies its involvement in the decidualization process as a part of the feto-maternal communication during the maintenance of canine pregnancy. Another interesting finding from our study, indicating a possible functional reciprocal effect between PRL and RLN, is the detection of RXFP1 and RXFP2 expression in PRL-secreting cells in the adenohypophysis of dogs (Nowak et al. 2018). This important and novel finding strongly implicates the involvement of RLN in regulating the production of PRL and other pituitary hormones, possibly contributing to the typically higher PRL levels observed during pregnancy. Clearly, further research is needed to test this hypothesis. As for other peptide hormones, leptin and its receptor (LepR) or GnRH and its receptor (GnRH-R), are also expressed in the canine placenta (Balogh et al. 2015; Schafer-Somi et al. 2015). Their localization in various cellular components of the canine placenta, including trophoblast cells, suggests multiple, yet-to-be-determined, paracrine- and/or autocrine functions.

7.1 Decidual Cells and Prepartum Luteolytic Cascade

More detailed description of the canine prepartum cascade has been provided recently (Kowalewski 2014; Kowalewski et al. 2015, 2020). The most important regulatory feature to be emphasized is the feto-maternal communication during the maintenance and termination of canine pregnancy (Fig. 5), which involves the maternal decidual cells and fetal trophoblast cells. Thus, maternal decidual cells are equipped with the nuclear PGR, and diminishing progesterone levels, and/or suppression of PGR function by an antigestagen, lead to the prepartum release of PGF2α from fetal trophoblast cells where the prostaglandin synthesizing machinery is strongly represented (i.e., COX2/PTGS2, PTGES, PGFS/AKR1C3). An alternative biosynthetic pathway was determined in canine placenta, indicating that the locally produced PGE2 acts as a substrate for the provision of prepartum PGF2α (Gram et al. 2014a). This finding is of clinical importance, since besides being possibly involved in cervical softening (Fuchs et al. 1984; Stys et al. 1981), PGE2 also seems to contribute actively to the production of luteolytic and myocontractile PGF2α in the dog. Interestingly, similar to the PGR, the OXTR is also only located in maternal decidual cells, and its expression increases during normal and antigestagen-induced luteolysis, clearly implying its contribution to prepartum prostaglandin production (Gram et al. 2014b). In contrast, the placental glucocorticoid receptor (GR/NR3C1) is localized in the fetal part of the placenta (cytotrophoblast) (Gram et al. 2016). Its expression is increased during natural, but not antigestagen-induced parturition, implying that the increased expression of the GR/NR3C1 is not mandatory for placental prostaglandin synthesis. Therefore, in contrast to the OXTR, which seems to be an upstream regulator of placental prostaglandin synthesis, it appears that the GR/NR3C1 acts downstream of the PGR. Analogous to the situation observed in humans, a role of the GR/NR3C1 as a local antigestagen has been proposed (Gram et al. 2016). This hypothesis could explain why parturition in dogs takes place at variably low progesterone concentrations, presumably depending on the local availability of the GR. It should be noted that because the prepartum cortisol increase in dogs is erratic (Nohr et al. 1993), its local effects may not reflect peripheral levels. Finally, as recently shown (details in (Nowak et al. 2019) and reviewed in (Kowalewski et al. 2020)), at the placental level, in association with increased PGF2α synthesis, prepartum luteolysis appears to be associated primarily with apoptotic events and disruption of placental vascularization, followed by an inflammatory response. The disruption of PGR signaling seems to play an important role in all of these processes and will be the subject of further research.

7.2 In Vitro Model of Canine Decidualization

Taking into account the importance of the decidualization process and the species-specific role of decidual cells in the maintenance of canine pregnancy, it becomes

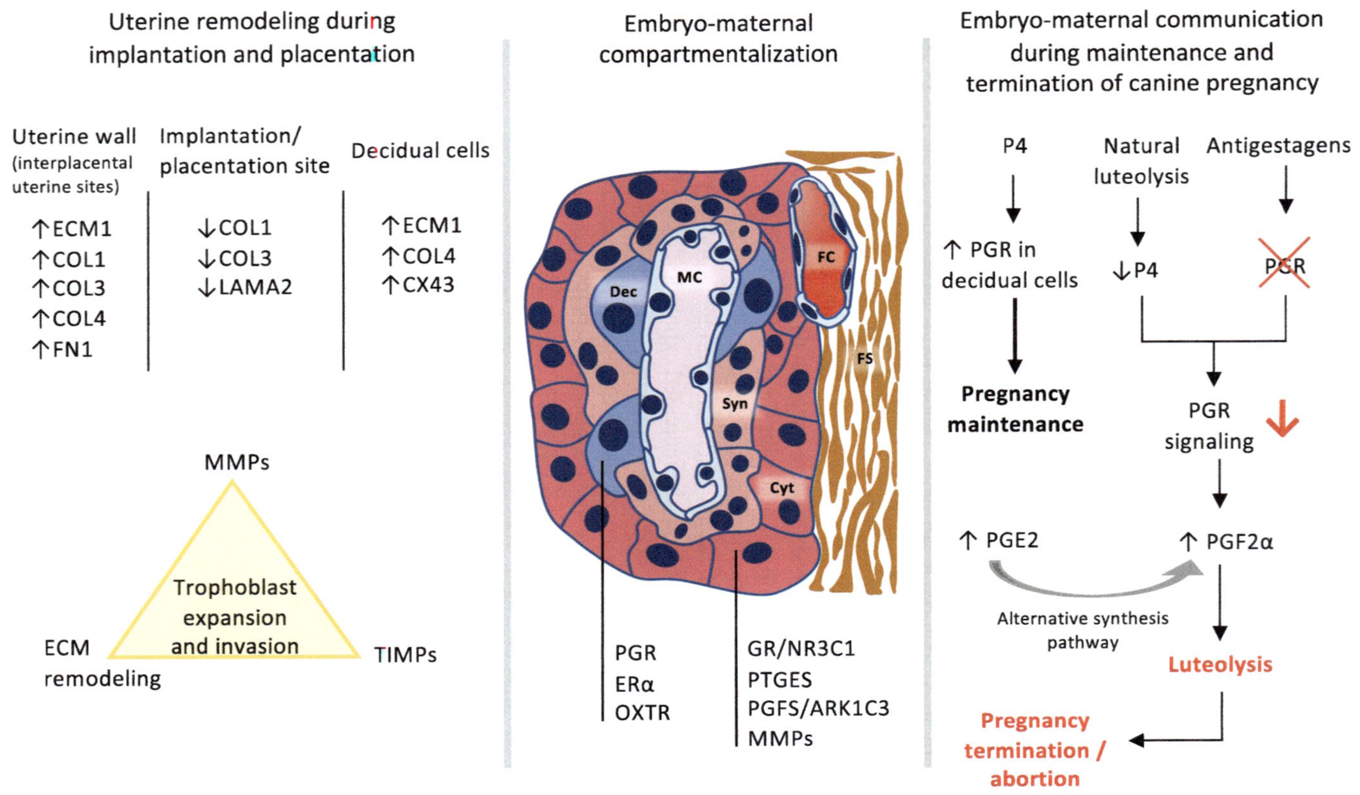

Fig. 5 Uterine remodeling during implantation and placentation and schematic representation of the placental feto-maternal prepartum interplay during induction of the luteolytic cascade

obvious that unveiling the process of decidual cell formation and their functionality is crucial to gaining an understanding of the regulation of canine reproduction, from both the physiological and clinical perspective. Consequently, to develop a deeper insight into this complex phenomenon in the dog, we have established and characterized a novel, species-specific in vitro model for decidualization by isolating canine primary uterine stromal cells from early diestric dogs (i.e., from naturally estrogenized females) (Kautz et al. 2015). The cAMP-mediated approach was successfully utilized and an immortalized cell line was established to serve as a unique model for basic research on canine decidualization (Graubner et al. 2017b). Besides morphological changes induced in decidualizing canine stromal cells, indicating the mesenchymal-epithelial transition associated with increased metabolic and secretory activity, upregulated COL4 and CX43 expression (Graubner et al. 2017b; Kautz et al. 2015), and concomitantly retaining the mesenchymal character (indicated by continuous vimentin and alphaSMA expression), several decidualization markers were induced in these cells, including PGR, PRLR, IGF1, or the PTGES/PTGER2/PTGER4 system. The decidualization capability of PGE2 and the involvement of the two cAMP-mediating receptors of PGE2 (PTGER2/-4) were also confirmed, providing an insight into the reciprocal functional interrelationship between PGE2 and PGR in mediating the decidualization process in the dog (Graubner et al. 2017b, 2020). Thus, while PGE2 upregulates the PGR and PRLR in a PTGER2/-4-dependent manner in decidualizing canine stromal cells in vitro, progesterone appears to be involved in regulating the availability of PGE2 receptors (Graubner et al. 2020). Most recently, we were able to show that cAMP, but not PGE2, is capable of inducing changes in the expression of extracellular matrix proteins, such as COL4 or ECM1, in decidualizing cells (Graubner et al. 2020). The same accounts for the expression of CX43. Thus, the mechanisms acting upstream of cAMP activity in regulating the morpho-functional transition of canine uterine stromal cells towards decidua formation need to be further investigated.

8 Summary and Outlook

Although not considered for a long time in basic and clinical research, the canine placenta has proven to be a dynamic and important endocrine organ, actively responding to circulating hormones of central (i.e., hypothalamo-hypophyseal), ovarian, as well as local, auto−/paracrine origin. The translational and comparative aspects for other species arise from the species-specific endotheliochorial character of the canine placenta, the shallow invasion of trophoblast, and also from underlying endocrine patterns of canine reproduction. As highlighted above, understanding the formation of canine decidua appears to be key to the improvement and optimization of clinical approaches for better control of canine reproduction. The unique placental feto-maternal communication between maternal decidual cells and fetal trophoblast, depending on ovarian steroid provision, appears to be one of the most interesting features and worth further study. It warrants future investigation into the biological nature of the underlying progesterone-mediated signaling important for sustaining canine pregnancy and regulating its length.

Acknowledgments The authors would like to thank Dr. Sharon Mortimer for careful editing of the manuscript.

Funding: Research supported by The Swiss National Science Foundation (SNSF); research grant number 31003A_182481.

References

Amoroso EC (1952) Placentation. In: Parkes AS (ed) Marshall's physiology of reproduction. Longmans Green, London, pp 127–316

Amoroso EC (1961) Histology of the placenta. Br Med Bull 17:81–90

Anderson JW (1969) Ultrastructure of the placenta and fetal membranes of the dog. I. the placental labyrinth. Anat Rec 165:15–35

Attupuram NM, Kumaresan A, Narayanan K, Kumar H (2016) Cellular and molecular mechanisms involved in placental separation in the bovine: a review. Mol Reprod Dev 83:287–297

Baan M, Taverne MA, Kooistra HS, de Gier J, Dieleman SJ, Okkens AC (2005) Induction of parturition in the bitch with the progesterone-receptor blocker aglepristone. Theriogenology 63:1958–1972

Balogh O, Staub LP, Gram A, Boos A, Kowalewski MP, Reichler IM (2015) Leptin in the canine uterus and placenta: possible implications in pregnancy. Reprod Biol Endocrinol 13:13

Barrau MD, Abel JH, Torbit CA, Tietz WJ (1975) Development of the implantation chamber in the pregnant bitch. Am J Anat 143:115–130

Beceriklisoy HB, Walter I, Schafer-Somi S, Miller I, Kanca H, Izgur H et al (2007) Matrix metalloproteinase (MMP)-2 and MMP-9 activity in the canine uterus before and during placentation. Reprod Domest Anim 42:654–659

Bukowska D, Kempisty B, Jackowska M, Wozna M, Antosik P, Piotrowska H et al (2011) Analysis of integrins and vascular endothelial growth factor isoforms mRNA expression in the canine uterus during perimplantation period. Pol J Vet Sci 14:253–258

Cervar-Zivkovic M, Dieber-Rotheneder M, Barth S, Hahn T, Kohnen G, Huppertz B et al (2011) Endothelin-1 stimulates proliferation of first-trimester trophoblasts via the A- and B-type receptor and invasion via the B-type receptor. J Clin Endocrinol Metab 96:3408–3415

Clark DE, Hurst PR, McLennan IS, Myers DB (1993) Immunolocalization of collagen type I and laminin in the uterus on days 5 to 8 of embryo implantation in the rat. Anat Rec 237:8–20

Concannon PW, Hansel W, Visek WJ (1975) The ovarian cycle of the bitch: plasma estrogen, LH and progesterone. Biol Reprod 13:112–121

Concannon PW, Weinstein P, Whaley S, Frank D (1987) Suppression of luteal function in dogs by luteinizing hormone antiserum and by bromocriptine. J Reprod Fertil 81:175–180

Concannon P, Tsutsui T, Shille V (2001) Embryo development, hormonal requirements and maternal responses during canine pregnancy. J Reprod Fertil Suppl 57:169–179

Enders AC, Carter AM (2004) What can comparative studies of placental structure tell us?—a review. Placenta 25(Suppl A):S3–S9

Fellows EJ, Hazzard TM, Kutzler MA (2012) Gene expression in pre-term, pre-labour and parturient canine placenta. Reprod Domest Anim 47(Suppl 6):182–185

Fuchs AR, Goeschen K, Rasmussen AB, Rehnstrom JV (1984) Cervical ripening and plasma prostaglandin levels. Comparison of endocervical and extra-amniotic PGE2. Prostaglandins 28:217–227

Fujimoto N, Terlizzi J, Aho S, Brittingham R, Fertala A, Oyama N et al (2006) Extracellular matrix protein 1 inhibits the activity of matrix metalloproteinase 9 through high-affinity protein/protein interactions. Exp Dermatol 15:300–307

Furukawa S, Kuroda Y, Sugiyama A (2014) A comparison of the histological structure of the placenta in experimental animals. J Toxicol Pathol 27:11–18

Gram A, Buchler U, Boos A, Hoffmann B, Kowalewski MP (2013) Biosynthesis and degradation of canine placental prostaglandins: prepartum changes in expression and function of prostaglandin F2alpha-synthase (PGFS, AKR1C3) and 15-hydroxyprostaglandin dehydrogenase (HPGD). Biol Reprod 89:2

Gram A, Fox B, Buchler U, Boos A, Hoffmann B, Kowalewski MP (2014a) Canine placental prostaglandin E2 synthase: expression, localization, and biological functions in providing substrates for Prepartum PGF2alpha synthesis. Biol Reprod 91:154

Gram A, Boos A, Kowalewski MP (2014b) Uterine and placental expression of canine oxytocin receptor during pregnancy and normal and induced parturition. Reprod Domest Anim 49(Suppl 2):41–49

Gram A, Hoffmann B, Boos A, Kowalewski MP (2015) Expression and localization of vascular endothelial growth factor a (VEGFA) and its two receptors (VEGFR1/FLT1 and VEGFR2/FLK1/KDR) in the canine corpus luteum and utero-placental compartments during pregnancy and at normal and induced parturition. Gen Comp Endocrinol 223:54–65

Gram A, Trachsel A, Boos A, Kowalewski MP (2016) Elevated utero/placental GR/NR3C1 is not required for the induction of parturition in the dog. Reproduction 152:303–311

Gram A, Boos A, Kowalewski MP (2017) Cellular localization, expression and functional implications of the utero-placental endothelin system during maintenance and termination of canine gestation. J Reprod Dev 63:235–245

Graubner FR, Gram A, Kautz E, Bauersachs S, Aslan S, Agaoglu AR et al (2017a) Uterine responses to early pre-attachment embryos in the domestic dog and comparisons with other domestic animal species. Biol Reprod 97:197–216

Graubner FR, Reichler IM, Rahman NA, Payan-Carreira R, Boos A, Kowalewski MP (2017b) Decidualization of the canine uterus: from early until late gestational in vivo morphological observations, and functional characterization of immortalized canine uterine stromal cell lines. Reprod Domest Anim 52(Suppl 2):137–147

Graubner FR, Boos A, Aslan S, Kucukaslan I, Kowalewski MP (2018) Uterine and placental distribution of selected extracellular matrix (ECM) components in the dog. Reproduction 155:403–421

Graubner FR, Pereira MT, Boos A, Kowalewski MP (2020) Canine decidualization in vitro: extracellular matrix modification, progesterone mediated effects and selective blocking of prostaglandin E2 receptors. J Reprod Dev

Grether BM, Friess AE (1993) The glandular compartments of the canine placenta—a scanning electron microscopic study. Schweiz Arch Tierheilkd 135:272–278

Grether BM, Friess AE, Stoffel MH (1998) The glandular chambers of the placenta of the bitch in the second third of pregnancy (day 30-44): an ultrastructural, ultrahistochemical and lectinhistochemical investigation. Anat Histol Embryol 27:95–103

Grummer R, Chwalisz K, Mulholland J, Traub O, Winterhager E (1994) Regulation of connexin26 and connexin43 expression in rat endometrium by ovarian steroid hormones. Biol Reprod 51:1109–1116

Grummer R, Reuss B, Winterhager E (1996) Expression pattern of different gap junction connexins is related to embryo implantation. Int J Dev Biol 40:361–367

Grummer R, Hewitt SW, Traub O, Korach KS, Winterhager E (2004) Different regulatory pathways of endometrial connexin expression: preimplantation hormonal-mediated pathway versus embryo implantation-initiated pathway. Biol Reprod 71:273–281

Hermanson JW, de Lahunta A, Evans HE (2020) Miller and Evan's anatomy of the dog, 5th edn. Elsevier, St. Louis, MO

Hoffmann B, Hoveler R, Nohr B, Hasan SH (1994) Investigations on hormonal changes around parturition in the dog and the occurrence of pregnancy-specific non conjugated oestrogens. Exp Clin Endocrinol 102:185–189

Hoffmann B, Busges F, Engel E, Kowalewski MP, Papa P (2004) Regulation of corpus luteum-function in the bitch. Reprod Domest Anim 39:232–240

Kautz E, Gram A, Aslan S, Ay SS, Selcuk M, Kanca H et al (2014) Expression of genes involved in the embryo-maternal interaction in the early-pregnant canine uterus. Reproduction 147:703–717

Kautz E, de Carvalho PP, Reichler IM, Gram A, Boos A, Kowalewski MP (2015) In vitro decidualisation of canine uterine stromal cells. Reprod Biol Endocrinol 13:85

Kehrer A (1973) Zur Entwicklung und Ausbildung des Chorions der Placenta zonaria bei Katze, Hund und Fuchs. Z Anat Entwicklungsgesch 143:24–42

Klein R, Schams D, Failing K, Hoffmann B (2003) Investigations on the re-establishment of the positive feedback of oestradiol during anoestrus in the bitch. Reprod Domest Anim 38:13–20

Kowalewski MP (2014) Luteal regression vs. prepartum luteolysis: regulatory mechanisms governing canine corpus luteum function. Reprod Biol 14:89–102

Kowalewski MP (2017) Regulation of Corpus luteum function in the domestic dog (*Canis familiaris*) and comparative aspects of luteal function in the domestic cat (*Felis catus*). In: Meidan R (ed) The life cycle of the Corpus luteum. Springer International, Cham

Kowalewski MP (2018) Selected comparative aspects of canine female reproductive physiology. In: Skinner Michael K (ed) Encyclopedia of reproduction, 2nd edn. Academic Press, Elsevier, London

Kowalewski MP, Beceriklisoy HB, Aslan S, Agaoglu AR, Hoffmann B (2009) Time related changes in luteal prostaglandin synthesis and steroidogenic capacity during pregnancy, normal and antiprogestin induced luteolysis in the bitch. Anim Reprod Sci 116:129–138

Kowalewski MP, Beceriklisoy HB, Pfarrer C, Aslan S, Kindahl H, Kucukaslan I et al (2010) Canine placenta: a source of prepartal prostaglandins during normal and antiprogestin-induced parturition. Reproduction 139:655–664

Kowalewski MP, Michel E, Gram A, Boos A, Guscetti F, Hoffmann B et al (2011) Luteal and placental function in the bitch: spatio-temporal changes in prolactin receptor (PRLr) expression at dioestrus, pregnancy and normal and induced parturition. Reprod Biol Endocrinol 9:109

Kowalewski MP, Gram A, Kautz E, Graubner FR (2015) The dog: nonconformist, not only in maternal recognition signaling. Adv Anat Embryol Cell Biol 216:215–237

Kowalewski MP, Tavares Pereira M, Kazemian A (2020) Canine conceptus-maternal communication during maintenance and termination of pregnancy, including the role of species-specific decidualization. Theriogenology 150:329–338

Krakowka S, Long D, Koestner A (1978) Influence of transplacentally acquired antibody on neonatal susceptibility to canine distemper virus in gnotobiotic dogs. J Infect Dis 137:605–608

Kutzler M, Sahlfeld L, Fellows E (2012) Who let the dogs in: a canine trophoblast invasion model for pre-eclampsia. Reprod Domest Anim 47(Suppl 6):186–189

Laws MJ, Taylor RN, Sidell N, DeMayo FJ, Lydon JP, Gutstein DE et al (2008) Gap junction communication between uterine stromal cells plays a critical role in pregnancy-associated neovascularization and embryo survival. Development 135:2659–2668

Leiser R, Kaufmann P (1994) Placental structure: in a comparative aspect. Exp Clin Endocrinol 102:122–134

Leiser R, Koob B (1993) Development and characteristics of placentation in a carnivore, the domestic cat. J Exp Zool 266:642–656

McGeady TA, Quinn PJ, Fitzpatrick ES, Ryan MT, Kilroy D, Lonergan P (2017) Veterinary embryology, 2nd edn. Wiley Blackwell, Hoboken, NJ

Miglino MA, Ambrosio CE, dos Santos MD, Wenceslau CV, Pfarrer C, Leiser R (2006) The carnivore pregnancy: the development of the embryo and fetal membranes. Theriogenology 66:1699–1702

Nohr B, Hoffmann B, Steinetz BE (1993) Investigation of the endocrine control of parturition in the dog by application of an antigestagen. J Reprod Fertil Suppl 47:542–543

Nowak M, Gram A, Boos A, Aslan S, Ay SS, Onyay F et al (2017) Functional implications of the utero-placental relaxin (RLN) system in the dog throughout pregnancy and at term. Reproduction 154:415–431

Nowak M, Boos A, Kowalewski MP (2018) Luteal and hypophyseal expression of the canine relaxin (RLN) system during pregnancy: implications for luteotropic function. PLoS One 13: e0191374

Nowak M, Rehrauer H, Ay SS, Findik M, Boos A, Kautz E et al (2019) Gene expression profiling of the canine placenta during normal and antigestagen-induced luteolysis. Gen Comp Endocrinol 282:113194
Papa PC, Hoffmann B (2011) The corpus luteum of the dog: source and target of steroid hormones? Reprod Domest Anim 46:750–756
Papa PC, Kowalewski MP (2020) Factors affecting the fate of the canine corpus luteum: potential contributors to pregnancy and non-pregnancy. Theriogenology 150:339–346
Payan-Carreira R, Santos C, Miranda S, Pereira RM, Santos D, Pires MA (2014) Temporal changes in neutral endopeptidase/CD10 immunoexpression in the cyclic and early pregnant canine endometrium. Theriogenology 82:815–826
Reynaud K, Saint-Dizier M, Tahir MZ, Havard T, Harichaux G, Labas V et al (2015) Progesterone plays a critical role in canine oocyte maturation and fertilization. Biol Reprod 93:87
Schafer-Somi S, Beceriklisoy HB, Budik S, Kanca H, Aksoy OA, Polat B et al (2008) Expression of genes in the canine pre-implantation uterus and embryo: implications for an active role of the embryo before and during invasion. Reprod Domest Anim 43:656–663
Schafer-Somi S, Kowalewski MP, Kanca H, Bozkurt MF, Gram A, Sabitzer S et al (2015) GnRH and its receptor (GnRH-R) are expressed in the canine placenta and uterus. Theriogenology 84:1482–1489
Schafer-Somi S, Gabriel C, Aslan S (2020) Embryo-maternal communication in dogs: immune system related factors. Theriogenology 150:382–387
Steinetz BG, Goldsmith LT, Lust G (1987) Plasma relaxin levels in pregnant and lactating dogs. Biol Reprod 37:719–725
Steinetz BG, Goldsmith LT, Harvey HJ, Lust G (1989) Serum relaxin and progesterone concentrations in pregnant, pseudopregnant, and ovariectomized, progestin-treated pregnant bitches: detection of relaxin as a marker of pregnancy. Am J Vet Res 50:68–71
Stoffel MH, Gille U, Friess AE (1998) Scanning electron microscopy of the canine placenta. Ital J Anat Embryol 103:291–300
Stoffel MH, Friess AE, Hartmann SH (2000) Ultrastructural evidence of transplacental transport of immunoglobulin G in bitches. J Reprod Fertil 118:315–326
Stys SJ, Dresser BL, Otte TE, Clark KE (1981) Effect of prostaglandin E2 on cervical compliance in pregnant ewes. Am J Obstet Gynecol 140:415–419
Tavares Pereira M, Nowaczyk R, Payan-Carreira R, Miranda S, Aslan S, Kaya D, Kowalewski MP (2021a) Selected uterine immune events associated with the establishment of pregnancy in the dog. Front Vet Sci 7:625921
Tavares Pereira M, Nowaczyk R, Aslan S, Ay SS, Kowalewski MP (2021b) Utero-placental immune milieu during normal and aglepristone-induced parturition in the dog. Animals (Basel) 11(12):3598
Verhaar MC, Strachan FE, Newby DE, Cruden NL, Koomans HA, Rabelink TJ et al (1998) Endothelin-a receptor antagonist-mediated vasodilatation is attenuated by inhibition of nitric oxide synthesis and by endothelin-B receptor blockade. Circulation 97:752–756
Vermeirsch H, Simoens P, Hellemans A, Coryn M, Lauwers H (2000a) Immunohistochemical detection of progesterone receptors in the canine uterus and their relation to sex steroid hormone levels. Theriogenology 53:773–788
Vermeirsch H, Simoens P, Lauwers H (2000b) Immunohistochemical detection of the estrogen receptor-alpha and progesterone receptor in the canine pregnant uterus and placental labyrinth. Anat Rec 260:42–50
Wynn RM, Corbett JR (1969) Ultrastructure of the canine placenta and amnion. Am J Obstet Gynecol 103:878–887
Yu J, Berga SL, Zou W, Sun HY, Johnston-MacAnanny E, Yalcinkaya T et al (2014) Gap junction blockade induces apoptosis in human endometrial stromal cells. Mol Reprod Dev 81:666–675
Yu J, Berga SL, Johnston-MacAnanny EB, Sidell N, Bagchi IC, Bagchi MK et al (2016) Endometrial stromal decidualization responds reversibly to hormone stimulation and withdrawal. Endocrinology 157:2432–2446

Placentation in the African Elephant (*Loxodonta africana*)

W. R. (Twink) Allen and Fiona J Stansfield

Contents

Abstract The female elephant shows a 3-week "follicular phase" to commence her 16-week estrous cycle at the end of which a second surge in pituitary luteinizing hormone (LH) release matures and ovulates an ovarian follicle in association with estrous behavior and mating, whereas the first LH surge at the start of the follicular phase causes luteinization of 3–5 partially developed follicles. The prolonged pregnancy of 22 months is supported by a zonary endotheliochorial placenta which secretes placental lactogen (ePL) from around 40 days of gestation in association with replacement of the lumenal epithelium of the endometrium by trophoblast and the development of large corpora lutea (CLs) in the maternal ovaries from the previously formed luteinized follicles in response to the first LH peak early in the follicular phase. The zonary placenta develops above, rather than within, the endometrium. The elephant placenta secretes neither estrogens nor progestagens

W. R. (Twink) Allen (✉)
Sharjah Equine Hospital, Sharjah, United Arab Emirates

Robinson College, University of Cambridge, Cambridge, UK

The Paul Mellon Laboratory of Equine Reproduction, 'Brunswick', Newmarket, Suffolk, UK

F. J. Stansfield
The Paul Mellon Laboratory of Equine Reproduction, 'Brunswick', Newmarket, Suffolk, UK

Department of Production Animal Studies, Faculty of Veterinary Science, Onderstepoort, University of Pretoria, Pretoria, South Africa

R. D. Geisert, T. E. Spencer (eds.), *Placentation in Mammals*, Advances in Anatomy, Embryology and Cell Biology 234, https://doi.org/10.1007/978-3-030-77360-1_9

throughout gestation, as pregnancy maintenance relies on 5α-dihyroprogesterone and other 5α reduced progestagens secreted by secondary CLs stimulated by ePL and the stromal tissue of the fetal gonads, which become extremely enlarged during the second half of the 22-month pregnancy. In female fetuses, this ovarian enlargement includes the development and subsequent regression of multiple primary and secondary follicles with a consequent substantial decline in primary follicle numbers at birth. During the next 8–9 years of pre-pubertal life, however, oocyte and primary follicle numbers recover to levels near those found in late gestation, which may be evidence of postnatal oogenesis occurring in the elephant.

Keywords Double LH peak · Zonary placenta · Placental lactogen · Secondary corpora lutea · Enlarged fetal gonads · Postnatal oogenesis

1 Introduction

Professor E. C. Amoroso FRS, or "Amo" as he was widely and affectionately known, was a true giant in the broad field of comparative placentation during the 1950s to 1990s. He, fittingly, coauthored with his junior protégé, John Perry, in the Proceedings of the Royal Society in 1964, a truly seminal paper entitled, "The foetal membranes and placenta of the African elephant (*Loxodonta africana*)." This proved the benchmark for many subsequent studies of elephant placentation, including those of the present authors, and it is therefore entirely fitting that a chapter entitled, "Placentation in the African Elephant" should be included in this book honouring Amo's great contribution to placentation in many species.

As the biggest land mammal in the world, it is perhaps not surprising that the adult female elephant shows a prolonged estrous cycle of around 16 weeks and an even longer gestation period of some 22 months during which the 130–140 kg newborn elephant calf has been supported in utero by a zonary endotheliochorial placenta that is structurally similar to, although a great deal bigger than, that in the dog and cat (Amoroso and Perry 1964; Short 1966; Laws 1969; Hanks and Short 1972; Perry 1974; Smith and Buss 1975; Kasputin et al. 1996; Allen et al. 2003; Wooding et al. 2005). The male elephant (Fig. 1a) also displays some unusual and unique sexual characteristics. He comes into a period of enhanced sexual excitement and activity known as "musth" for some 6–8 weeks, usually only once per year, during which his intra-abdominally situated testes secrete testosterone in greatly increased amounts, which raise his peripheral blood testosterone concentrations by 100-fold or more (Buss and Johnson 1967; McNeilly et al. 1983). This "testosterone flush" causes him to dribble urine constantly from his penis (Fig. 1b), and this increase in testosterone also causes enlargement of the temporal glands on each side of his head and activates them to secrete increased quantities of a pungent fluid more or less constantly (Fig. 1c). The hyperactivity of these temporal glands clearly causes discomfort which makes the musth male repeatedly drive his tusks into the ground

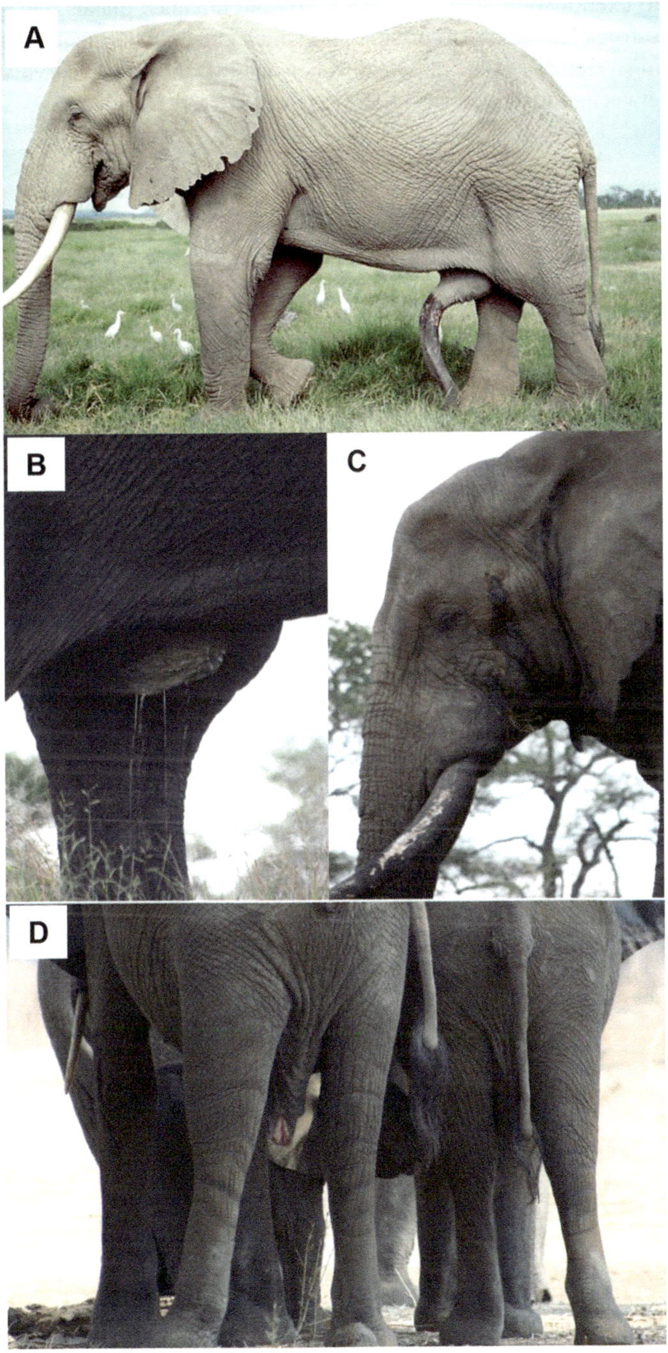

Fig. 1 (**a**) An adult male African elephant with his penis partially erect. (**b**) Musth elephant dribbling urine from his penis. (**c**) Musth elephant with swollen and painful temporal glands. He drives his tusks into the ground to create back pressure on the temporal glands to ease his discomfort. (**d**) A female elephant exhibiting her vulval opening low down between her back legs

(Fig. 1c) to exert a soothing back pressure on the inflamed glands. He goes off in search of estrous females, intromission into which can only be accomplished by the female using a cremaster type muscle to "open and lift" the lips of her vulva which is situated between, not above, her back legs (Fig. 1d).

2 Anatomy of the Uterus and Ovaries

Similarities and contrasts exist between the uterus of the horse and the elephant. Although both are bicornuate, the body of the elephant uterus (Fig. 2Ab) is shorter than that of the horse (Fig. 2Aa) and the two horns, joined to each other by a short intercornuate ligament, run parallel to each other for a much greater distance before diverging laterally away towards the ovaries (Fig. 2Ab). A great amount of fat is

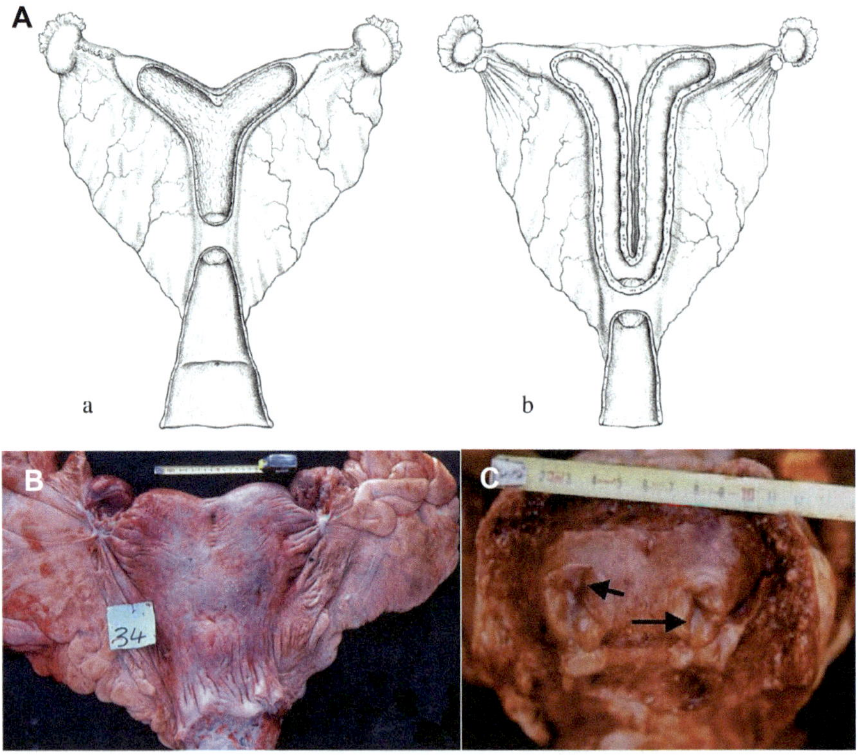

Fig. 2 (**A**) Diagram comparing the basic structure of the horse (**Aa**) and elephant (**Ab**) uteri. Note the shorter body and longer horns of the elephant uterus. (**B**) Elephant uterus showing the considerable fat deposition in the broad ligament and puckering of the ligament to increase the strength of its attachment to the dorsal abdomen. (**C**) Cross section of the uterine horns of an elephant showing the "star-shaped" arrangement of the endometrial folds (arrows). (Reprinted with permission from Allen et al. 2003)

deposited in the broad ligament (Fig. 2B) making it significantly thicker and stronger than its equine equivalent and resulting in a puckered and rope-like appearance as it passes anterio-laterally to become incorporated in the ovarian and lateral uterine ligaments that are anchored to the dorso-lateral wall of the abdomen (Fig. 2B; Allen et al. 2003). Since the gravid uterus weighs in excess of 150 kg during the final stages of gestation (Laws 1969; Craig 1984) the strength afforded by the broad ligament at its attachment is no doubt vital for a successful pregnancy. A cross section across the uterus posterior to the lateral divergence of the two horns shows the folding of the endometrium to form the star-shaped uterine lumen (Fig. 2C) as described originally by Perry (1953) and Amoroso and Perry (1964).

In relation to the overall body size of the female elephant, the adult elephant ovary is surprisingly small. A distinct serosal pouch completely envelops the ovary which is composed of the infundibulum of the oviduct and its incorporation into the ovarian bursar (Fig. 3a; Perry 1953). In nonpregnant elephants, groups of small follicles (0.3–0.8 mm diameter) are frequently seen in the ovaries, but towards the end of the 3-week follicular phase of the estrous cycle single, large, apparently, pre-ovulatory follicles of around 1.5 cm diameter are observed. In pregnant elephants, on the other hand, and even those in the very early stages of gestation, some 3–6 large CLs measuring 3–5 cm in diameter are always present, usually, although not invariably, bunched together on the ovary situated ipsilateral to the gravid uterine horn (Fig. 3b; Lueders et al. 2012).

Many uteri in older females, both pregnant and nonpregnant, exhibit multiple papillomatous outgrowths on the outer uterine serosal surface. Measuring 0.5–2.0 cm in diameter and raised 0.5–1.0 cm above the surface, they tend to be clustered near the tip of the uterine horn (Fig. 4a). While not dissimilar in appearance to the explants of refluxed menstrual endometrium that grow on the serosal surfaces of the uterus and neighboring intestines in women suffering from endometriosis (Leyendecker et al. 1998), the lumps are, in fact, no more than simple exfoliative outgrowths of the uterine serosa, with no evidence of any mucosal components (Fig. 4b).

3 Hormones of the Estrous Cycle and Early Pregnancy

The elephant estrous cycle is unusual in that it commences with an 18–21 day "follicular growth period" which is characterized by 2 surges of LH release from the anterior pituitary gland. The first surge occurs during the initial 3–5 days which stimulates growth and partial luteinization of a wave of some 5–8 follicles followed by a second peak 14–18 days later. The second LH surge is accompanied by estrous behavior and mating prior to proper ovulation of a single, mature follicle which results in the development of a typical progesterone-secreting CL, while the earlier wave of luteinized follicles still lie dormant in the ovaries (Fig. 5a; Yamamoto et al. 2011; Lueders et al. 2010). Progestagens secreted by the newly formed CL take the female out of estrous and into a classical progestagen-dominated diestrus during

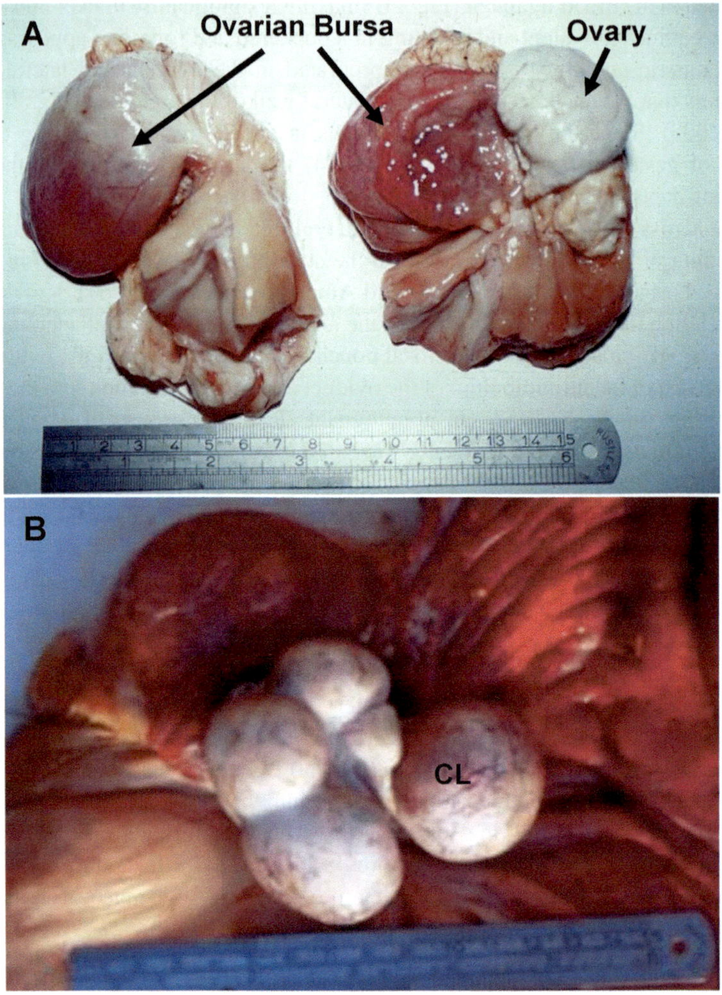

Fig. 3 (**a**) A pair of elephant ovaries, one of which remains enclosed in its ovarian bursa. The other ovary has been extruded from the bursa to show its rather uniform and uninteresting structure. (**b**) The ipsilateral ovary from a pregnant elephant exhibiting multiple accessory corpora lutea (CL). (**a**; Reprinted with permission from Allen et al. 2003)

which serum levels of 5α-dihydroprogesterone and other 5α-progestagens rise to a peak at around Week 6 after ovulation. 5α-dihydroprogesterone concentrations decline again steadily to reach baseline at the onset of a new follicular phase, some 14 weeks after the previous ovulation (Fig. 5a). Uniquely, native progesterone is hardly secreted at all by either the ovaries or the fetoplacental unit in the elephant (Hodges et al. 1997; Meyer et al. 1997; Allen et al. 2002).

In the event of conception following mating in association with ovulation of one of the second wave of follicles, serum progestagen concentrations again begin to

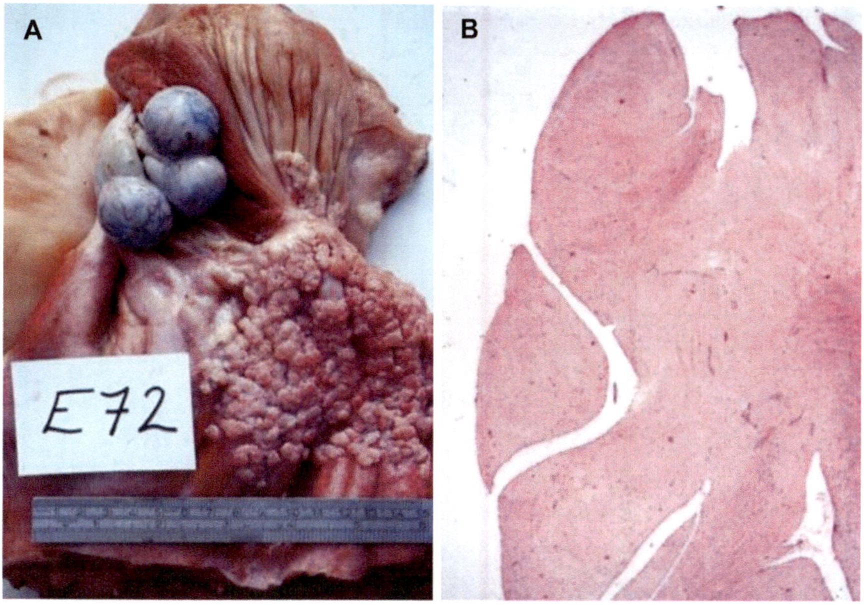

Fig. 4 (**a**) Uterus from an aged female elephant showing the right horn covered in wart-like growths. (**b**) Histological section of a uterine "wart." (**a**; Reprinted with permission from Allen et al. 2003)

decline 4–6 weeks later, but then show a sudden and sharp secondary rise around an estimated Day 45 (Fig. 5b). The increase in progestagen is coincident with initial attachment of the allantochorion to the endometrium at the commencement of placentation and the associated onset of secretion of placental lactogen (ePL) by the implanting trophoblast (Allen et al. 2003; Sakamoto et al. 2012; Yamamoto et al. 2012). This, in turn, causes considerable enlargement of the still-present luteinized follicles which formed early in the previous pre-mating follicular phase and their conversion into fully active accessory CLs which are clustered frequently, but not always, on the ovary ipsilateral to the gravid uterine horn (Fig. 3b; Hodges et al. 1997; Allen et al. 2002; Stansfield and Allen 2012). These secondary or accessory CLs persist throughout gestation and secrete progestagens to maintain the pregnancy state, with additional help from the fetal gonads during the second half of gestation.

4 Gross Development of the Placenta

It remains unknown whether fertilization takes place in the ovarian bursa or the convoluted fallopian tube in the elephant; the time taken for tubal transport of the embryo to the uterus is also unknown. However, the young elephant conceptus lodges at the base of one of the uterine horns where it is held stationary by a dramatic

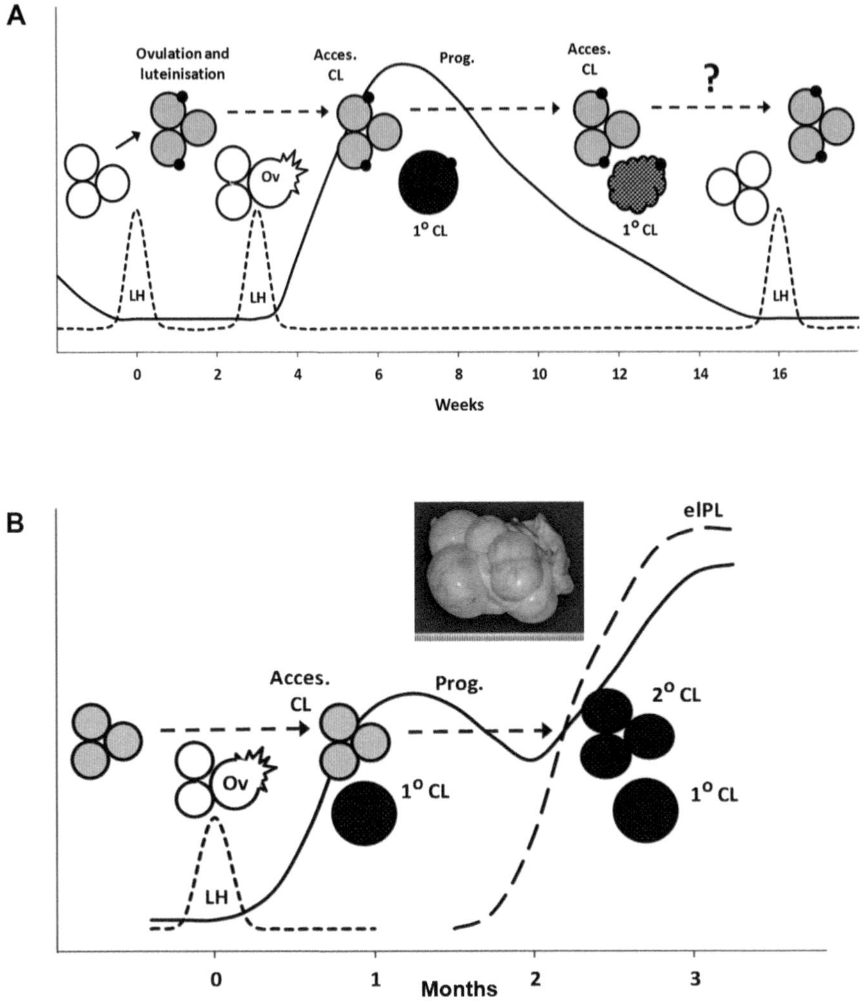

Fig. 5 (**a**) Graphical representation of the elephant estrous cycle. Two LH surges occur, one at the beginning and the other at the end of the 21 day "follicular growth period." The first stimulates the development of a number of persisting luteinized follicles and the second stimulates ovulation of a mature follicle, the corpus luteum (CL) which persists for 14 weeks before regressing to enable the next follicular growth period to commence. (Adapted with permission from Lueders et al. 2010). (**b**) Graphical representation of the hormone changes during early pregnancy in the elephant. Progesterone concentrations decline steadily as occurs during diestrus in the estrous cycle, but progesterone concentrations sharply increase again between 40 and 50 days after ovulation following development of secondary CL on the maternal ovaries. Formation of the secondary CL is stimulated by the action of trophoblast production of placental lactogen which stimulate the luteinized follicles persisting from the previous follicular phase that led to the pregnancy. (Adapted with permission from Lueders et al. 2012)

increase in the tonicity of the myometrium. The gravid uterus literally clamps down upon the elliptical conceptus to fix the position in the uterine horn allowing the central region of the trophoblast to make intimate physical contact with the lumenal epithelium of the endometrium.

The conceptus bulge becomes apparent during the first 2–3 months of gestation as a discrete, swelling at the point of reflection of the gravid uterine horn (Fig. 6a). The ovoid conceptus shows the developing placental band as a pale ribbon of thickening in its mid-section (Fig. 6b). With advancing gestation, the conceptus bulge enlarges steadily to fill the whole of the gravid horn, although it never expands into the uterine body or non-gravid horn (Fig. 6c, d). The reddish-brown placental band broadens as gestation progresses, usually forming a complete band of tissue with brown-colored lateral edges that result from the formation of the hematogenous zone in this region (Fig. 6d). In some animals, however, the placental band can narrow in some regions and may even show distinct gaps in its structure (Fig. 6e). The allantochorion on either side of the placental zone is a pale and slightly roughened beneath which the surface of the allantois is covered in small pustule-like round plaques (Fig. 6f). Another unusual feature of the elephant placenta is the clear division of the allantois into four conjoined compartments within the chorion.

Later in pregnancy, the placental band shows large areas of reddish-brown tissue consisting of materno-fetal lamellae through which run the larger blood vessels. The lateral edges of the band are dark brown in color due to leakage of maternal blood to form the hematogenous zones (Fig. 7a). Another unusual feature is the relative narrowness (2–4 cm) of the placental hilus that provides the only attachment of the conceptus to the endometrium (Fig. 7b). This structure carries the dense plexus of maternal blood vessels that vascularize the placenta (Fig. 7b). However, despite the great growth and expansion of the placenta throughout gestation, the original materno-placental hilus gains little additional width, remaining a narrow channel that carries an increasing concentration of enlarging uterine blood vessels. The band is the only structure that holds the conceptus within the uterus and it can be severed easily (Fig. 7b) to allow the conceptus at term to simply fall out of the uterus. Indeed, at normal parturition which takes place while the mother remains standing, the 120 kg calf does indeed fall vertically down the long maternal vagina while still attached to the umbilical cord, thereby "dragging" the placenta off its heavily vascularized maternal placental hilus. This results in considerable hemorrhage into the uterine lumen (Allen 2006) where the resulting large hematoma (Fig. 7c) takes 1–1.5 years to be fully resolved and resorbed away, thereby functioning as a "temporary contraceptive" and contributing significantly to the 4-yearly inter-calving interval exhibited by the elephant which is still lactating heavily when she does eventually return to estrous and is re-mated some 2 years after birth of the calf (Allen et al. 2003).

The umbilical cord of the elephant fetus initially attaches mesometrially. Near term it measures 60–100 cm in length and 4–6 cm in width. It usually remains untwisted and it contains 3 or 4 stout (1 cm width) blood vessels due to occasional branching of one vessel within the cord. It is not long enough to be delivered intact

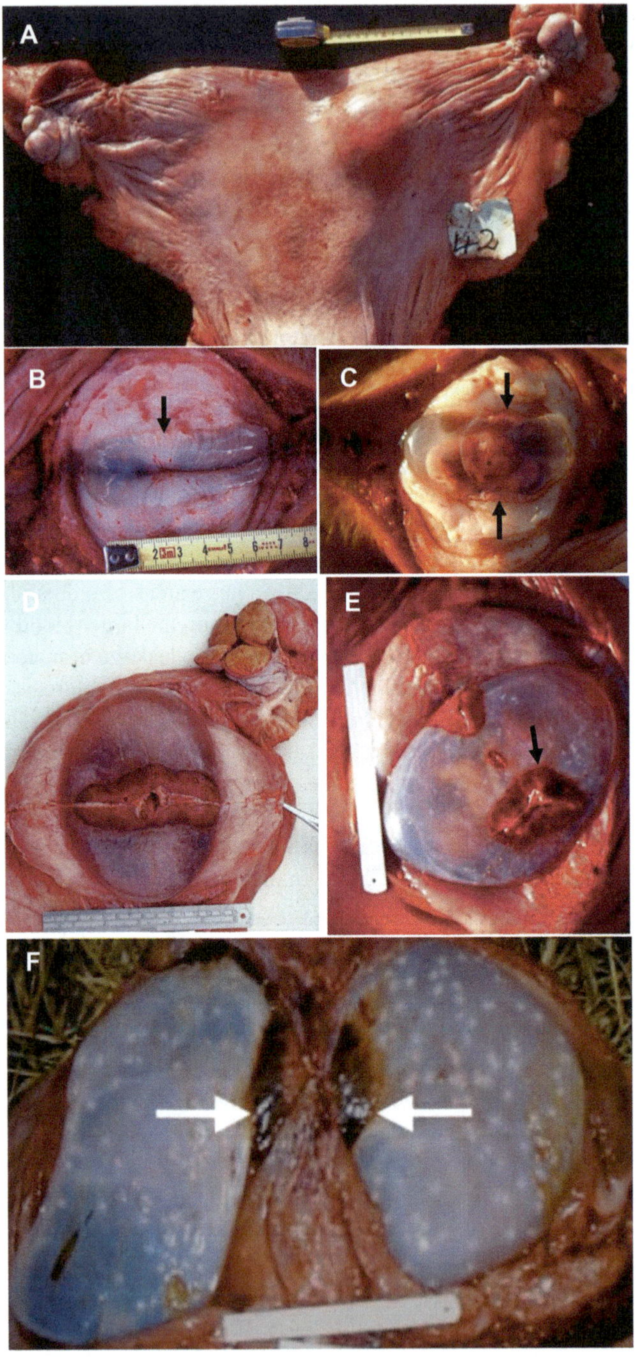

Fig. 6 (**a**) Elephant uterus at approximately 4.5 months of pregnancy showing the conceptus bulge at the base of one uterine horn. (**b**) Uterus at approximately 4 months of gestation opened to indicate the intense myometrial tonicity surrounding the conceptus. The edge of the thin, pale placental band (arrowed) can just be discerned in the conceptus membranes. (**c**) A 4.5-month conceptus opened to

with the calf at term so it usually remains in utero with the placenta unless, as occurs occasionally, the placenta is delivered with the calf.

5 Histological Development of the Placenta

At an estimated 30–45 days of gestation, the trophoblast at the equatorial region of the conceptus is closely apposed to the endometrium and is extruding finger-like outgrowths that work their way beneath the lumenal epithelium of the endometrium and lift it off the basement membrane (Fig. 8a). In some areas a considerable proportion of the lumenal epithelium has been replaced by trophoblast which then passes a short distance down the lumenae of the endometrial glands, forcing the epithelium off its basement membrane as it does so (Fig. 8b). And in other places, the trophoblast invested endometrial stroma is starting to produce blunt, capillary-filled upgrowths from the surface (Figs. 8c and 9a) as the progenitors of the very long fingers of stroma that will eventually comprise the maternal component of the placental band. Thus, a transition from attachment to, erosion of, and, finally, complete replacement of the endometrial epithelium by trophoblast can be distinguished, and these changes then lead to early growth of the trophoblast-invested upgrowths of endometrial stroma and associated vasculature (Fig. 8d). Interestingly, there is little sign of any accumulation of maternal lymphocytes or other immune cells in response to this erosion and attachment process by the trophoblast.

By 4 months of gestation, the placental band is now clearly visible as a 3 cm wide circumferential thickening. Histologically, a line of simple protrusions of endometrial stroma, each invested by a single layer of trophoblast cells, rise from the endometrial surface (Figs. 8e and 9b). They are beginning to become branched at their tips and, as such, show a likeness to the branching microcotyledons on the surface of the early 50–60-day equine placenta (Fig. 9c). An appreciable thickness of endometrial stroma still persists between the trophoblast cells and endothelium of the maternal blood vessels within the core of each lamellum (Figs. 8f and 9c). Their

Fig. 6 (continued) reveal the fetus extruded by the persisting considerable tonicity of the myometrium. The cut edges of the placental band are now seen more clearly above and below the head of the fetus (arrowed). (**d**) A 6.6-month pregnant uterus opened to expose the conceptus in situ. The placental band with its darker lateral hemophagous edges is seen passing around the circumference of the middle of the rugby ball-shaped conceptus with the thin, pale endometrial attachment band running along its central region. (**e**) Elephant placenta carrying a 240 g fetus (estimated gestation age 6.7 months) and showing distinct gaps in the placental band with a skewed longitudinal, rather than latitudinal, arrangement of one piece of the placental band (arrowed) in the equatorial region of the ovate conceptus. (**f**) Pregnant elephant uterus carrying a 1.9 kg fetus (8.9 months). Note the dark brown coloration of the hemophagous zone (arrowed) at the lateral edges of the placental band and the numerous pustules scattered over the surface of the allantoic sacs beneath the outermost trophoblast. (**a**, **b**, **d**, **e** and **f**; Reprinted with permission from Allen et al. 2003)

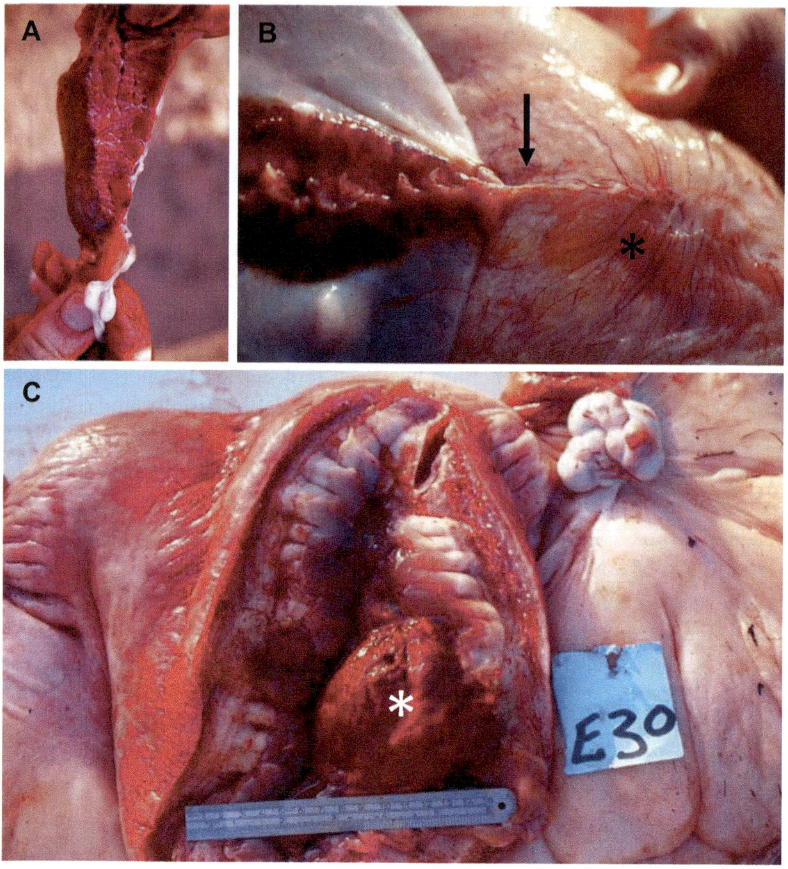

Fig. 7 (**a**) Cross section of one side of the placental band of a 98 kg fetus (estimated gestational age = 20.6 months). Note its similar appearance to liver tissue and the brown coloration on the ventral surface due to leakage of maternal blood in this hemophagous zone. (**b**) Close-up view of the maternal placental hilus (arrow) being severed from the endometrium in the uterus carrying a conceptus with the 240 g fetus (6.7 months). Note its relative narrowness and the concentration of endometrial blood vessels feeding into it (asterisk). (**c**) A previously gravid uterine horn opened approximately 6 months post calving to show a large resorbing blood clot (asterisk) attached to the endometrial surface and occluding the uterine lumen. (**a** and **b**; Reprinted with permission from Allen et al. 2003)

stromal lamellae become increasingly elongated and more branched with advancing gestation thereby allowing closer apposition of fetal trophoblast to the maternal endothelium of the capillaries within the lamellum (Figs. 8g, h and 9d). And, at each outermost edge of the placental band, the lamellae bend laterally towards the endometrium, to form a blind cleft between the endometrial and placental surfaces (Fig. 9d). The lumenal epithelium within these lateral clefts becomes increasingly convoluted and the cells become tall and columnar, with blunt pseudopodia protruding from their apical surfaces. In the placental band apposed to the uterine

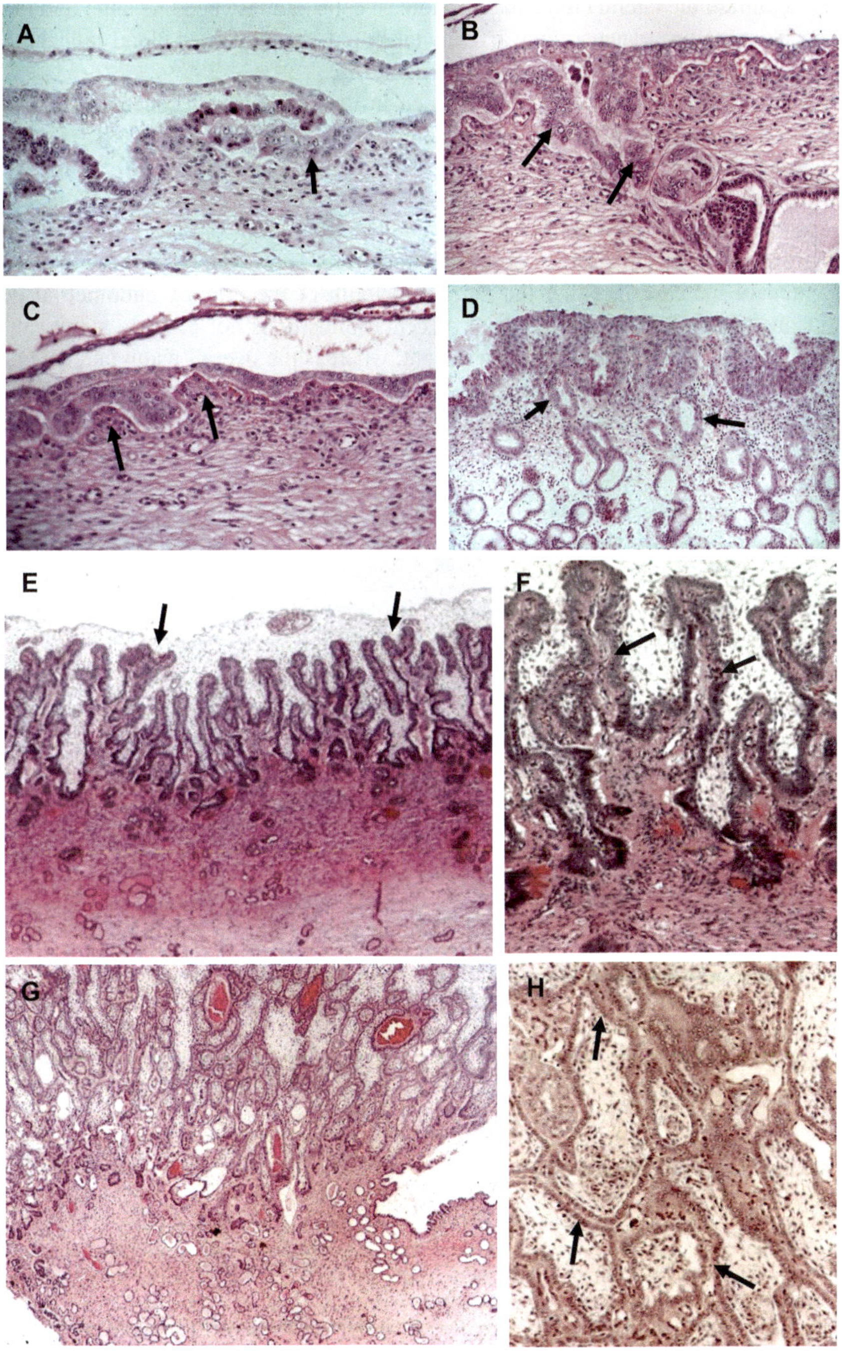

Fig. 8 (**a**) Histological section of the endometrium at the conceptus bulge in an elephant estimated to be at approximately 40 days of gestation showing a "finger" of trophoblast cells passing beneath the lumenal epithelium and lifting it off the surface (arrow). (**b**) Adjacent "finger" of trophoblast is seen heading down the mouth of an endometrial gland (arrows). (**c**) At the same site in the 40-day

epithelium in the lateral clefts, the trophoblast cells likewise become taller and more flocculent in appearance as they actively take up the maternal red blood cells that fill the core of the lamellae in this region (Fig. 9d).

With advancing gestation, and as seen in the placentae of conceptuses of 5--14 months estimated gestational ages, the trophoblast-covered stromal lamellae increase steadily in length, width, and branching complexity (Fig. 9d). In this way, the overall size and complexity of the interface stems solely from elongation, secondary and tertiary branching, and then tight folding of the original stromal upgrowths within their envelope of allantochorion (Fig. 9d), rather than from any significant increase in either the number or bulk of the primary endometrial protrusions. The width of the maternal placental hilus does not increase significantly after the fetus has reached 10 kg in weight, whereas the overall width and development of the whole placenta increases continuously to term (Fig. 9b–d). There is no sign of any syncytial formation in the trophoblast layer, and the reduction of stroma in each upgrowth continues to bring the trophoblast cells into increasingly intimate contact with the endothelium of the maternal blood vessels to increase the efficiency of placental exchange. And in the lateral hematogenous clefts, the leakage and take-up of maternal red blood cells also continues to increase steadily with advancing gestation (Fig. 9d).

In summarizing essential steps during placentation in the pregnant elephant, lifting and displacement of the lumenal epithelium of the endometrium by protrusions of trophoblast cells, the complete replacement of large areas of lumenal epithelium by trophoblast in the progenitor region of the placental band, and the beginning of trophoblast-covered stromal upgrowths are all important features at the very beginning (Allen 2006). They show marked similarity to the removal of the lumenal and glandular epithelia of the mare's endometrium by the specialized invasive trophoblast cells of the chorionic girdle region of the equine fetal membranes to form the chorionic gonadotrophin (eCG)-secreting endometrial cups (Allen et al. 1973; Hamilton et al. 1973). There are, however, marked differences between the horse and elephant as to what happens when the trophoblast has lifted

---

Fig. 8 (continued) pregnant elephant, the trophoblast layer is seen lying atop of the epithelium-denuded endometrial stroma with the very first signs of upward growth of short, blunt villi of stroma (arrows). (**d**) Section of the placental region in another elephant at an estimated 60 days of gestation showing blunt villi of trophoblast-invested stroma growing above the surface of the endometrium to commence the development of the zonary placenta (asterisk). Note the larger trophoblast cells in the epithelial lining of the apical portions of the endometrial glands (arrows). (**e**) Low power and, (**f**) higher power, sections of the developing placental band in a pregnant elephant carrying a 1.6 g fetus and therefore estimated to be at 4 months of gestation. The stroma-covered trophoblast stromal villi are growing straight up from the surface of the endometrium and are now just beginning to branch. Each villus still has a wide core of endometrial stroma containing the maternal blood capillaries (arrows). (**g**) Low and (**h**) higher power sections at the base of the placental hilus in an elephant at an estimated 11 months of gestation. The stromal-covered villi are now much taller and branched and the proportion of endometrial stroma in each villus has lessened so as to bring the trophoblast into closer contact with the maternal blood capillaries (arrows)

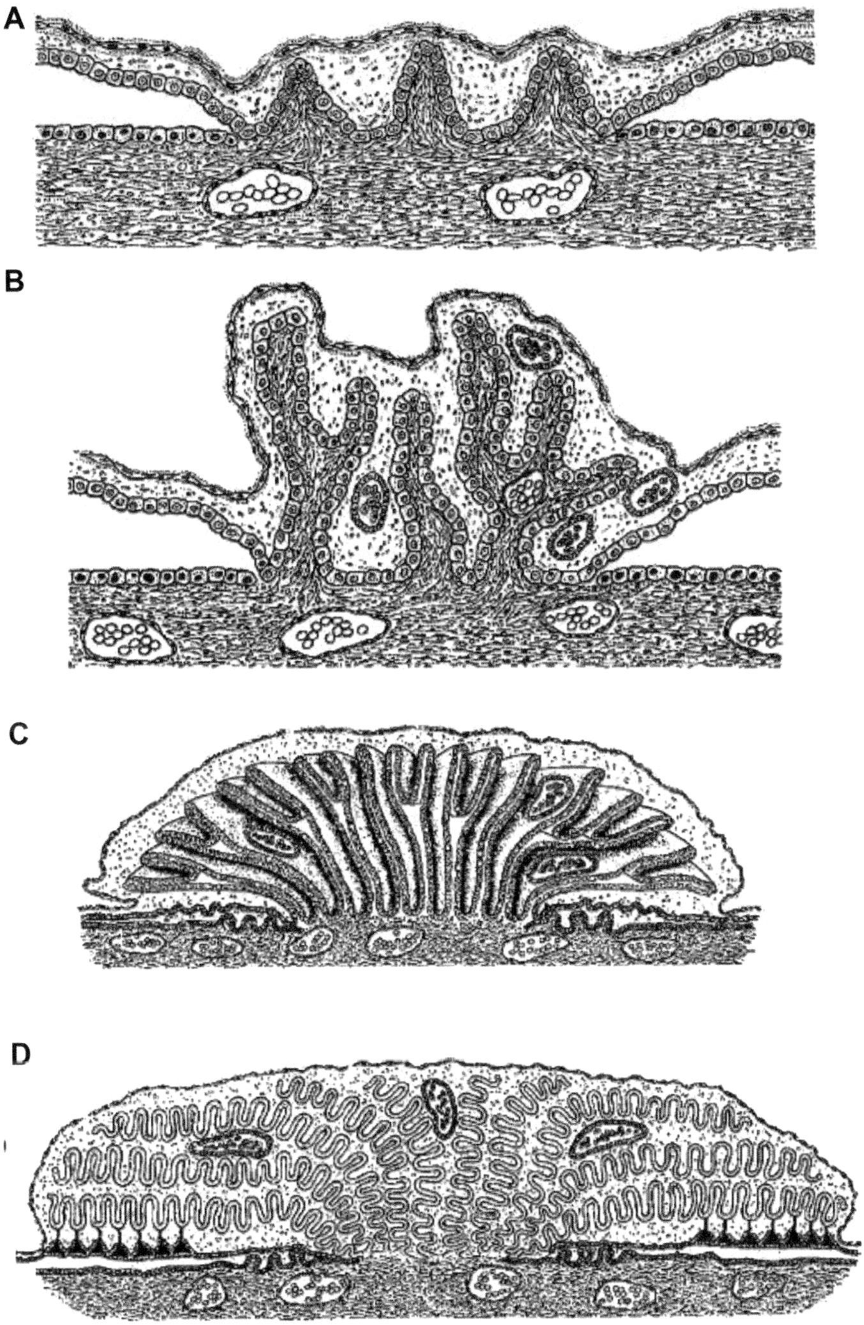

Fig. 9 Diagrammatic representation of significant morphological changes that occur during development of the zonary placental band on the elephant conceptus. (**a**) Commencing upgrowth of the trophoblast covered, capillary-filled stromal villi above the surface of the endometrium; (**b**) Elongation and branching of the trophoblast-covered lamellae, each containing maternal blood vessels that are still surrounded by adequate amounts of stromal tissue. Fetal blood cells containing

the epithelial cells off the basement membrane and bundled them out of the way. In the horse, the binucleate girdle cells eventually pass through the lumenal and glandular basement membranes and stream out into the endometrial stroma where they enlarge and become tightly packed together to form the endometrial cups (Allen et al. 1973). In the elephant, on the other hand, the trophoblast cells do not invade the endometrium so deeply and they only penetrate into, and remove the epithelial cells from the mouths and apical necks of, the endometrial glands (Fig. 8b, d, e). This lack of invasiveness shown by the elephant trophoblast may be due to a greater cohesiveness between cells, or a lack of enzymes necessary to promote deep penetration of the endometrium. The initially subepithelial capillaries and their associated stromal cells are clearly hyperstimulated by the presence of the trophoblast to begin the process of rapid upward growth. Thus, instead of a conventional invasion of fetal tissue into the maternal layer, protrusions of the latter, led by the proliferation of the blood vessels, grow up dramatically to supply the demands of an expanding trophoblast and its steadily growing dependent fetus (Figs. 10 and 11). It seems likely that the trophoblast cells would secrete sufficient quantities of the vascular endothelial growth factors (VEGF) family (Gnosh et al. 2000; Charnock-Jones et al. 2001), together with mitogens such as insulin-like growth factor 2 (IGF-2), hepatocyte growth factor-scatter factor (HGF-SF), and epidermal growth factor (EGF), to achieve this rapid growth and accompanying vascularization (Stewart et al. 1995; Lennard et al. 1995, 1998).

6 Involvement of the Fetal Gonads

Beyond mid-gestation, and in parallel to the similar situation in the pregnant mare, the gonads of both male and female fetuses begin to enlarge greatly so that they come to occupy a sizeable proportion of the fetal abdomen by 16–18 months of gestation (Fig. 12a; Allen et al. 2005). And as in the mare, much of this growth of the elephant fetal gonads comes from multiplication of the interstitial cells which stain strongly for the 3β-hydroxysteroid dehydrogenase (3β-HSD) enzyme (Fig. 12b), thereby indicating their secretion of progestagens. And again, like the pregnant mare, these progestagens consist of 5α-DHP and other 5α-progestagens rather than native progesterone. But in sharp contrast to the situation in equine pregnancy, but in

Fig. 9 (continued) nuclei are scattered throughout the mesoderm separating adjacent lamellae; (**c**) As the branched stromal lamellae continue to elongate, those at the lateral edge of the band lay over towards the maternal endometrium to create the blind ending cleft in which the hematogenous zone will form; (**d**) In the maternal placental band the now very elongated trophoblast stromal lamellae become increasingly folded or pleated so as to maximize the available surface area of contact for placental exchange. Leakage of maternal blood into the lateral clefts of the placental band creates the hematogenous zones in which the morphologically differentiated trophoblast cells take up blood components. (Reprinted with permission from Allen et al. 2003)

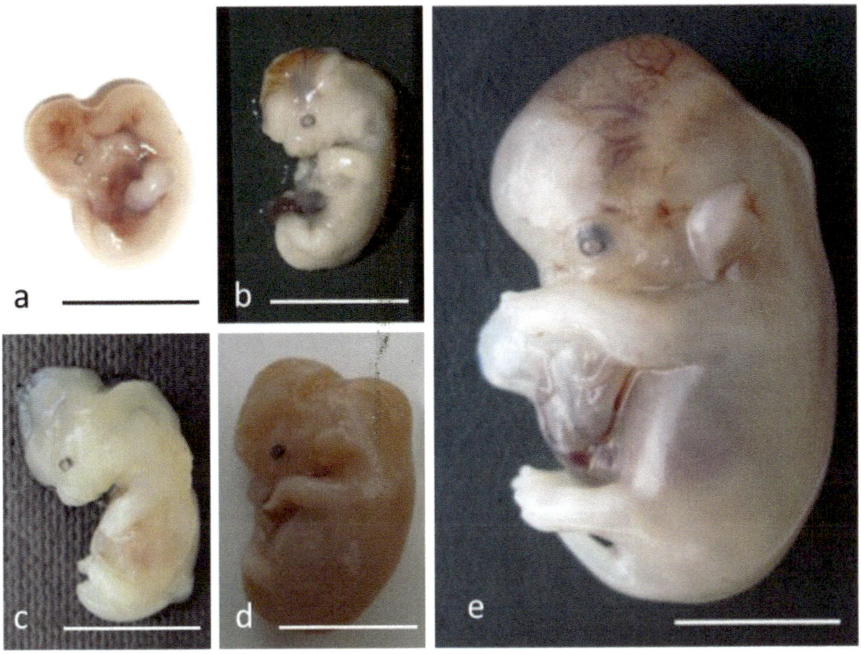

Fig. 10 Five elephant embryos aged on the basis of their weight using the formula developed by Craig (1984). (**a**) 0.47 g = 76 days; (**b**) 0.50 g = 78 days; (**c**) 0.55 g = 87 days; (**d**) 0.75 g = 89 days; (**e**) 3.0 g = 96 days (Scale bar = 10 mm). (Reprinted with permission from Stansfield et al. 2012a)

parallel to that in the dog placenta (Hoffmann et al. 1994; Nishiyama et al. 1999), the elephant placenta secretes neither progestagens nor estrogens (Allen et al. 2002). A major difference between the enlarged fetal ovaries in the elephant versus the horse is growth and subsequent regression of multiple follicles (Fig. 12c) throughout the second half of gestation in the former. This leads to considerable wastage of oocytes such that, at birth, a female newborn elephant calf retains approximately 561,000 small follicles compared to the much higher figure of 4,500,000 small follicles found at 18 months of gestation (Fig. 12d; Stansfield et al. 2012b). Very surprisingly, however, during the first few years of postnatal life, the prepubertal ovaries of the female elephant calf show small clusters of cells in their ovarian stroma that stain positively for a number of stem cell markers, including Stella, Oct-4, Nanog, and Lin 28 (Fig. 13; WR Allen and F Stansfield unpublished finding). Furthermore, counting small follicle numbers in the ovaries of young prepubertal calves shows a remarkable recovery in follicle numbers, back to approximately 1,262,000 at 6–8 years of age (Fig. 14; Stansfield et al. 2012b). Thus, it seems the female elephant calf shows good evidence of postnatal oogenesis!

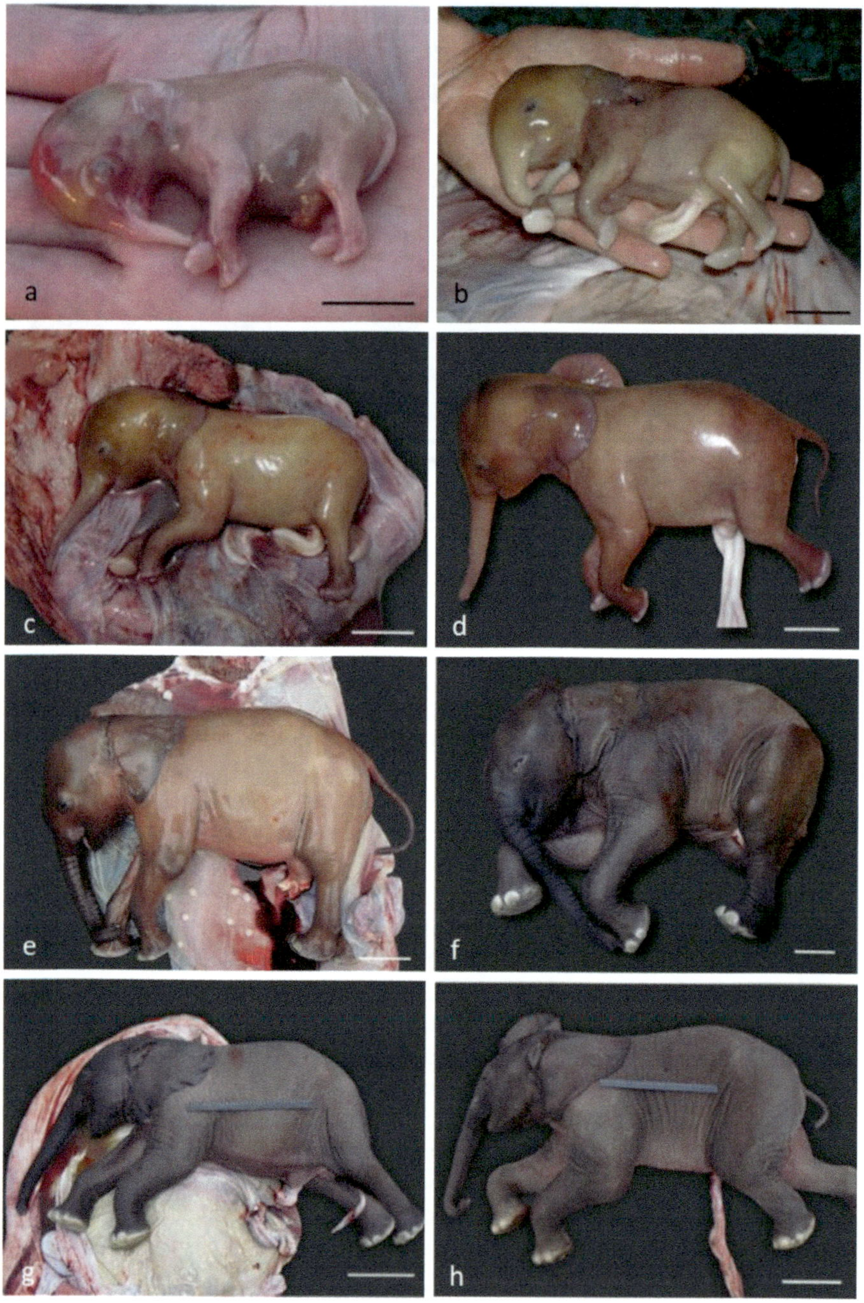

Fig. 11 Eight elephant fetuses aged on the basis of their weight using the formula developed by Craig (1984). (**a**) 36 g = 4.1 months; (**b**) 215 g = 5.3 months; (**c**) 463 g = 5.9 months; (**d**) 2.0 kg = 8 months; (**e**) 7 kg = 11 months; (**f**) 47 kg = 17 months; (**g**) 85 kg = 20 months; (**h**) 139 kg = 22 months (Scale bars: **a**–**c** = 2 cm; **d** = 3 cm; **e** = 5 cm; **f** = 10 cm; **g** and **h** = 20 cm). (Reprinted with permission from Stansfield 2012)

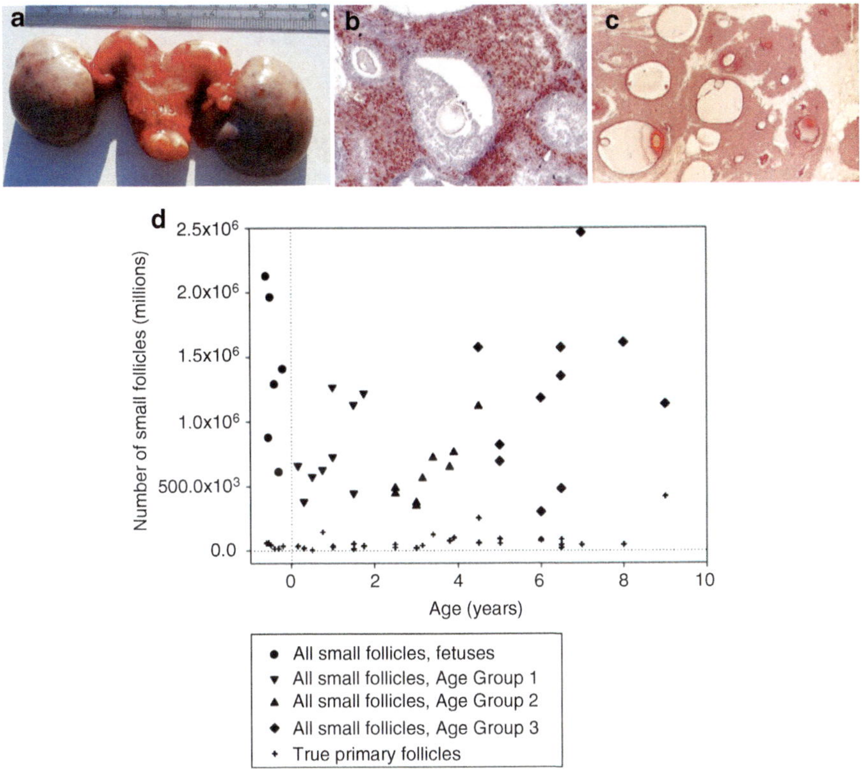

Fig. 12 (**a**) The reproductive tract of a female elephant fetus at 18 months of gestation showing the small, underdeveloped uterus and the very enlarged fetal ovaries. (**b**) Higher power section of the stroma of the fetal ovary stained with a 3β-hydroxysteroid dehydrogenase antiserum. The interstitial cells stain strongly to show their secretion of progestagens to support the long-lived corpora lutea in the maternal ovaries to maintain the pregnancy state. (**c**) A histological section of one ovary showing enhanced primary and secondary follicles throughout the ovarian stroma (×100). (**d**) Dot plot showing the numbers of small follicles counted in the ovaries of late-stage female fetuses and prepubertal female calves. Note the dramatic fall in numbers during the final months of pregnancy and the significantly large mean numbers again during the first 8 years of postnatal life

7 Postpartum Uterine Involution

Although the band of allantochorion and apposing uterine endometrium involved in the initial formation of the elephant zonary endotheliochorial placenta in early gestation is a significant proportion (a quarter to a third) of the whole conceptus at that time, as gestation advances and the placenta grows and increases exponentially in size, the original attachment band through which all the maternal blood vessels pass to vascularize the placenta remains unchanged and is now remarkably small and narrow (Fig. 7b) compared to the relatively enormous overlying placenta it nourishes (Fig. 7a). Parturition at term occurs with the mother standing, which means the

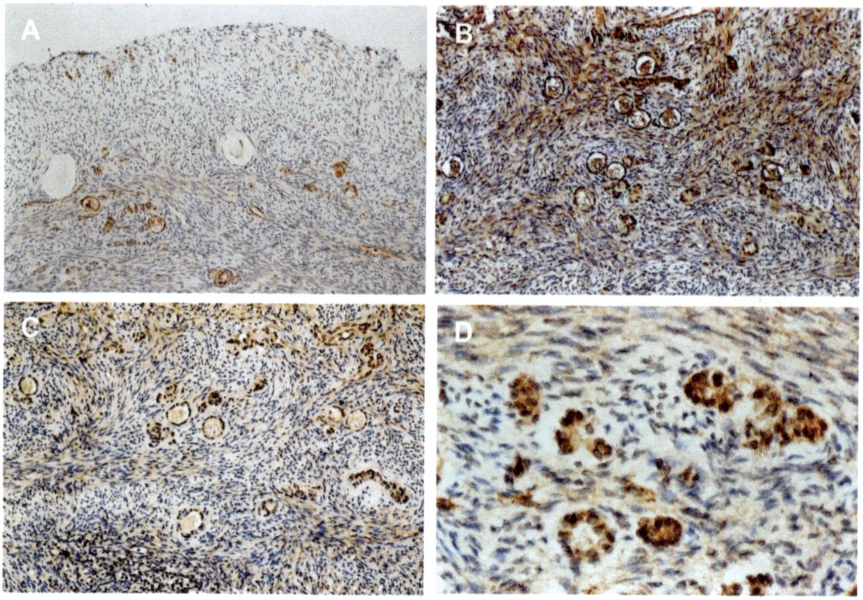

Fig. 13 Sections of the ovaries and testes of young elephant calves between 6 months and 4.5 years of age showing positive staining of individual or groups of cells with the stem cell markers (**a**) Stella, (**b**) Oct 4, (**c**) Nanog, and (**d**) Lin28

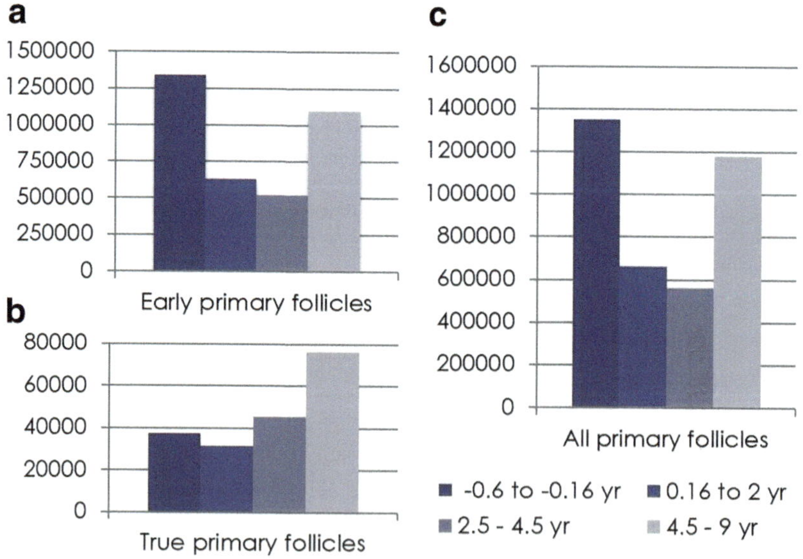

Fig. 14 Mean numbers of (**a**) early primary, (**b**) true primary, and (**c**) all primary follicles counted in the ovaries of fetuses between 10 months of gestation and term and in female calves from birth to 9 years of age

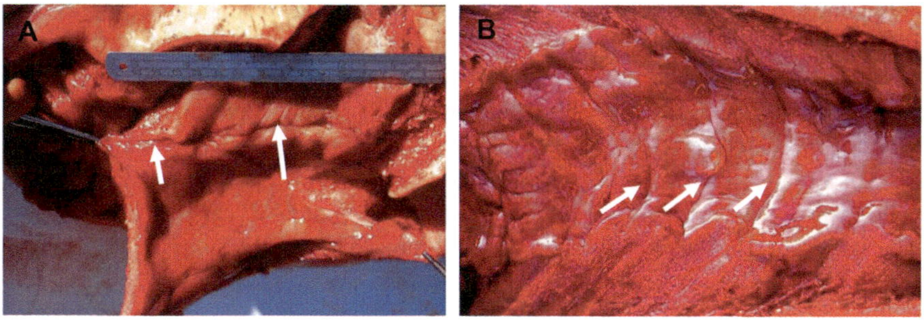

Fig. 15 (**a**) Opened uterus of a female elephant 10 months after calving showing the pronounced endometrial scar persisting from the previous placental attachment (arrows). (**b**) Opened uterus of 60-year-old female elephant showing multiple circumferential scars in the endometrium persisting from several previous pregnancies (arrows). (Reprinted with permission from Allen et al. 2003)

newborn calf rather "plops" out onto the ground dragging the allantochorion and zonary placenta with it. As a result, the narrow attachment band is literally torn off the endometrial surface which, in turn, causes considerable hemorrhage into the uterine lumen (Fig. 7c). This then takes some months to be fully resorbed, thereby likely acting as a temporary "contraceptive" to prevent the possibility of a further pregnancy before complete uterine involution has occurred. Thereafter, a narrow, but deep and pronounced, endometrial scar passing round the inner circumference of the uterine horn persists to mark the presence of the previous pregnancy (Fig. 15a) and by counting these evenly spaced scars in the uteri of adult aging elephants (Fig. 15b) it is possible to determine accurately the number of previous gestations each has had as has been demonstrated previously in the bitch by Orfanou et al. (2009).

8 Conclusions

Three features of pregnancy in the elephant show an interesting similarity to equine pregnancy. First, initial attachment of the conceptus at the base of one or other of the uterine horns. Second, development of secondary or accessory CLs in the maternal ovaries at around 40 days of gestation in response to the commencing secretion of a luteotropic hormone by the placenta at that time, which is eCG in the mare versus ePL in the elephant. And third, tremendous growth and enlargement of the fetal gonads during the second half of gestation and their secretion of steroid hormones to help support the pregnancy state, which is estrogen precursors in the mare and progestagens in the elephant. But apart from these few shared aspects, elephant pregnancy includes some fascinating and apparently unique features. For example, a prolonged follicular phase before mating and conception, which includes not one, but two, well-spaced waves of pituitary LH release and resulting follicular growth. A group of luteinized follicles develops from the first of these LH pulses which, after "hanging around" in the ovaries for some weeks, suddenly enlarge greatly and become converted into highly productive and long-lived accessory CLs by the action of ePL secreted by the attaching trophoblast. These then function to maintain the pregnancy state in the face of the steroidogenically inactive placenta.

Then, perhaps the most fascinating aspect of elephant pregnancy is the dramatic loss of oocytes from the enlarged gonads in the female fetus in late gestation followed by their apparent replacement during the first 6–8 years of postnatal life. The elephant is indeed a wondrous and unusual animal, both physically and physiologically!

Acknowledgements The authors express their sincere appreciation to Dr. Susanna Stout and the skilled members of the elephant culling teams in Kruger National Park, South Africa, and The Savé Valley Conservancy in Zimbabwe for their kind assistance in obtaining the elephant postmortem organs and tissues described in this chapter. Dr. Sandra Wilsher gave invaluable help with the preparation of the manuscript.

References

Allen WR (2006) Ovulation, pregnancy, placentation and husbandry in the African elephant (*Loxodonta africana*). Philos Trans R Soc B 361:821–834

Allen WR, Hamilton DW, Moor RM (1973) The origin of the equine endometrial cups. II. Invasion of the endometrium by trophoblast. Anat Rec 177:485–502

Allen WR, Mathias SS, Wooding FB, Skidmore JA, van Aarde RJ (2002) Placentation in the African elephant (*Loxodonta africana*). I. Endocrinological aspects. Reprod Suppl 60:105–116

Allen WR, Mathias S, Wooding FBP, van Aarde RJ (2003) Placentation in the African elephant (*Loxodonta africana*). II. Morphological changes in the uterus and placenta throughout gestation. Placenta 24:596–617

Allen WR, Stout S, Frost M (2005) Placentation in the African elephant, *Loxodonta africana*. IV. Growth and organisation of the fetal gonads. Reproduction 130:713–720

Amoroso EC, Perry JS (1964) The foetal membranes and placenta of the African elephant (*Loxodonta africana*). Proc R Soc B 248:1–34

Buss IO, Johnson OW (1967) Relationship of Leydig cell characteristics and intratesticular testosterone levels to sexual activity in the African elephant. Anat Rec 157:191–196

Charnock-Jones DS, Clarke DE, Licence D, Day K, Wooding FB, Smith SK (2001) Distribution of vascular endothelial growth factor (VEGF) and its binding sites at the maternal-fetal interface during gestation in pigs. Reproduction 122:753–760

Craig JC (1984) Fetal mass and date of conception in African elephants: a revised formula. Afr J Sci 80:512–516

Gnosh HD, Starky AN, Charnock-Jones DS, Dhawan L, Dhara S, Smith SK, Sengupta J (2000) Expression of vascular endothelial growth factor (VEGF) and placental growth factor (PlGF) in conceptuses and endometrium during implantation in the rhesus monkey. Mol Hum Reprod 6:935–941

Hamilton DW, Allen WR, Moor RM (1973) The origin of the equine endometrial cups III. Light and electron microscopic study of fully developed equine endometrial cups. Anat Rec 177:485–502

Hanks J, Short RV (1972) The formation and function of the corpus luteum in the African elephant, *Loxodonta Africana*. J Reprod Fertil 29:79–89

Hodges JK, Heisterman M, Beard A, van Aarde RJ (1997) Concentrations of the 5-alpha-reduced progestins, five-alpha-pregnane-3, 20 dione and 5 alpha pregnane-3 alpha ol 20-one in luteal tissue and circulating blood and their relationship to luteal function in the African elephant, *Loxodonta africana*. Biol Reprod 56:640–646

Hoffmann B, Höveler R, Nohr B, Hasan SH (1994) Investigations on hormonal changes around parturition in the dog and the occurrence of pregnancy-specific non-conjugated oestrogens. Exp Clin Endocrinol 102:185–189

Kasputin N, Critzer J, Olso D, Malvern PV (1996) Non-luteal oestrous cycles of 3-week duration are initiated by anovualtory luteinising hormone peaks in African elephants. Biol Reprod 55:1147–1154

Laws RM (1969) Aspects of reproduction in the African elephant, *Loxodonta Africana*. J Reprod Fertil Suppl 6:193–217

Lennard SN, Stewart F, Allen WR (1995) Transforming growth factor beta1 (TGFbeta1) expression in the endometrium of the mare during placentation. Mol Reprod Dev 42:131–140

Lennard SN, Gerstenberg C, Allen WR, Stewart F (1998) Expression of epidermal growth factor and its receptor in equine placental tissues. J Reprod Fertil 112:49–57

Leyendecker G, Kung G, Herbertz M, Mall G (1998) Endometriosis; a dysfunction and disease of the archimetra. Hum Reprod Update 4:752–762

Lueders I, Niemeuller C, Gray C, Rich P, Hildebrandt TB (2010) Luteogenesis during the oestrous cycle in Asian elephants (*Elephas maximus*). Reproduction 140:77–786

Lueders I, Niemuller C, Rich P, Gray C, Hermes R, Goeritz F, Hildebrandt TB (2012) Gestating for 22 months: luteal development and pregnancy maintenance in elephants. Proc R Soc B Biol Sci 279:3687–3696

McNeilly AJ, Martin RD, Hodges JK, Smuts GL (1983) Blood concentrations of gonadotrophins, prolactin and gonadal steroids in males and in non-pregnant and pregnant female African elephants (*Loxodonta africana*). J Reprod Fertil 67:113–120

Meyer HH, Jewgenower K, Hodges JK (1997) Binding activity of 5-alpha gestagens to the progestin receptor from African elephants (*Loxodonta africana*). Gen Comp Endocrinol 105:164–167

Nishiyama T, Tsumagari S, Ito M, Kimura J, Watanabe G, Taya K, Takeisehi M (1999) Immuno-histochemical study of steroidogenic enzymes in the ovary and placenta during pregnancy in the dog. Anat Histol Embryol 28:125–129

Orfanou DC, Ververidis HN, Pourlis A, Fragkou IA, Kokoli AN, Boscos CM, Taitzoglou IA, Tzora A, Nerou CM, Athanasiou L, Fthenakis GC (2009) Post-partum involution of the canine uterus—gross anatomical and histological features. Reprod Domest Anim 2009(Suppl 2):152–155

Perry JS (1953) The reproduction of the African elephant (*Loxodonta africana*). Philos Trans R Soc B 237:93–149

Perry JS (1974) Implantation, foetal development and early placentation of the African elephant (*Loxodonta africana*). Proc R Soc B 269:109–135

Sakamoto S, Kaewmanee S, Nagaoka K, Watanabe G, Taya K (2012) The secretory pattern and source of immunoreactive prolactin in the pregnant African (*Loxodonta africana*) and Asian (*Elephas maximus*) elephants. J Reprod Dev 58:105–111

Short RV (1966) Oestrous behaviour, ovulation and the formation of the corpus luteum in the African elephant, *Loxodonta africana*. East Afr Wildl J 4:56–68

Smith SK, Buss IO (1975) Formation, function and persistence of the corpora lutea of the African elephant, *Loxodonta africana*. J Mammology 56:30–43

Stansfield FJ (2012) The ontogeny of the ovarian follicular reserve in the African elephant (*Loxodonta Africana*). PhD thesis. Department of Production Animal Studies, University of Pretoria

Stansfield FJ, Allen WR (2012) Luteal maintenance of pregnancy in the African elephant (*Loxodonta africana*). Reproduction 143:845–854

Stansfield FJ, Nothling JO, Allen WR (2012a) Development of the germinal ridge and ovary in the African elephant (*Loxodonta Africana*). Reproduction 144:1–12

Stansfield FJ, Nothling JO, Allen WR (2012b) Growth and development of the ovary and small follicle pool from mid-foetal life to pre-puberty in the African elephant (*Loxodonta africana*). BMC Vet Res 23:119–127

Stewart F, Lennard SN, Allen WR (1995) Mechanisms controlling formation of the equine chorionic girdle. Biol Reprod Monogr Ser 1:151–159

Wooding FBP, Stewart F, Mathias S, Allen WR (2005) Placentation in the African elephant. III. Ultrastructural and functional features of the placenta. Placenta 26:449–470

Yamamoto T, Kewmanee S, Niemuller C, Gray C, Streitch WJ, Hildebrandt TB (2011) Role of the double luteinising hormone peak, luteinising follicles and the secretion of inhibin for dominant follicle selection in Asian elephants (*Elephas maximus*). Biol Reprod 85:714–720

Yamamoto Y, Yamamoto T, Taya K, Watanabe G, Stansfield FJ, Allen WR (2012) Placentation in the African elephant (*Loxodonta africana*). V. The trophoblast secretes placental lactogen. Placenta 32:506–512

Development of the Mouse Placenta

Sourav Panja and Bibhash C. Paria

Contents

Abstract Placenta forms as a momentary organ inside the uterus with a slew of activities only when the woman is pregnant. It is a discoid-shaped hybrid structure consisting of maternal and embryonic components. It develops in the mesometrial side of the uterus following blastocyst implantation to keep the two genetically different entities, the mother and embryo, separated but connected. The beginning and progression of placental formation and development following blastocyst implantation coincides with the chronological developmental stages of the embryo. It gradually acquires the ability to perform the vascular, respiratory, hepatic, renal, endocrine, gastrointestinal, immune, and physical barrier functions synchronously that are vital for fetal development, growth, and safety inside the maternal environment. The uterus ejects the placenta when its embryonic growth and survival supportive roles are finished; that is usually the birth of the baby. Despite its irreplaceable role in fetal development and survival over the post-implantation progression of pregnancy, it still remains unclear how it forms, matures, performs all of its activities, and starts to fail functioning. Thus, a detailed understanding about

S. Panja · B. C. Paria (✉)
Division of Neonatology, Department of Pediatrics, Vanderbilt University Medical Center, Nashville, TN, USA
e-mail: sourav.panja@vumc.org; bc.paria@vumc.org

R. D. Geisert, T. E. Spencer (eds.), *Placentation in Mammals*, Advances in Anatomy, Embryology and Cell Biology 234, https://doi.org/10.1007/978-3-030-77360-1_10

normal developmental, structural, and functional aspects of the placenta may lead to avoid pregnancy problems that arise with the placenta.

Keywords Blastocyst · Decidua · Implantation · Labyrinth · Placenta

Abbreviations

Al	Allantois
Am	Amnion
AM	Antimesometrial side
Amc	Amniotic cavity
AVE	Anterior visceral endoderm
BC	Blastocoel
Bl	Blastocyst
Ch	Chorion
ChC	Chorionic cavity
CM	Circular muscle
cTGCs	Canal-associated trophoblast giant cells
DC	Decidua
DVE	Distal visceral endoderm
EC	Endothelial cell
ECC	Exocoelomic cavity
EEE	Extra-embryonic ectoderm
EEEn	Extra-embryonic endoderm
Em	Embryo
EP	Epiblast
EPC	Ectoplacental cone
EpcC	Ectoplacental cavity
ExEM	Extra-embryonic mesoderm
GlyT	Glycogen trophoblast
ICM	Inner cell mass
IGF-II	Insulin-like growth factor II
JZ	Junctional zone
LE	Luminal epithelium
LM	Longitudinal muscle
M	Mesometrial side
MBS	Maternal blood sinusoid
PaE	Parietal endoderm
PDZ	Primary decidual zone
PE	Primitive endoderm
Pro-AmC	Pro-amniotic cavity
PS	Primitive streak
p-TGCs	Parietal giant cells

SDZ	Secondary decidual zone
SpA-TGC	Spiral artery-associated trophoblast giant cells
SpT	Spongiotrophoblast
S-TGCs	Sinusoidal trophoblast giant cells
SynT-I	Syncytotrophoblast I
SynT-II	Syncytotrophoblast II
TGCs	Trophoblast giant cells
Tr	Trophectoderm
ViE	Visceral endoderm
ZP	Zona pellucida

1 Introduction

Placenta is a vital organ for the development and growth of the fetus from the time of implantation to birth. It is responsible for fetal respiratory, nutritional, excretory, endocrine, and immunological functions. It also has an extraordinary ability to adapt to adverse maternal and environmental cues and diminishes their impact on the fetus. It starts to form and grow at the interface of the uterine endometrium and blastocyst following a successful institution of the blastocyst–uterine contact, known as the implantation process. Placenta development, growth and maturation are directedly linked to the blastocyst's morphogenetic developmental advances to the stages of first the egg cylinder, then the gastrula, and lastly the fetus (Rossant and Cross 2001).

Placental developmental, positional, and functional defects are threats to pregnancy complications such as early pregnancy failure, stillbirth, fetal growth restriction, preeclampsia, and preterm birth. Some pregnancy complications that arise late in pregnancy such as preeclampsia and preterm birth may reflect errors that occur during early in placental development (Burton et al. 2016; Hemberger et al. 2020). There are several major placental abnormalities based on the site of implantation, degree of trophoblast invasion, and functions of the placenta. The placenta normally forms onto the upper part of the uterus in humans. However, when it attaches too deeply it causes problems such as the placenta accreta (firm attachment with the myometrium), increta (penetration inside the myometrium), and percreta (breach through the myometrium to nearby organs) (Bauer and Bonanno 2009; Oyelese and Smulian 2006). On the contrary, inadequate trophoblast invasion has been implicated in the pathophysiology of preeclampsia (Phipps et al. 2019). Placental previa is a condition when the placenta develops in the lower part of the uterus and partially or completely covers the cervix (Kollmann et al. 2016). Placental insufficiency is a disorder marked by insufficient blood flow to the placenta (Brown and Hay 2016; Browne 1963). Placental abruption occurs when the placenta separates from the uterus prior to the birth of the baby (Oyelese and Ananth 2006). Placental insufficiency, placental abruption, and preeclampsia may share a common pathologic mechanism involving poor placentation and placental ischemia.

Pregnancy problems due to placental defects are difficult to fix, but the fetal growth and birth complications can be managed if it is early diagnosed. By far, the placenta is an understudied organ and, hence, a least understood structure. A better understanding of the molecular mechanisms responsible for placenta formation, maturation, and functions will have a great impact on the development of treatment tools to mitigate pregnancy complications due to placental flaws. Researchers are relentlessly seeking to reveal what a normal placenta is and how it forms and performs its various roles. A brief review of previous as well as more recent works in revealing the origins of the placental cell lineages and the chronology of placental developmental events during the post-implantation period in mice is described in this article.

2 The Gist of Preimplantation Embryonic Developmental Stages

The emergence of placental cell types initiates at the preimplantation embryonic stage when the totipotent blastomeres start segregating to form trophectoderm (Tr) and inner cell mass (ICM) lineages (Guo et al. 2020). A haploid female gamete or ovum when becomes diploid (zygote) as a result of a union with a haploid male gamete (sperm) through a process called fertilization achieves the ability to undergo cleavage division and differentiation to create a live organism (Ikawa et al. 2010). Fertilization of an ovum occurs at the fimbriated end of the oviduct in mice. Mouse embryos take about 3 days to develop from the zygotic stage to the blastocyst stage (Fig. 1). The zygote begins 4 days of travel down the oviduct to the uterus. After three rounds of mitotic division inside the oviduct, the zygote progresses through the 2-cell stage and 4-cell stage to the 8-cell stage. Individual cell in each stage of these embryos is referred to as blastomere. Blastomeres are considered totipotent (ability to generate all cell-types required to form an organism) until the early 8-cell stage. Totipotent to pluripotent transition conceivably begins at this stage when the 8-cell stage embryo undergoes a process called compaction during which blastomere to blastomere contact is maximized by polarization and tight adhesion. Thus, compaction marks the beginning of first event of morphogenic and blastomere's differentiation into two cell lineages that are found in the blastocyst. Subsequent two mitotic divisions give rise to 16-cell and 32-cell stage embryos, called the morula and blastocyst, respectively. As the 8-cell embryo transitions to 16-cell, most of the blastomeres at this stage are positioned outside encircling a few interior blastomeres. This embryonic stage is termed as the morula. The development of the morula-stage embryo in mice takes place inside the oviduct. The development of the blastocyst stage from the morula stage occurs either inside the oviduct or the uterus. There is evidence that mitotic cleavage stages of blastocyst formation vary among species. The pig and hamster embryos initiate blastocyst formation at the fourth cleavage stage (16-cell stage), a cleavage stage ahead of other species such as mice, rats,

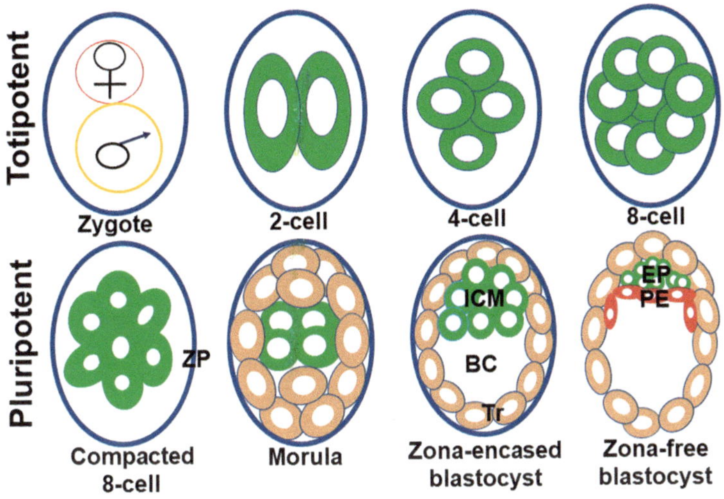

Fig. 1 Preimplantation stages of the mouse embryo. Blastocoel (BC), Epiblast (EP), Inner cell mass (ICM), Zona pellucida (ZP), Primitive endoderm (PE), Trophectoderm (Tr)

sheep, bovine, humans, and marsupials where blastocyst formation begins at or over the fifth cleavage stage (≥32-cell stage). As the morula transitions to the blastocyst stage, a fluid-filled cavity known as blastocoel (BC) appears inside the embryo when Tr cells begin to pump fluid into intracellular spaces, and further differentiation of blastomeres begins. While a small population of blastomeres that remain as a cluster inside the blastocyst is named the ICM, a layer of blastomeres that surrounds the BC and the ICM is called the Tr (Fig. 1) (Chazaud and Yamanaka 2016; Posfai et al. 2019; Rossant and Tam 2009). The number of Tr cells in an early blastocyst is always higher than ICM cells.

All stages of the preimplantation embryos (1-cell stage to the blastocyst stage) remain covered by a glycoprotein envelope called the zona pellucida (ZP) (Fig. 1) and remain buoyant inside the lumens of the oviduct and uterus. The energy required for zygotic development to the 2-cell stage is derived from the nutrient stored inside the ovum. In order to continue to develop past the 2-cell stage, the embryo activates its own genome to produce energy from the prestored nutrients or nutrients sequestered from the luminal fluid of the oviduct and uterus. Recent studies suggest that embryonic gene activation may begin at the mid 1-cell embryo and dramatically increases during the 2-cell stage (Abe et al. 2018). As the blastocyst matures inside the uterine lumen, it hatches from the ZP and achieves the attachment capacity to the uterine endometrium (Yoshinaga 2013). The ICM of the blastocyst contains precursor cells for the epiblast (EP) and primitive endoderm (PE) or hypoblast. Just prior to implantation, the ICM differentiates to the EP and PE. The EP are precursors of all the cells of the embryo and extraembryonic mesoderm (Fig. 2). The PE lines the blastocoel surface of the epiblast and later generates extraembryonic endoderms (EEEn) such as the parietal and visceral endoderms (PaE and ViE) (Fig. 2). The location of the ICM or EP within the vesicular blastocyst determines the embryonic

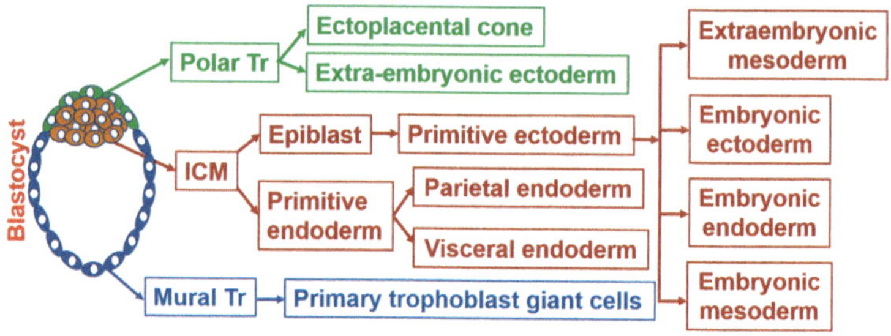

Fig. 2 The emergence of cell lineages from the blastocyst. Inner cell mass (ICM) and trophectoderm (Tr)

pole of the blastocyst. The domain of the blastocyst opposite to the embryonic pole is known as the abembryonic pole. Trophoblast cells that reside over the ICM and later EP are known as polar trophoblasts (Polar Tr) and the rest of trophoblast cells that surround the blastocoel are termed mural trophoblasts (Mural Tr). While polar Tr cells of the blastocyst give rise to extraembryonic ectoderm (EEE) and ectoplacental cone (EEC) following implantation, mural Tr cells generate trophoblast giant cells (TGCs) (Fig. 2) (Rossant and Cross 2001). The further development of the zona-escaped blastocyst to the fetal stage only happens when the blastocyst makes direct contact with the uterus through a process called implantation. Defects in any developmental processes of preimplantation embryos due to abnormal cell division, cell death, cell fate, and differentiation are likely causes of preimplantation embryonic death, poor quality of blastocysts, complete implantation failure, or abnormal implantation leading to placentation defects, miscarriage, preterm birth, and retarded growth of the fetus.

3 A Gist of the Blastocyst Implantation Process

Blastocyst implantation is a landmark event of pregnancy in which a floating blastocyst makes attachment with the uterine endometrium (Figs. 3 and 4) to remain united with it until the birth. Through this process of implantation, a semiallogenic blastocyst rewards with a shelter, nutrients, oxygen, and immunological protection for its further development to the egg cylinder stage and beyond. In exchange for these services, the uterus achieves its one and only reproductive function of carrying a pregnancy in the preservation of life.

Successful implantation of the blastocyst requires an implantation-competent blastocyst, a receptive uterus, and a mutualistic interaction between the blastocyst and the uterus. Blastocyst implantation process is divided into three steps: apposition, adhesion, and invasion (Carson et al. 2000; Zhang et al. 2013). An anomaly in any of these stages is considered to be the root cause of instantaneous pregnancy failure or placental developmental defects leading to later pregnancy complications

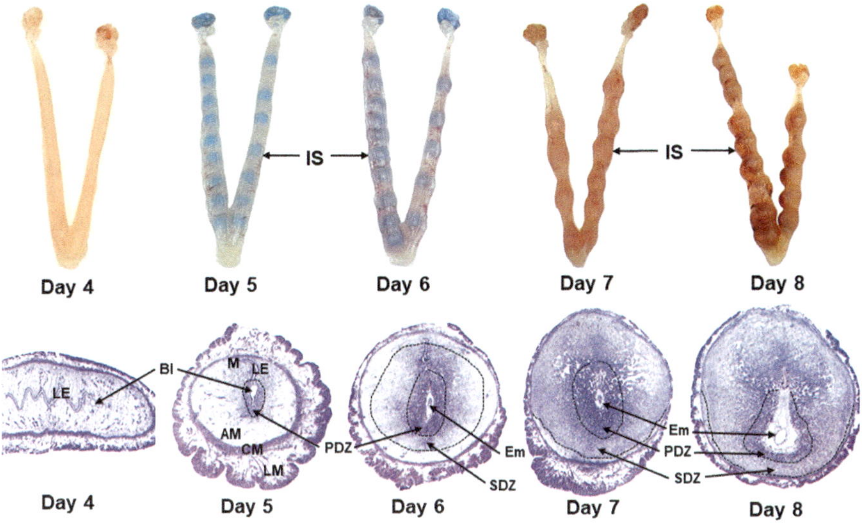

Fig. 3 Upper panel: Photographs of the preimplantation day 4 receptive uterus and post-implantation uteri with implantation sites from days 5 to 8 of pregnancy. Blastocyst implantation sites (IS) on days 5 and 6 were visualized by intravenous injection of Chicago Blue dye. Lower panel: Photomicrographs (20×) of Hematoxylin- and Eosin-stained sections of the day 4 preimplantation uterus (longitudinal section) and implantation sites (cross sections) of days 5–8 of pregnancy. Antimesometrial (AM) side, Blastocyst (Bl), Circular muscle (CM), Embryo (Em), Longitudinal muscle (LM), Luminal epithelium (LE), Mesometrial side (M); Primary decidual zone (PDZ), Secondary decidual zone (SDZ)

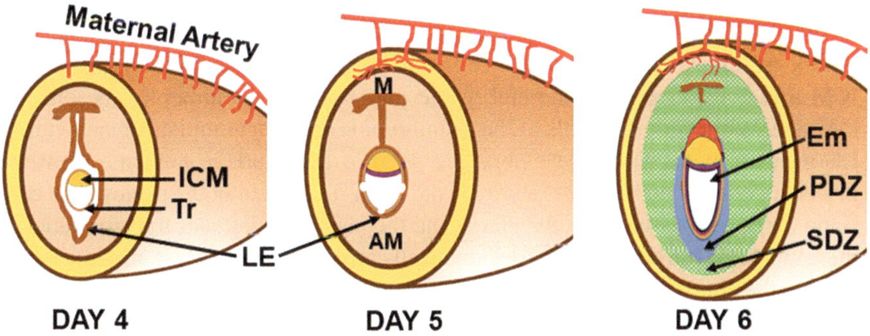

Fig. 4 Schematic drawings of uterine and embryonic changes during early pregnancy (days 4–6 of pregnancy). Antimesometrial (AM) side, Embryo (Em), Inner cell mass (ICM), Luminal epithelium (LE), Mesometrial (M) side, Primary decidual zone (PDZ), Secondary decidual zone (SDZ), Trophectoderm (Tr)

such as placental insufficiency, preeclampsia, fetal developmental disorders, and preterm birth (Cha et al. 2012).

In general, the uterine endometrium remains receptive to blastocyst implantation for a short period which is termed "implantation window" (Paria et al. 1993). In humans, the uterus achieves receptivity in the mid-secretory phase (20–26 days) coinciding with differentiation of endometrial stromal cells into specialized decidual cells of each 28-day menstrual cycle with or without an occurrence of copulation and the presence of a blastocyst. Uterine exposure to high levels of progesterone during the luteal phase is considered critical for changes in the uterus that are needed for uterine receptivity (Simon et al. 2003). In contrast to the human, the mouse uterus exposed to only progesterone does not attain receptivity either in the lack or excess of estrogen. A successful mating during the cycle requires to attain receptivity. The preimplantation mouse uterus exposed to the augmented level of luteal progesterone achieves receptivity only when it is exposed to a small rise of circulating ovarian estrogen level around the noon of day 4 of pregnancy (Paria et al. 1993). An ideal uterine window of implantation in mice lasts only for a very short period from midnight of day 4 to early morning of day 5 (Das et al. 1994). Like uterine receptive state, blastocyst implantation-competent stage is also a determinant factor for initiation of implantation. A study using the delayed implantation model in mice has shown that activated blastocyst functions as a proinflammatory entity at the time of implantation (He et al. 2019).

The types of implantation vary from species to species based on the degree of invasion. The primates and guinea pigs exhibit interstitial type of implantation in which the blastocyst reaches the stroma disregarding the luminal epithelium (LE). In cattle, sheep, and pigs, implantation is a superficial type in which embryonic trophoblasts simply remain attached to the uterine epithelium (no penetration). In rodents such as mice and rats, implantation is an eccentric type in which a mouse blastocyst placed itself inside a cup-shaped endometrial crypt (implantation chamber). These crypts are formed on day 4 of pregnancy as a result of occasional inward folding of epithelial layer primarily in the antimesotrial (AM) side of the receptive uterus (Figs. 3 and 4) (Cha et al. 2014). Once this cocooning process ends, the blastocyst initiates contact with LE cells using some of its mural Tr cells positioning ICM and polar Tr cells of the embryonic pole towards the lumen or mesometrial (M) side. While mural Tr cells of the abembryonic pole begin initial contact with the LE in mice, polar Tr of the embryonic pole do this work in humans. However, placenta is developed exclusively from the polar Tr cells in both species. The early stage of the contact between the uterus and the blastocyst initiates transformation of the endometrial stromal cells to decidual cells only at the blastocyst–uterine contact site. As the contact and direct communication between the uterus and blastocyst gets stronger, the uterine LE cells at the site of blastocyst contact undergo apoptosis and/or entosis by the contact-making trophoblast cells to confirm direct contact with the underneath stromal/decidual cells (Fig. 4) (Li et al. 2015; Zhang and Paria 2006). Following implantation, the round blastocyst undergoes transformation to form an elongated "egg cylinder." This occurs as a result of rapid proliferation of cells of the EP, PE, and polar Tr.

4 A Gist of Placentation

Placenta formation begins with blastocyst implantation and keeps on growing (Figs. 3, 5 and 6). After implantation occurs, the blastocyst implantation site progressively grows in size from day 5 to day 8 of pregnancy mainly due to development of the decidua (Fig. 3). Later in pregnancy, the rapid growth of the

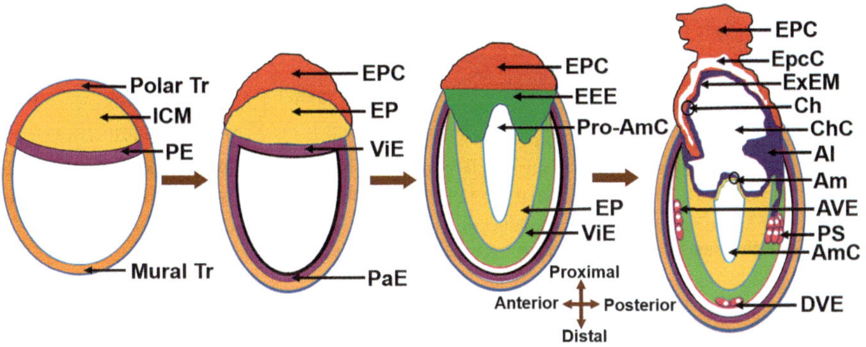

Fig. 5 Schematic drawing of the synchronous development of the placenta and the embryo following blastocyst implantation. Allantois (Al), Amnion (Am), Amniotic cavity (AmC), Anterior visceral endoderm (AVE), Chorion (Ch), Chorionic cavity (ChC), Distal visceral endoderm (DVE), Ectoplacental cone (EPC), Ectoplacental cavity (EpcC), Extra-embryonic ectoderm (EEE), Epiblast (EP), Extra-embryonic mesoderm (ExEM), Inner cell mass (ICM), Parietal endoderm (PaE), Primitive endoderm (PE), Primitive streak (PS), Pro-amniotic cavity (Pro-AmC), Visceral endoderm (ViE), and Trophectoderm (Tr)

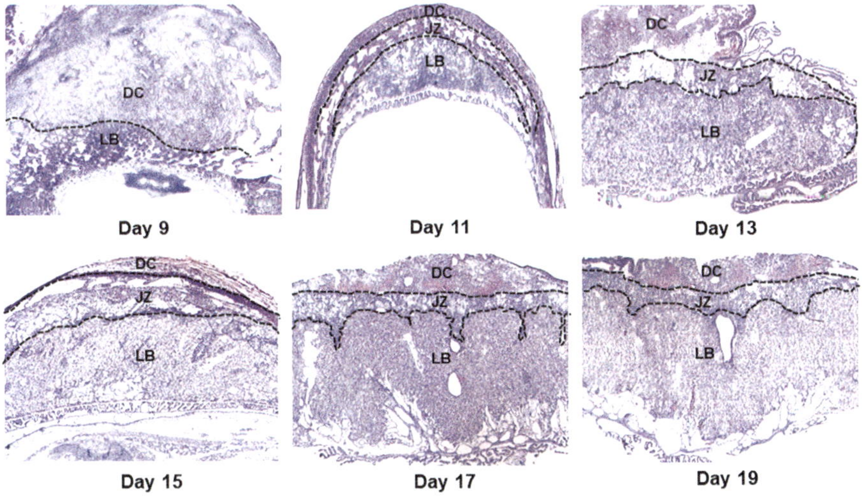

Fig. 6 Photomicrographs of hematoxylin- and eosin-stained histological sections of placentae from days 9–19 of pregnancy. Decidua (DC), Junctional zone (JZ), Labyrinth (LB)

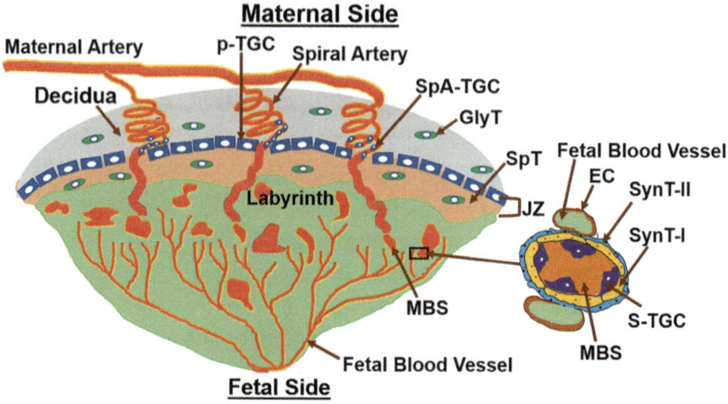

Fig. 7 A schema of a mouse matured placenta. Endothelial cell (EC), Glycogen trophoblast (GlyT), Junctional zone (JZ), Maternal blood sinusoid (MBS), Parietal TGC (p-TGC), Sinusoidal TGC (S-TGC), Spiral artery-associated TGC (SpA-TGC), Spongiotrophoblast (SpT), Syncytotrophoblast I (SynT-I), Syncytotrophoblast II (SynT-II)

placenta and the embryo governs the size of the implantation site. A definite structure of the placenta in mice is evident by gestational days 10–11. Its development finishes by day 15 of pregnancy and maximum volume reaches by day 17 (Coan et al. 2004). It is expelled upon birth of the fetus. It has both maternal uterine and fetal components. It develops between the uterus and the fetus and acts as a caretaker of pregnancy and the fetus. Although both the humans and mice have a hemochorial type of placenta (Hafez 2017), the differences between the human and mouse placentation are considerable. Unlike the human villous placenta, the site of placental exchange in rodents in the labyrinth region has a complex arrangement of maternal and vascular channels (Hemberger et al. 2020). The mouse placenta comprises of three structures: the maternal decidua, the junctional zone (JZ), and the labyrinth (Figs. 6 and 7). The maternal decidual zone contains uterine decidual cells, maternal vasculature, glycogen trophoblast (GlyT) cells, and spiral artery-associated trophoblast giant cells (SpA-TGC). The JZ contains a p-TGC layer and a spongiotrophoblast layer consisting of spongiotrophoblasts (SpT) and GlyT cells. The labyrinth contains chorion-derived epithelial cells and allantois-derived vascular cells (Fig. 7). Here we review the current understanding of the post-implantation embryonic events that outline the operational organization of the mouse placenta.

4.1 Uterine Decidua Formation for Placentation

Blastocyst attachment onto the uterine luminal epithelium in mice stimulates uterine stromal cell proliferation. Transformation of these proliferating stromal cells to decidual cells initiates shortly after the blastocyst–uterine attachment reaction (Figs. 3 and 4). It first occurs in antimesometrial (AM) stromal cells that are closest to the implantation chamber. This avascular and densely packed decidual cell region that is termed as the primary decidual zone (PDZ) is considered the first protective scaffold of the implanted embryo. The PDZ formation is fully established by day 6 (Figs. 3 and 4). Later, all other proliferating stromal cells surrounding the PDZ both at the AM and mesometrial (M) sides transform into decidual cells, forming a secondary decidual zone (SDZ) (Figs. 3 and 4). While AM decidual cells are polyploid, M decidual cells are not (Das 2009, 2010). As the pregnancy progresses from day 5 to 8, the AM decidua degenerates making space for the growing embryo, while the M decidua gradually thins out to make room for placentation, eventually forming the decidua basalis. Defective decidua formation may result in infertility or later onset of pregnancy complications such as preeclampsia, recurrent abortion, preterm birth, and intrauterine fetal growth restriction.

For a pregnancy to continue further, a solid anchoring of other placental features onto the decidua basalis and the alteration of maternal spiral arteries into soft vessels are necessary. Uterine artery at the site of placentation branches into several spiral arteries by angiogenesis (Figs. 4 and 7). Spiral arteries begin as high-resistance, low-capacity arteries and remodel to low-resistance, high-capacity structures. The key regulators of these arterial changes are leukocytes, specifically those localized in the decidua basalis (Mori et al. 2016). This vascular transformation allows the increase in maternal oxygenated blood flow needed to sustain the fetus. These spiral arteries later converge into large arterial canals that cross the JZ to the base of the chorionic plate. From there, blood flows within small maternal blood sinusoids (MBS) in the labyrinth towards the junctional zone where they converge in larger channels that return deoxygenated maternal blood. Arterial canals lack smooth muscle cells and converge at the junction to the giant cells where they open into the trophoblast-lined canals that enter the labyrinth (Fig. 7). The veins are large, thin walled, without trophoblast cells. The umbilical artery branches into arterioles that carry deoxygenated blood from the fetus to the labyrinth side closest to the junctional zone. There, they branch into capillaries. The fetal deoxygenated blood with these capillaries flows in countercurrent with the oxygenated blood from the mother.

4.2 *Ectoplacental Cone (EPC) and Egg Cylinder Formation*

At implantation, mural Tr cells stop proliferating, but endoreduplicate to form primary TGCs which help in the attachment of the blastocyst initially with the uterine LE and later with the stromal compartment. In contrast, polar Tr cells begin proliferating to create extra- EEE and EPC (Fig. 2). Initially, the polar Tr grows towards the EP forming the EEE, which shoves the EP in the direction of the blastocoel. Later, outside cells of the EEE proliferate and grow to form an EPC towards the uterine lumen and mesometrial side of the uterus. EPC resembles like a cap on top of the EEE (Fig. 5). When the outside cells of the EPC interlock with the uterine LE, narcosis of the epithelial cells begins, leaving Tr cells to make direct contact with the uterine decidual cells. These cells also often surround the decidual cells by sending projections. These trophoblast cells are called parietal giant cells (p-TGCs) that have the largest polyploid nuclei and highest content of nuclei among all cell-types in the EPC (Bevilacqua and Abrahamsohn 1988, 1989). P-TGCs that pass through the decidual cells establish contact with the endothelial cells and displace the cells from uterine capillary walls. On day 8 of pregnancy, the EPC fully developed (Woods et al. 2018).

Upon implantation, the EP grows, elongates, and transforms into a U-shaped columnar epithelium (primitive ectoderm) that surrounds the pro-amniotic cavity (Pro-Amc) (Fig. 5). The formation of the Pro-Amc is accomplished by cavitation. While EPC is forming, the PE that lines the luminal surface of the EP proliferates and expands to form the PiE and ViE. The cells of these two endoderms later migrate along the mural trophoblast cells. The PiE lines the luminal surface of the mural trophoblasts and the ViE envelops the EP. In a later stage of embryonic development, the ViE below the EP thickens and identifies as distal visceral endoderm (DVE). Cells from the DVE later move to the anterior region of the embryo where they become known as anterior visceral endoderm (AVE). A second subpopulation of DVE migrates and assembles at the posterior side of the embryo and form the primitive streak (PS) (Rivera-Perez et al. 2003; Rivera-Perez and Hadjantonakis 2014).

4.3 Junctional Zone (JZ) Formation

The JZ arises from the cells of the core of the EPC. It is situated between the labyrinth and the maternal decidua (Figs. 6 and 7) and functions as a source of hormones, growth factors, and energy that are required for normal growth of the placenta and the embryo. The JZ developmental defects are considered a leading cause of placental as well as fetal growth retardations. The JZ consists of two layers: A p-TGC layer that directly borders the decidua and a spongiotrophoblast (SpT) layer which consists of SpT and GlyT cells (Fig. 7). While SpT cells are dense, compact, and nonmigratory, GlyT cells are vacuolated and often migrate to the decidua through p-TGC layer (Simmons and Cross 2005). Studies have suggested that SpT and GlyT cells originate from unique cell-types of the EPC. SpT and GlyT cells also surround the central canals that pass through the labyrinth and canal-associated trophoblast giant cells (cTGCs) which lines the canal (Fig. 7). GlyT cells serve as an energy source as they store glycogen and are also source of insulin-like growth factor II (IGF-II). SpT cells along with TGCs are the main source of hormones and growth factors (Woods et al. 2018).

4.4 Labyrinth and Gastrulation Formation

Labyrinth is the largest structure of the placenta and is formed next to the JZ (Figs. 6 and 7). It is a vital but a complex structure by organization and function. The construction of labyrinth involves interactions among three structures: the EPC, the chorion, and the allantois (Fig. 5). The EPC and chorion are formed from trophoblast cells, whereas the allantois is derived from extraembryonic mesoderm which originates from the epiblast.

The PS formation is the first morphological landmark of gastrulation (E 6.5) and labyrinth construction (Fig. 5). At gastrulation, the PS forms on the side of the embryo opposite to the AVE, denoting AVE as the anterior pole of the body axis and PS as the posterior pole of the embryo (Rossant and Tam 2009). The PS is the site where cells undergo an epithelial-to-mesenchymal transition. As the PS elongates, epiblast cells that are on the inside of the cup ingress up in the streak. The EP cells that move through the streak become mesoendoderm which are the precursors not only for embryonic mesoderm and extraembryonic mesoderm (ExEM) of the chorion (Ch), amnion (Am), yok sac and allantois (Al) but also for amniotic ectoderm as well as embryonic ectoderm and endoderm (Fig. 2). After they exit the primitive streak, the cells create a wave of mesoendoderm that expands between the visceral endoderm and epiblast (Woods et al. 2018).

Chorion Formation

During gastrulation, the EEE folds inside the Pro-AmC from both sites and fuses together forming the ectoplacental cavity (EpcC) (Fig. 5). At the same time, a chorionic or exocoelomic cavity (ChC) is formed when extraembryonic mesoderm (ExEM) grew from the ventral and dorsal regions of the allantois and reaches the anterior margin of the egg cylinder through the Pro-AmC. The allantois initially arises from the PS as a bud. Later it grows and expands within the exocoelomic cavity. The cavity that remains in contact with the EP is called the amniotic cavity (AmC). Proximal ExEM normally grows along the base of the EEE of the ectoplacental cavity and together form the chorion (ChC). The area that contains these two layers is called the chorionic plate. Distal ExEM when it comes in contact with a layer of embryonic ectoderm form the amnion (Am) that lies in between the ChC and AmC. At day 7 of pregnancy, extraembryonic mesoderm cells of the allantois proliferate and grow inside the chorionic cavity reaching the chorion. This eliminates the EpcC and brings the chorionic plate in contact with the EEC. Chorionic plate now is a multilayered structure and becomes known as the chorio-allantoic plate (Simmons et al. 2008).

Allantois (Al) and Yolk Sac

The AL forms the fetal vascular compartment of the placenta and umbilical cord in mice (Arora and Papaioannou 2012). It arises as a bud of mesoderm from the primitive streak and grows toward the chorion through the exocoelomic cavity. The growing allantois consists of an outer layer of mesothelium that envelops a mesenchymal core (Fig. 5). Endothelial cell (EC) generation starts in the allantois from the mesenchyme. Whether hematopoietic cells arise from the endothelium or its mesenchyme is still a matter of debate. However, studies suggest that hematopoietic cells in placenta emerge via endothelial to hematopoietic transition. The ECs initially form a primary vascular plexus and connect with the dorsal aortae of the embryo prior to chorio-allantois fusion. The primary plexus then remodels to form the central vessel by a process called vasculogenesis and later the umbilical artery and vein at day 10 of pregnancy (Azevedo and Pelajo-Machado 2018). The artery and vein then invade chorion by angiogenesis forming the fetal vascular components of the labyrinth. After chorio-allantoic attachment, the chorion begins to form the villi across the chorionic surface creating spaces into which the fetal blood vessels grow from the allantois (Downs 1998; Downs and Gardner 1995).

The maternal blood spaces within the labyrinth are separated by three layers of trophoblast cells and a layer of fetal endothelial cells. The three layers of trophoblast cells are a layer of mononuclear sinusoidal trophoblast giant cells (S-TGCs) that surround maternal blood sinusoids followed by two layers of syncytiotrophoblasts, SynT-I, and SynT-II in contact with fetal vessels on the opposite site (Fig. 7). Syncytiotrophoblast cell layers are multinucleated and form as a result of trophoblast cell–cell fusion. SynT-II remains in contact with fetal endothelial cells. Thus, maternal blood comes in direct contact with trophoblast cells rather than endothelial cells of the vasculature in mice (Watson and Cross 2005).

Yolk sac is composed of two layers: The ViE and the ExEM. The difference between the mouse and human yolk sacks is that while the mouse yolk sac envelops the embryo, human yolk sac remains attached to, but does not envelop the embryo (Freyer and Renfree 2009). It has been shown that mouse yolk sac contains in vivo colony forming cells capable of producing granulocytic, megakaryocytic, and erythroid spleen colonies (Garcia and Larina 2014; Palis and Koniski 2005; Yamane 2018). Thus, yolk sac may be an exporter of hematopoietic stem cells which subsequently colonize other hematopoietic sites. It has been long known that yolk sac produces cholesterol, an essential component of the membrane and steroid synthesis (Woollett 2011). Furthermore, the endoderm layer of the yolk sac expresses growth factors and transcription factors that are required for vasculogenesis and angiogenesis (Freyer and Renfree 2009).

5 Conclusion

As the placenta is a principal transient organ of nutrient supply to the growing embryo during pregnancy, adequate placental function is instrumental for developmental progression of the intrauterine embryo. Studies have established that defects or delay in implantation, uterine stromal cell decidualization or uterine vascular remodeling at the implantation site lead to a spectrum of pregnancy complications such as poor placentation, embryonic demise, or abnormal embryonic growth and development.

The mouse is the leading model system for studying and understanding the basic structural and functional aspects of the placenta due to ethical constraints and inaccessibility of very early human implantation tissues. In this regard, gene expression and manipulation studies in mice have provided a wealth of information. The cell lineage tracing studies established that the majority defining structures of the placenta is originated from trophoblast cells of the blastocyst, but endothelial cells that make the fetal placental vasculature is aroused from the ICM of the blastocyst. Phenotypic analysis of global as well as conditional gene knockout mice has provided further information about key roles of genes and gene networks behind the complex morphological and molecular events that occur during placenta formation. More recently stem cell technologies have also begun to shed new lights on the origin of placenta cell lineages. So far thousands of genes that cause either embryonic lethality and/or placental defects have been identified. However, it is also possible that there are hundreds if not thousands of genes that influence placental developmental and functional defects remain to be identified. Taken together, gaining insights into genes and gene networks that affect implantation, decidualization, and placental structural and functional defects using the mouse model will promote the study of human pregnancy complication involving placental and embryonic developmental defects.

Acknowledgement Research in the authors' laboratory is supported by R01 HD094946 from the Eunice Kennedy Shriver National Institute of Child Health and Human Development/National Institute of Health.

References

Abe KI, Funaya S, Tsukioka D, Kawamura M, Suzuki Y, Suzuki MG, Schultz RM, Aoki F (2018) Minor zygotic gene activation is essential for mouse preimplantation development. Proc Natl Acad Sci U S A 115(29):E6780–E6788. Available from: PM:29967139

Arora R, Papaioannou VE (2012) The murine allantois: a model system for the study of blood vessel formation. Blood 120(13):2562–2572. Available from: PM:22855605

Azevedo PN, Pelajo-Machado M (2018) Mechanism of hematopoiesis and vasculogenesis in mouse placenta. Placenta 69:140–145. Available from: PM:29680159

Bauer ST, Bonanno C (2009) Abnormal placentation. Semin Perinatol 33(2):88–96. Available from: PM:19324237

Bevilacqua EM, Abrahamsohn PA (1988) Ultrastructure of trophoblast giant cell transformation during the invasive stage of implantation of the mouse embryo. J Morphol 198(3):341–351. Available from: PM:3221406

Bevilacqua EM, Abrahamsohn PA (1989) Trophoblast invasion during implantation of the mouse embryo. Arch Biol Med Exp (Santiago) 22(2):107–118. Available from: PM:2619314

Brown LD, Hay WW Jr (2016) Impact of placental insufficiency on fetal skeletal muscle growth. Mol Cell Endocrinol 435:69–77. Available from: PM:26994511

Browne JC (1963) Placental insufficiency. Scott Med J 8:459–465. Available from: PM:14089221

Burton GJ, Fowden AL, Thornburg KL (2016) Placental origins of chronic disease. Physiol Rev 96 (4):1509–1565. Available from: PM:27604528

Carson DD, Bagchi I, Dey SK, Enders AC, Fazleabas AT, Lessey BA, Yoshinaga K (2000) Embryo implantation. Dev Biol 223(2):217–237. Available from: PM:10882512

Cha J, Sun X, Dey SK (2012) Mechanisms of implantation: strategies for successful pregnancy. Nat Med 18(12):1754–1767. Available from: PM:23223073

Cha J, Bartos A, Park C, Sun X, Li Y, Cha SW, Ajima R, Ho HY, Yamaguchi TP, Dey SK (2014) Appropriate crypt formation in the uterus for embryo homing and implantation requires Wnt5a-ROR signaling. Cell Rep 8(2):382–392. Available from: PM:25043182

Chazaud C, Yamanaka Y (2016) Lineage specification in the mouse preimplantation embryo. Development 143(7):1063–1074. Available from: PM:27048685

Coan PM, Ferguson-Smith AC, Burton GJ (2004) Developmental dynamics of the definitive mouse placenta assessed by stereology. Biol Reprod 70(6):1806–1813. Available from: PM:14973263

Das SK (2009) Cell cycle regulatory control for uterine stromal cell decidualization in implantation. Reproduction 137(6):889–899. Available from: PM:19307426

Das SK (2010) Regional development of uterine decidualization: molecular signaling by Hoxa-10. Mol Reprod Dev 77(5):387–396. Available from: PM:19921737

Das SK, Wang XN, Paria BC, Damm D, Abraham JA, Klagsbrun M, Andrews GK, Dey SK (1994) Heparin-binding EGF-like growth factor gene is induced in the mouse uterus temporally by the blastocyst solely at the site of its apposition: a possible ligand for interaction with blastocyst EGF-receptor in implantation. Development 120(5):1071–1083. Available from: PM:8026321

Downs KM (1998) The murine allantois. Curr Top Dev Biol 39:1–33. Available from: PM:9475996

Downs KM, Gardner RL (1995) An investigation into early placental ontogeny: allantoic attachment to the chorion is selective and developmentally regulated. Development 121(2):407–416. Available from: PM:7768182

Freyer C, Renfree MB (2009) The mammalian yolk sac placenta. J Exp Zool B Mol Dev Evol 312 (6):545–554. Available from: PM:18985616

Garcia MD, Larina IV (2014) Vascular development and hemodynamic force in the mouse yolk sac. Front Physiol 5:308. Available from: PM:25191274

Guo S, Cui X, Jiang X, Duo S, Li S, Gao F, Wang H (2020) Tracing the origin of the placental trophoblast cells in mouse embryo developmentdagger. Biol Reprod 102(3):598–606. Available from: PM:31621828

Hafez S (2017) Comparative placental anatomy: divergent structures serving a common purpose. Prog Mol Biol Transl Sci 145:1–28. Available from: PM:28110748

He B, Zhang H, Wang J, Liu M, Sun Y, Guo C, Lu J, Wang H, Kong S (2019) Blastocyst activation engenders transcriptome reprogram affecting X-chromosome reactivation and inflammatory trigger of implantation. Proc Natl Acad Sci U S A 116(33):16621–16630. Available from: PM:31346081

Hemberger M, Hanna CW, Dean W (2020) Mechanisms of early placental development in mouse and humans. Nat Rev Genet 21(1):27–43. Available from: PM:31534202

Ikawa M, Inoue N, Benham AM, Okabe M (2010) Fertilization: a sperm's journey to and interaction with the oocyte. J Clin Invest 120(4):984–994. Available from: PM:20364096

Kollmann M, Gaulhofer J, Lang U, Klaritsch P (2016) Placenta praevia: incidence, risk factors and outcome. J Matern Fetal Neonatal Med 29(9):1395–1398. Available from: PM:26043298

Li Y, Sun X, Dey SK (2015) Entosis allows timely elimination of the luminal epithelial barrier for embryo implantation. Cell Rep 11(3):358–365. Available from: PM:25865893

Mori M, Bogdan A, Balassa T, Csabai T, Szekeres-Bartho J (2016) The decidua-the maternal bed embracing the embryo-maintains the pregnancy. Semin Immunopathol 38(6):635–649. Available from: PM:27287066

Oyelese Y, Ananth CV (2006) Placental abruption. Obstet Gynecol 108(4):1005–1016. Available from: PM:17012465

Oyelese Y, Smulian JC (2006) Placenta previa, placenta accreta, and vasa previa. Obstet Gynecol 107(4):927–941. Available from: PM:16582134

Palis J, Koniski A (2005) Analysis of hematopoietic progenitors in the mouse embryo. Methods Mol Med 105:289–302. Available from: PM:15492402

Paria BC, Huet-Hudson YM, Dey SK (1993) Blastocyst's state of activity determines the "window" of implantation in the receptive mouse uterus. Proc Natl Acad Sci U S A 90(21):10159–10162. Available from: PM:8234270

Phipps EA, Thadhani R, Benzing T, Karumanchi SA (2019) Pre-eclampsia: pathogenesis, novel diagnostics and therapies. Nat Rev Nephrol 15(5):275–289. Available from: PM:30792480

Posfai E, Rovic I, Jurisicova A (2019) The mammalian embryo's first agenda: making trophectoderm. Int J Dev Biol 63(3-4-5):157–170. Available from: PM:31058294

Rivera-Perez JA, Hadjantonakis AK (2014) The dynamics of morphogenesis in the early mouse embryo. Cold Spring Harb Perspect Biol 7(11):a015867. Available from: PM:24968703

Rivera-Perez JA, Mager J, Magnuson T (2003) Dynamic morphogenetic events characterize the mouse visceral endoderm. Dev Biol 261(2):470–487. Available from: PM:14499654

Rossant J, Cross JC (2001) Placental development: lessons from mouse mutants. Nat Rev Genet 2 (7):538–548. Available from: PM:11433360

Rossant J, Tam PP (2009) Blastocyst lineage formation, early embryonic asymmetries and axis patterning in the mouse. Development 136(5):701–713. Available from: PM:19201946

Simmons DG, Cross JC (2005) Determinants of trophoblast lineage and cell subtype specification in the mouse placenta. Dev Biol 284(1):12–24. Available from: PM:15963972

Simmons DG, Natale DR, Begay V, Hughes M, Leutz A, Cross JC (2008) Early patterning of the chorion leads to the trilaminar trophoblast cell structure in the placental labyrinth. Development 135(12):2083–2091. Available from: PM:18448564

Simon C, Dominguez F, Valbuena D, Pellicer A (2003) The role of estrogen in uterine receptivity and blastocyst implantation. Trends Endocrinol Metab 14(5):197–199. Available from: PM:12826321

Watson ED, Cross JC (2005) Development of structures and transport functions in the mouse placenta. Physiology (Bethesda) 20:180–193. Available from: PM:15888575

Woods L, Perez-Garcia V, Hemberger M (2018) Regulation of placental development and its impact on fetal growth-new insights from mouse models. Front Endocrinol (Lausanne) 9:570. Available from: PM:30319550

Woollett LA (2011) Review: transport of maternal cholesterol to the fetal circulation. Placenta 32 (Suppl 2):S218–S221. Available from: PM:21300403

Yamane T (2018) Mouse yolk sac hematopoiesis. Front Cell Dev Biol 6:80. Available from: PM:30079337

Yoshinaga K (2013) A sequence of events in the uterus prior to implantation in the mouse. J Assist Reprod Genet 30(8):1017–1022. Available from: PM:24052329

Zhang Q, Paria BC (2006) Importance of uterine cell death, renewal, and their hormonal regulation in hamsters that show progesterone-dependent implantation. Endocrinology 147(5):2215–2227. Available from: PM:16469810

Zhang S, Lin H, Kong S, Wang S, Wang H, Wang H, Armant DR (2013) Physiological and molecular determinants of embryo implantation. Mol Aspects Med 34(5):939–980. Available from: PM:23290997

Placentation in the Human and Higher Primates

Graham J. Burton and Eric Jauniaux

Contents

Abstract Placentation in humans is precocious and highly invasive compared to other mammals. Implantation is interstitial, with the conceptus becoming completely embedded within the endometrium towards the end of the second week post-fertilization. Villi initially form over the entire surface of the chorionic sac, stimulated by histotrophic secretions from the endometrial glands. The secondary

G. J. Burton (✉)
Centre for Trophoblast Research, Department of Physiology, Development and Neuroscience, University of Cambridge, Cambridge, UK
e-mail: gjb2@cam.ac.uk

E. Jauniaux
Faculty of Population Health Sciences, EGA Institute for Women's Health, University College London, London, UK
e-mail: e.jauniaux@ucl.ac.uk

R. D. Geisert, T. E. Spencer (eds.), *Placentation in Mammals*, Advances in Anatomy, Embryology and Cell Biology 234, https://doi.org/10.1007/978-3-030-77360-1_11

yolk sac never makes contact with the chorion, and a choriovitelline placenta is never established. However, recent morphological and transcriptomic analyses suggest that the yolk sac plays an important role in the uptake of nutrients from the coelomic fluid. Measurements performed in vivo demonstrate that early development takes place in a physiological, low-oxygen environment that protects against teratogenic free radicals and maintains stem cells in a multipotent state. The maternal arterial circulation to the placenta is only fully established around 10–12 weeks of gestation. By then, villi have regressed over the superficial, abembryonic pole, leaving the definitive discoid placenta, which is of the villous, hemochorial type. Remodeling of the maternal spiral arteries is essential to ensure a high-volume but low-velocity inflow into the mature placenta. Extravillous trophoblast cells migrate from anchoring villi and surround the arteries. Their interactions with maternal immune cells release cytokines and proteases that are key to remodeling, and a successful pregnancy.

Keywords Human · Trophoblast · Histotroph · Secondary yolk sac · Cytotrophoblastic shell · Spiral arterial remodeling

1 Introduction

Placentation in humans has long been considered almost unique amongst mammals, involving a precocious and highly invasive form of implantation only seen amongst the great apes. The embedding of the conceptus into the uterine wall was assumed to lead to early establishment of the maternal circulation, which in turn enabled the evolution of our enlarged brain. Over the last two decades these views have been challenged from various directions. Firstly, molecular phylogenetics has revealed that the hemochorial state is likely the ancestral form of placenta and not the derived form once thought (Wildman et al. 2006; Carter and Mess 2007; Martin 2008; Elliot and Crespi 2009). Secondly, despite major morphological differences it has become apparent that many aspects of human placentation are more equivalent to those in other species than previously appreciated. This is particularly true during early pregnancy, for it is now recognized that humans display an extended period of histotrophic nutrition that continues throughout organogenesis, and that the yolk sac is likely not the vestigial organ it is often portrayed to be. In this chapter we focus on the development of the human placenta, and make comparisons with closely related higher primates.

2 Origin of the Trophoblast Lineage

Although the trophoblast is fundamental to the placenta, defining the lineage in molecular terms is problematic, for as yet no exclusive marker has been identified. Instead, it has been proposed that a panel of four criteria be used, including (1) selected protein markers such as cytokeratin 7 (KRT7), TF activator protein-2 gamma (TFAP2C), and *GATA*-binding protein 3 (GATA3), (2) expression of microRNAs from the C19 cluster, (3) hypomethylation of the promoter of the transcription factor ELF5, and (4) absence of expression of the HLA class I molecules HLA-A and HLA-B (Lee et al. 2016). Individual components of this panel may feature in different cell types, but collectively they serve to uniquely define the trophoblast.

Activation of the embryonic genome occurs between the 4- and 8-cell stage of development (Braude et al. 1988), and blastocyst formation is usually seen by day 4.5 post-fertilization (p-f). The epigenetic and transcriptional networks that regulate differentiation into the trophoblast lineage are broadly conserved between the mouse and the human (Hanna et al. 2018), although primate development is protracted compared to that of rodents (Niakan and Eggan 2013). There are important differences, however. For example, while CDX2, TEAD4, ELF5, and TFAP2C are common to both species, SOX2, ERSB, and EOMES are absent in the human (Soncin et al. 2018; Hemberger et al. 2020). Single-cell sequencing of blastocysts has revealed that two subpopulations of trophectoderm cells can be identified by days 6–7 p-f (Petropoulos et al. 2016). The differences in gene expression are mostly related to cell–cell signaling, and consistent with a more differentiated subpopulation localized using immunofluorescence for the chemokine receptor 7 (CCR7) protein to the polar region of the blastocyst overlying the inner cell mass. The receptor for fibroblast growth factor, FGFR1, which is initially expressed on all the trophoblast cells becomes localized to the same region as the blastocyst develops (Niakan and Eggan 2013).

3 Implantation

Implantation takes place at ~5–6 days p-f, and as in other species requires a synchronized dialogue between a receptive endometrium and an activated embryo (Teh et al. 2016). Attachment of the blastocyst to the uterine luminal epithelium occurs at a point equidistant between openings of the uterine glands. The polar trophoblast overlying the inner cell mass adheres preferentially, presumably reflecting the different properties of this subpopulation. Malrotation of the blastocyst at this stage will lead to the connecting stalk, the forerunner of the umbilical cord, not being centralized over the developing placenta, resulting in an eccentric, or even a velamentous cord insertion. Heparin-binding epidermal growth factor–like growth factor (HB-EGF) expressed by the uterine luminal epithelium has been identified as a

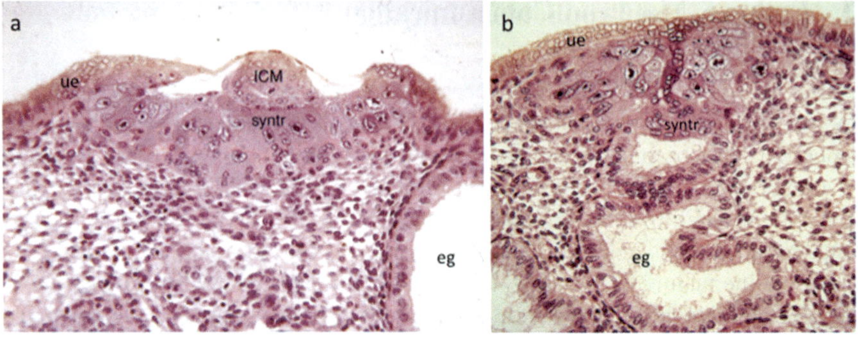

Fig. 1 Human blastocyst implanting at Carnegie stage 5A, approximately 7 days p-f. (**a**) The trophoblast in contact with the endometrium has differentiated into a primary syncytiotrophoblast (syntr) that contains nuclei of different sizes and chromatin patterns, and a population of underlying uninucleate cytotrophoblast progenitor cells. The mural trophectoderm of the abembryonic pole of the blastocyst is draped over the inner cell mass (ICM) and is still exposed to the uterine lumen as the uterine epithelium (ue) has not yet reformed. eg, endometrial gland. (**b**) A section towards the margin of the implantation site showing the primary syncytiotrophoblast (syntr) eroding into the wall of an endometrial gland (eg), making the histotroph available. (Photomicrograph courtesy of Prof. AC Enders and the Carnegie Collection)

key factor in the initial attachment, binding to members of the EGF family of receptors present on the surface of the trophoblast (Cha et al. 2012). Later, integrins, L-selectin ligands and selectin oligosaccharides are also involved.

Our knowledge of events during and immediately following implantation is extremely limited due to the inaccessibility of the tissues. Attempts have been made to mimic the process in vitro, culturing blastocysts on monolayers of endometrial cells (Lindenberg et al. 1989; Bentin-Ley et al. 2000). These have confirmed that implantation is of the intrusive type, with protrusions of trophoblast penetrating between and displacing the uterine cells. Where the trophoblast cells make contact with the epithelium they undergo proliferation and differentiation, fusing with neighbors to form small areas of multinucleated syncytial trophoblast. The same phenomenon occurs when blastocysts are cultured on plastic, suggesting it does not require biological signals from another tissue (Shahbazi et al. 2016). Mitotic figures are not seen within the syncytial masses, and the deeper trophoblast cells that formed the original wall of the blastocyst remain uninucleate and represent a progenitor population, the cytotrophoblast cells. Some of these cells most likely represent true trophoblast stem cells, but as yet molecular markers that characterize such cells unequivocally have to be identified.

The most extensive collection of human implantation material is the Carnegie Collection. The earliest specimen available (Carnegie stage 5a) shows a partially embedded conceptus; the uterine epithelium has been penetrated but not yet grown over the surface of the sac (Fig. 1). A flattened mass of syncytial trophoblast is apposed to the endometrium and performs the initial invasion (Enders 1989), and hence this is referred to as the trophoblastic plate stage of development. Nuclei within the syncytial masses are of different sizes and chromatin appearances, possibly reflecting either fusion between the trophoblast and maternal cells, or a progression towards polyploidy (Enders 2001). The mural trophoblast on the opposite pole of the gestational sac remains unicellular and projects into the uterine lumen. There is little evidence of transformation of the endometrial stromal cells into decidua.

Although human implantation is of the interstitial type and described as invasive, this terminology may create a false impression, for there is evidence that up growth of the endometrial cells plays an important role in encapsulating the conceptus. Thus, the original descriptions of human early implantation sites describe mitotic figures in the endometrial stromal cells close to the conceptus (Hertig and Rock 1941). More recent experimental data also support the concept of trophoblast-induced proliferation and subsequent migration around, and over, the conceptus to form the decidua capsularis (Gellersen et al. 2010). This view is compatible with high-resolution ultrasound scans showing the implantation site raised above the general level of the endometrial surface, bulging into the uterine lumen. The conceptus therefore always lies in the superficial layer of the endometrium, and it is regression of the underlying endometrium, by then converted into the decidua basalis, that allows it to sink deeper into the uterine wall (Hempstock et al. 2004).

4 The Lacunar Stage of Placental Development

With the reestablishment of the uterine epithelium at stage 5b, (~day 8 p-f), the syncytial trophoblast expands to cover the entire chorionic sac (Fig. 2). A key step in placental development occurs at the same time when fluid-filled spaces, referred to as lacunae, appear in the syncytial mass. As these enlarge and coalesce they subdivide the mass into two layers, one in contact with the endometrium, the forerunner of the basal plate, and one in contact with the cytotrophoblast cells of the original blastocyst wall, the forerunner of the chorionic plate. This separation establishes the boundaries of the placenta, with a series of pillars of syncytial trophoblast, the trabeculae, extending between the two and representing forerunners of the anchoring villi.

During this process, a new type of syncytial trophoblast arises that resembles the syncytiotrophoblast of the definitive placenta. It forms a unilaminar microvillous polarized epithelium that lines the lacunae, and represents a transition from the early invasive form to one suited for absorption and transport processes (Enders 1989).

At its outer surface, the original syncytial trophoblast erodes into the maternal capillaries within the superficial endometrium, which are dilated in the vicinity of the implantation site. Consequently, erythrocytes are released into the lacunae, although they are surprisingly few in number and have pale-staining characteristics (Hertig et al. 1956; Hamilton and Boyd 1960). Movement, if any, through the lacunae must be like a slow venous ebb and flow (Hamilton and Boyd 1960), and does not represent an effective maternal arterial circulation. The early syncytial trophoblast

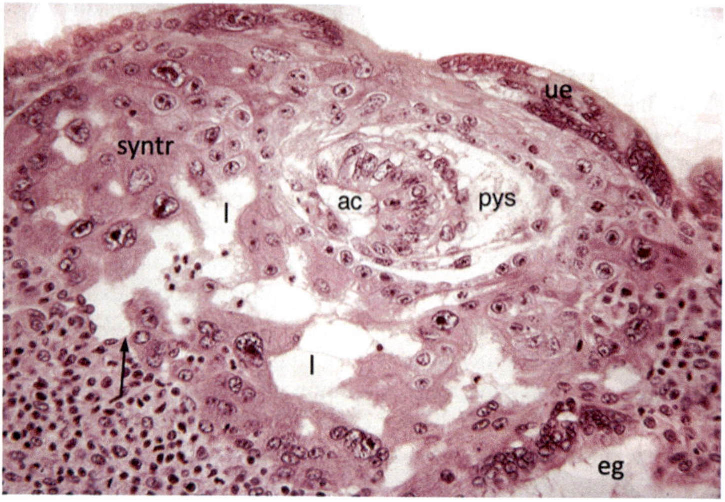

Fig. 2 Photomicrograph of a human conceptus at Carnegie stage 5B. The uterine epithelium (ue) has almost completely regrown over the implantation site. Fluid-filled spaces, the lacunae (l), have developed in the syncytiotrophoblast (syntr), and contain occasional erythrocytes released from an endometrial capillary eroded by the trophoblast (arrow). The syncytiotrophoblast also appears to be eroding into an endometrial gland (eg). The inner cell mass has differentiated into an epiblast layer that is continuous with the amnion, enclosing the amniotic cavity (ac), and hypoblast layer that grows around the inside of the cytotrophoblast cells (asterisk) to form the primary yolk sac (pys). (Photomicrograph courtesy of Prof. AC Enders and the Carnegie Collection)

also surrounds neighboring uterine glands (Hertig and Rock 1941; Enders 1989). The gland epithelial cells appear degenerate at these sites, establishing communications with the lacunae and enabling the continuance of histotrophic nutrition despite the interstitial form of implantation.

During this stage of development, cells derived from the hypoblast layer of the embryonic germ disc grow round the inner surface of the cytotrophophoblast cells, forming the primary yolk sac. Shortly after, extraembryonic mesoderm differentiates between the two tissues. The origin of the extraembryonic mesoderm in the human is still uncertain. Early reports suggested that it arises by delamination from the inner surface of the trophoblast (Hertig and Rock 1941), but the fact that the trophoblast rests on a well-defined basement membrane (Knoth and Larsen 1972) renders this unlikely. An alternative suggestion was that the mesoderm results from gastrulation that occurs at the most caudal end of the primitive streak before the main streak is established, the cells migrating over the inner surface of the trophoblast rather than penetrating the germ disc (Luckett 1978). Studies of macaques indicated an origin from the hypoblast layer as a further possibility (Enders and King 1988). Recent support for this observation comes from molecular analyses demonstrating that the two tissues share a number of common markers, in particular high expression of *GATA4* and *GATA6* (Nakamura et al. 2016). Although the precise origin remains uncertain, present evidence indicates that the extraembryonic mesoderm arises from the hypoblast layer or jointly from the epiblast and hypoblast layers (Boss et al. 2018). Once differentiated, extraembryonic mesoderm rapidly extends over the inner surface of the trophoblast. Henceforth, the two layers are referred to as the chorion, and the blastocyst as the chorionic sac.

Around day 12 p-f, the cytotrophoblast cells lying beneath the syncytiotrophoblast proliferate and penetrate into the trabeculae as cellular columns. Approximately 2 days later the cells reach the tips of the trabeculae, whereupon they extend through the syncytiotrophoblast and spread laterally, coalescing with neighbors to form a new layer interposed between the syncytiotrophoblast and the endometrium, the cytotrophoblastic shell (Boyd and Hamilton 1970; Burton and Jauniaux 2017). The importance of the shell will be considered later.

5 Early Villus Development

Continued cytotrophoblast cell proliferation results in side branches forming on the trabeculae and protruding into the lacunae. These branches represent primary villi, and are composed of a cytotrophoblast cell core with a covering of syncytiotrophoblast. Their presence marks the beginning of the villous stages of placentation. Further proliferative activity, with branching of the primary villi, initiates the development of primitive villous trees. As the trees are derived from the former trabeculae, they are always continuous with the developing chorionic plate. At the same time, the lacunar system is, by definition, transformed into the intervillous space.

After a further 2 days, around day 15 p-f, mesenchymal cells derived from the extraembryonic mesoderm lining the developing chorionic plate invade the primary villi, transforming them into secondary villi. Within a few days, the mesenchyme extends towards the villous tips, but never reaches as far as the cytotrophoblastic shell. Rather, the distal segments of what are now referred to as anchoring villi remain as cytotrophoblast cell columns. These columns are largely devoid of a covering of syncytiotrophoblast. A stem/progenitor cell niche persists at the proximal end of each column adjacent to the mesenchymal core where the cytotrophoblast cells continue to proliferate (Hemberger et al. 2010; Lee et al. 2018). Daughter cells differentiate under the influence of Notch signaling (Haider et al. 2016), and lose their mitotic potential as they move distally along the column. Ultimately they feed into the shell. The length of the columns reduces as pregnancy advances, reflecting the decline in proliferative potential of the cytotrophoblast cells after the end of the first trimester (Hemberger et al. 2010; Soncin et al. 2018).

Beginning between days 18 and 20 p-f, the first fetal capillaries can be observed within the mesenchyme of the secondary villi and the developing chorionic plate. They form in situ from hemangioblastic progenitor cell clusters, which in turn differentiate from the mesenchyme (Dempsey 1972; Demir et al. 1989; Robin et al. 2009; Aplin et al. 2015). The same progenitor cells give rise to groups of hematopoietic stem cells that form nucleated erythrocytes surrounded by the

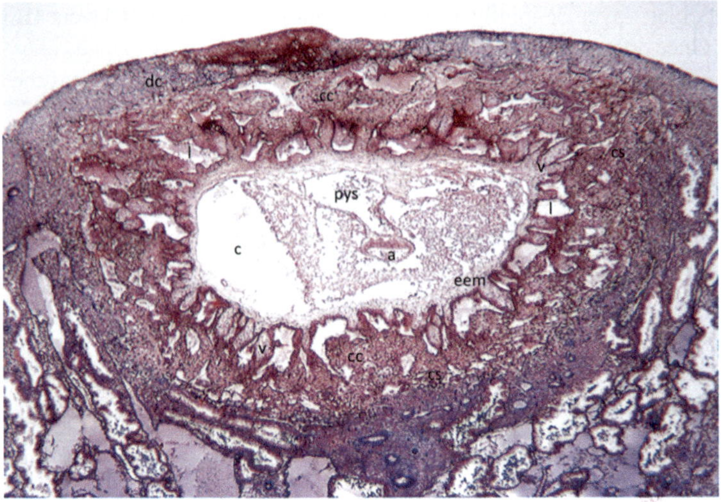

Fig. 3 Photomicrograph of a human conceptus at Carnegie stage 7, approximately 15–17 days p-f. The decidua capsularis (dc) now covers the conceptus which lies in the superficial endometrium. Villi (v) cover the entire surface of the chorionic sac. The villi are at the secondary stage, having a covering of trophoblast and a core of extra-embryonic mesenchyme (eem). They are separated by the lacunae (l). At their tips the villi are continued as a cytotrophoblast cell column (cc) of extravillous trophoblast cells, and these merge with neighbors at the maternal–fetal interface to create the cytotrophoblast shell (cs). By comparison, the embryo within the coelomic cavity (c) is relatively poorly developed, comprising a columnar epiblast layer that is continuous with the amnion (a) and the hypoblast that is continuous with the primary yolk sac (pys). (Photomicrograph courtesy of Prof. AC Enders and the Carnegie Collection)

developing endothelium. The appearance of capillaries in the villous mesenchymal stroma marks the development of the first tertiary villi.

Initially, villi form over the complete surface of the chorionic sac creating the chorion frondosum (Fig. 3), and by the end of this period the essential framework of the placenta is established.

6 The Cytotrophoblastic Shell

The cytotrophoblastic shell is best developed in early pregnancy when it is several cells thick and provides a means for rapid circumferential expansion of the implantation site (Hamilton and Boyd 1960). The cells have a distinctive phenotype, being rounded and containing large amounts of glycogen (Figs. 3 and 4). The shell helps create a unique microenvironment in which placental development is stimulated, while at the same time protecting the embryo against the potentially teratogenic effects of reactive oxygen species during the critical phase of organogenesis (Jauniaux et al. 2003a). In particular, when the outer surface of the shell encounters

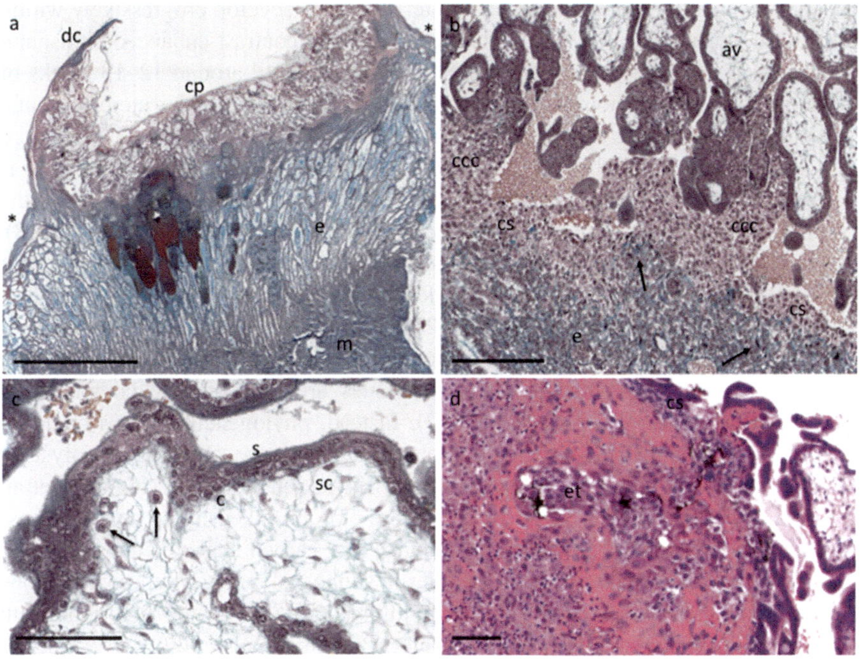

Fig. 4 Photomicrographs of a developing placenta at approximately 4 weeks p-f. (**a**) The chorionic sac is covered by the decidua capsularis (dc) and projects into the uterine lumen above the general line of the endometrium (e) which is asterisked. The chorionic sac has been opened on the superficial aspect and the embryo removed. Villi can be seen arising from the developing chorionic plate (cp). There has been hemorrhage into some of the endometrial glands. m, myometrium. (**b**) The mesenchymal stromal core does not extend to the tips of the anchoring villi (av) which terminate as the cytotrophoblast cell columns (ccc). Cytotrophoblast cells proliferate at the proximal ends of the columns, differentiate into extravillous trophoblast as they move down the column, and merge with neighbors to form the cytotrophoblastic shell (cs) at the maternal–fetal interface. At the maternal surface of the shell, cells undergo an epithelial-mesenchymal transition to form darkly staining spindle-shaped interstitial trophoblast (arrowed) that invades into the endometrium (e). (**c**) The epithelial covering of the villi comprises an outer multinucleated layer of syncytiotrophoblast (s) and an underlying layer of uninucleate villous cytotrophoblast progenitor cells (c). The mesenchymal cells of the core are loosely packed and sail like processes form fluid-filled stromal channels (sc) in which villous macrophages (arrowed) are often observed. (**d**) Specimen at 9 weeks p-f showing a spiral artery outlined by deposition of pink-staining fibrinoid plugged by endovascular trophoblast (et). Ink injected into the uterine artery can be seen within the intercellular spaces reaching the cytotrophoblastic shell (cs), indicating the possibility of a slow seepage of plasma into the intervillous space. Scale bars; 5 mm, 250 μm, 100 μm, and 100 μm respectively. Stain **a–c** Masson's trichrome, **d** hematoxylin, and eosin

the tip of a maternal spiral artery, cells migrate down the lumen of the artery as endovascular trophoblast. The volume of this migration is such that the trophoblast effectively "plugs" the mouths of the spiral arteries, restricting any inflow into the placenta to a slow seepage of plasma and largely excluding erythrocytes (Fig. 4d) (Hamilton and Boyd 1960; Hustin et al. 1988; Burton et al. 1999; Saghian et al.

2019). Towards the end of the first trimester, channels develop progressively within the plugs (Burton et al. 1999; Roberts et al. 2017), but contrast-enhanced ultrasound has revealed that flow does not significantly increase until around 12–13 weeks of gestation. Thus, it is thought that more proximal segments of the utero-placental vasculature may be rate limiting, in particular the radial arteries that do not undergo remodeling until the end of the first trimester (Roberts et al. 2017). This view is supported by computational models of the utero-placental vasculature that reveal the radial arteries to be a site of major resistance within the network (Clark et al. 2018). Overall, the effect is that placental development from implantation until the end of the first trimester occurs in a physiological low-oxygen environment of approximately 20 mmHg (Jauniaux et al. 2000, 2001).

This environment should not be considered hypoxic as has sometimes been the case, for there is no evidence that the tissues are energetically compromised (Cindrova-Davies et al. 2015; Gorr 2017). Instead, phylogenetically old carbohydrate metabolic pathways involving polyols are highly active, allowing glycolysis to continue at a high rate without reliance on fermentation to lactate (Jauniaux et al. 2005; Burton et al. 2017). These pathways are closely linked to the pentose phosphate pathway, and have the added advantage that carbon skeletons are preserved and can be used for synthesis of nucleotides in support of cell proliferation. The one requirement is a rich supply of glucose, and this is provided by the histotroph from the uterine glands, as will be discussed later. Polyols are also powerful osmolytes and may facilitate the drawing of water across the placenta to expand the coelomic and amniotic cavities.

Plugging of the spiral arteries during early pregnancy appears essential, and has also been observed in the rhesus macaque (Ramsey and Donner 1980). Poor development of the shell is associated with early onset of maternal blood into extensive areas of the placenta and spontaneous miscarriage, irrespective of the embryonic karyotype (Hustin et al. 1990; Jauniaux et al. 2003b). High levels of oxidative stress are observed in the placental tissues along with widespread degeneration of the syncytiotrophoblast (Hempstock et al. 2003b). Mechanical disruption of the maternal–fetal interface through intrauterine subchorionic hematomas may also lead to focal oxidative stress and weakening of the placental membranes (Johns et al. 2006; Burton and Jauniaux 2017).

7 Placental Development During the First Trimester

Once the main framework of the placenta is established, there is a period of rapid growth, far in excess of that of the embryo. New villi are formed from the lateral aspects of the anchoring villi as a result of villous sprouts that arise from the syncytiotrophoblast (Burton and Jones 2009). Initially, these club-shaped projections contain only syncytioplasm, but may contain aggregates of euchromatic nuclei in the expanded head. The proximal end is invaded by first cytotrophoblast and then by mesenchymal cells so that they undergo the same primary, secondary and tertiary stages seen following implantation. The original anchoring villi form the stem villi of the definitive placenta, and the new villi form progressively smaller branches, generations of intermediate villi that are free-floating. Terminal villi start to differentiate only after 20 weeks of gestation (Jackson et al. 1992), and so are not seen at this stage.

Villi during the first trimester have a low surface area–to-volume ratio, are relatively poorly vascularized, and so not well adapted to facilitate exchange (Fig. 4c). The trophoblast epithelium is two layered, with an outer layer of syncytiotrophoblast and a complete underlying layer of cytotrophoblast cells. The majority of these cells immunostain positively for CDX2, a marker of trophoblast stemness, and for Ki67, a proliferation marker, suggesting they are undergoing active division (Burton et al. 2020). This prolific expansion may be facilitated by the low oxygen concentration, which maintains embryonic stem cells in a pluripotent state (Lees et al. 2017), and stimulates trophoblast proliferation in first trimester primary cultures (Genbacev et al. 1996) and explant models (Caniggia et al. 2000).

The mesenchymal cells of the villous core are quite sparse, and have long sail-like processes that unite with neighbors to form a meshwork enclosing fluid-filled channels (Martinoli et al. 1984; Burton 1987). At the proximal end of the stem villi these channels appear to be in free communication with the coelomic cavity of the chorionic sac, and hence may facilitate diffusion of oxygen and nutrients into the cavity in the absence of an effective chorionic circulation. Villous macrophages, also referred to eponymously as Hofbauer cells, are frequently observed within the channels (Fig. 4c) (Castellucci et al. 1980). They are thought to arise initially by differentiation in situ from the hemangioblastic cultures (Demir et al. 1989), but once the feto-chorionic circulation is established at around 8–10 weeks of pregnancy they may be supplemented by bone marrow-derived monocytes. Semi-quantitative estimates indicate they represent approximately 40% of all stromal cells, both in early pregnancy and at term (Goldstein et al. 1988). They can be easily recognized by their large size, 10–35 μm in diameter, and rounded appearance with a highly vacuolated cytoplasm. It is thought they perform immune surveillance (Reyes et al. 2017), but may also have an importance in regulating placental morphogenesis. Later in pregnancy they are closely approximated to vascular structures and the trophoblast basement membrane, and are immunopositive for a number of growth factors (Benirschke et al. 2012).

During the first trimester the endometrial stromal cells undergo the decidual transformation. While this change may occur during the secretory phase of the non-pregnant cycle, it is very limited in its extent. In the early weeks of a pregnancy the transformation becomes more extensive, involving the endometrium around the entire uterine wall. The region beneath the developing placenta is henceforth referred to as the decidua basalis, and the remainder of the uterine lining as the decidua parietalis. The thin layer overlying the implantation site is the decidua capsularis. Decidual cells are characterized by their enlarged size and rounded shape, and secrete a number of factors, including prolactin, IGF-binding protein, and relaxin. The decidua plays an important role in modulating trophoblast invasion as evidenced by the deep, unregulated invasion seen at sites of ectopic implantation, such as in the fallopian tube, where decidua is absent. The decidua is also essential for supplying histotrophic nutrition to the conceptus during early pregnancy.

8 Histotrophic Nutrition

The communications between the uterine glands and the placenta seen at the time of implantation persist throughout the first trimester (Hamilton and Boyd 1960; Burton et al. 2002; Moser et al. 2015), enabling an extended period of histotrophic nutrition prior to onset of the maternal placental circulation. The epithelial cells lining the glands undergo a characteristic morphological change during early pregnancy, referred to as the Arias-Stella reaction (Arias-Stella 2002). They adopt a hypersecretory phenotype that appears to be endocrine mediated as it occurs even in ectopic pregnancies. Glycogen accumulates within the apical cytoplasm of the cells, and may be released through a combination of apocrine secretion and conversion into glucose through the actions of glycogen phosphorylase (Demir et al. 2002; Jones et al. 2015). Lipid droplets are also prominent in the secretions (Hempstock et al. 2004). The secretions are delivered into the intervillous space through openings in the cytotrophoblastic shell, and later the developing basal plate (Burton et al. 2002: #157; Hempstock et al. 2004). They disperse within the intervillous space and immunofluorescence studies confirm uptake by the syncytiotrophoblast where they co-localize with the lysosomal pathway (Hempstock et al. 2004). Breakdown of maternal proteins may provide the amino acids and elements required for anabolic pathways within the trophoblast similar to that seen in the rodent yolk sac (Brent and Fawcett 1998), but some, such as glycodelin-A, pass across the placenta intact and accumulate within the coelomic (chorionic) and amniotic cavities (Seppälä et al. 1992; Jauniaux and Gulbis 2000).

The glands are also an important source of growth factors, including epidermal growth factor (EGF), vascular endothelial growth factor, and leukemia inhibitory factor (Hempstock et al. 2004). Application of EGF to first trimester villous explants stimulates proliferation of the cytotrophoblast cells (Maruo et al. 1992), and hence it has been proposed that the histotroph plays an important role in stimulating placental proliferation and differentiation as in domestic species (Filant and Spencer 2014; Burton 2018; Burton et al. 2020). Obtaining proof of an equivalent servo-mechanism by which hormones from the placenta upregulate expression of growth factors within the glands is difficult in the human, but the presence of the Arias-Stella reaction provides support. In addition, the recent demonstration that early pregnancy hormones, such as human chorionic gonadotropin and prolactin, upregulate secretion of glycodelin-A and osteopontin from uterine gland organoids is strongly indicative (Turco et al. 2017). Prolactin is not secreted by the trophoblast in the human as in other species (Carter 2012), but by the decidua. Correct decidualization and subsequent paracrine signaling to the glands may therefore be essential for placental development and a successful pregnancy (Conrad et al. 2017).

9 The Amnion

The amnion arises from the epiblast layer of the inner cell mass (Fig. 3). The cells form a cluster, or rosette, polarize, and then a central cavity emerges (Deglincerti et al. 2016; Shahbazi et al. 2016), in a similar fashion to that observed in the rhesus monkey (Enders et al. 1986). The cells of the epiblast remain columnar, while those of the amnion are more squamous. The sac gradually expands as gestation advances, extending on to the ventral surface of the embryo following folding of the latter, and sheathing the connecting stalk to form the umbilical cord. The amnion finally makes contact with the chorionic plate at the end of the third month, obliterating the coelom.

10 Formation and Role of the Yolk Sac

The primary yolk sac is formed by hypoblast cells growing around the inner surface of the trophoblast, but later undergoes extensive remodeling when the layer of extraembryonic mesoderm interposed between the two vacuolates and splits into two layers. The outer layer lines the trophoblast and contributes to the chorion as described earlier, and the inner layer covers the yolk sac, forming its outer mesothelial layer. The space between the two layers is the coelomic cavity. At the same time, the yolk sac reduces in size as the more peripheral portions are nipped off, forming the secondary yolk sac. This sac reaches a maximum diameter of 6–7 mm between the sixth and tenth weeks of gestation, after which it decreases.

The secondary yolk sac is connected to the gut tube of the embryo by the vitelline duct, and is surrounded by the coelomic fluid. Morphologically it consists of two epithelial layers, an outer mesothelial layer and an inner endodermal layer, separated by a small amount of mesenchyme in which the vitelline vessels differentiate. The mesothelial layer is formed of flattened cells that display all the features of an absorptive epithelium, with a dense covering of microvilli, coated pits, and pinocytotic vesicles (Gonzalez-Crussi and Roth 1976; Jones and Jauniaux 1995). By contrast, the endodermal cells are more columnar and contain large quantities of rough endoplasmic reticulum, Golgi bodies and secretory droplets, indicative of a synthetic function. It is notable that the vitelline capillary plexus lies closely approximated to mesothelial epithelium.

The secondary human yolk sac fails to contact the chorion (Fig. 5), and a choriovitelline placenta is never formed as in other species. Hence, the yolk sac has largely been considered vestigial, although it is recognized as an important early site of hematopoiesis in the embryo (Pereda and Niimi 2008). It is also known to synthesize several key serum proteins, including alpha-fetoprotein, alpha$_1$-antitrypsin, albumin, and transferrin, prior to the fetal liver having differentiated sufficiently to take on this role (Buffe et al. 1993; Jones and Jauniaux 1995). Recently, the potential importance of the yolk sac for transfer of nutrients has been

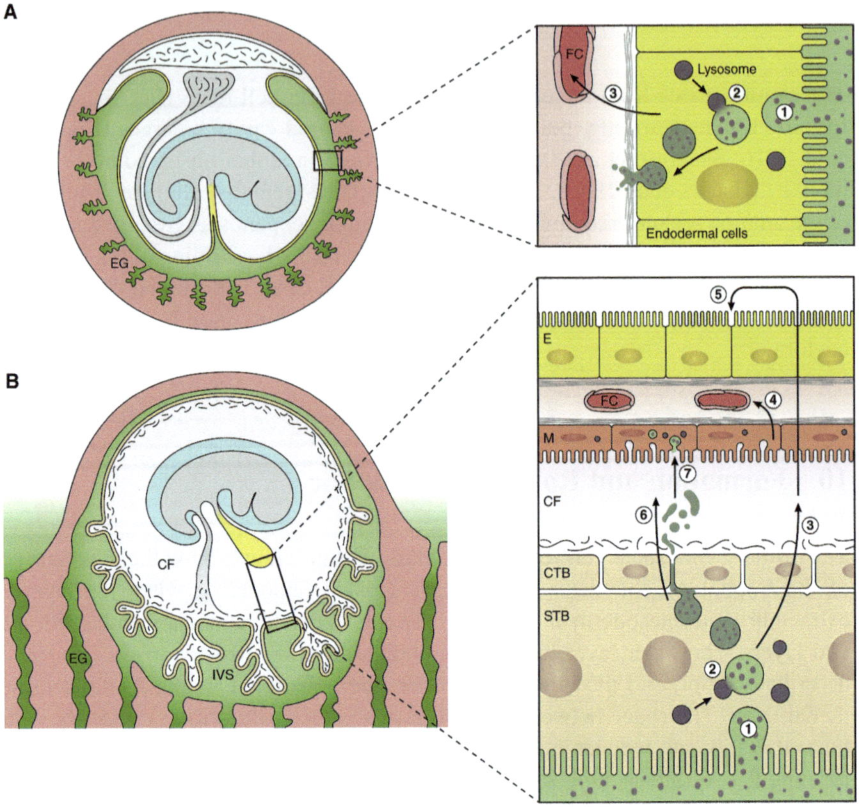

Fig. 5 Diagrammatic comparison of the nutrient pathway during early pregnancy in the mouse (**a**), and the speculated pathway in the human (**b**). In the mouse, histotroph from the endometrial glands (EG) is phagocytosed (1) by the endodermal cells (E) of the visceral layer of the inverted yolk sac (YS). Following fusion with lysosomes (2), digestion of maternal proteins leads to release of amino acids that are transported (3) to the fetal circulation (FC). In the human, histotroph released into the intervillous space (IVS) is phagocytosed (1) by the syncytiotrophoblast (STB). Following digestion by lysosomal enzymes (2), free amino acids may be transported (3) by efflux transporters to the coelomic fluid (CF) where they accumulate. Nutrients in the CF may be taken up by the mesothelial cells (M) of the yolk sac and transported (4) into the fetal circulation (FC). Alternatively, they may diffuse into the cavity of the yolk sac and be taken up by the endodermal cells (5). Some intact maternal proteins may also be released into the CF by exocytosis of residual bodies (6), and be engulfed by the mesothelial cells (7). *CTB* cytotrophoblast cells. (Reproduced with permission from (Cindrova-Davies et al. 2017))

highlighted by RNAseq data that confirm the presence of transcripts encoding a broad array of transporter proteins for lipids, amino acids, metal ions, and other micronutrients (Cindrova-Davies et al. 2017). The conservation of the majority of these transcripts across the mouse and chicken where there is strong experimental evidence of the role of their encoded proteins in transport indicates that the human yolk sac may be more active than previously appreciated. In addition, immunohistochemistry confirms the presence of transporter proteins, such as tocopherol transfer protein, GLUT1, and transferrin, which are primarily localized to the outer mesothelial layer (Jauniaux et al. 2004; Benirschke et al. 2012). The presence of the ligands for many of the transporters in the coelomic fluid bathing the epithelium adds further support for a potential role of the yolk sac in maternal–fetal transport (Cindrova-Davies et al. 2017).

It has been proposed, therefore, that the human yolk sac contributes to a physiological choriovitelline placenta during early pregnancy (Fig. 5), with the coelomic cavity interposed between the trophoblast and the yolk sac acting as an intermediary nutrient reservoir (Cindrova-Davies et al. 2017). From the evidence available it would appear to be most important for the handling and metabolism of cholesterol and lipids, which are essential as signaling intermediates and for formation of cell and organelle membranes (Woollett 2008). By the ninth week of pregnancy the yolk sac begins to show morphological evidence of a decline in function, and when the coelomic cavity is obliterated at around 15 weeks the remnants of the yolk sac become incorporated into the umbilical cord. Abnormal development of the yolk sac has been linked with miscarriage (Nogales et al. 1993; Freyer and Renfree 2009), but whether this is cause or effect is unknown at present.

11 The Extravillous Trophoblast

As the name implies, extravillous trophoblast (EVT) extends beyond the confines of the villous trees, invading into the decidua where it plays an essential role in remodeling of the maternal spiral arteries. The EVT originates from the cytotrophoblastic cell columns and the cytotrophoblastic shell as previously described. Cells on the outer surface of the shell undergo an epithelial-mesenchymal transition (EMT), changing shape from rounded profiles to pleiomorphic spindle-shaped cells that stain darkly with hematoxylin and eosin and express placenta-specific protein 8, PLAC8, that promotes invasiveness (Fig. 4c) (Chang et al. 2018). The stimulus for the transition is not known, but cells in the distal part of the column are immunopositive for CDCP1, which represses epithelial and promotes mesenchymal states (Wong et al. 2019). The transcription factor ZEB2 has recently been shown to potentially play a key role in regulating the transition, for overexpression in BeWo and JEG3 choriocarcinoma cell lines stimulates a transcript profile indicative of an EMT and increases cell invasiveness (DaSilva-Arnold et al. 2019). Whether expression of ZEB2 is responsive to the increased concentration of oxygen in the decidua or factors secreted by the decidual cells is not known at present. Measurements of DNA suggest that invasive extravillous trophoblast cells undergo endoreduplication and are tetraploid (Zybina et al. 2002, 2004; Velicky et al. 2018), and at the same time start to express markers of senescence.

Extravillous trophoblast cells migrate via two routes; the endovascular down the lumen of the arteries and the interstitial through the endometrial stroma where they cluster around the spiral arteries and the endometrial glands. Invasion of the interstitial trophoblast is extensive, and in a normal pregnancy they reach the inner third of the myometrium where they fuse with neighbours to form multinucleated placental bed giant cells (Al-Lamki et al. 1999). No active function has been attributed to these cells, although they are immunopositive for human placental lactogen, subunits of hCG (Al-Lamki et al. 1999), and members of the angiopoietin and vascular endothelial growth factor families (Schiessl et al. 2009). Potentially, they may promote regeneration of the endometrium following delivery.

The regulation of the invasion is complex, and numerous factors have been shown to influence invasion positively and negatively in vitro (Benirschke et al. 2012). In reality, it will depend on the local concentration of cytokines, glycoproteins, and possibly oxygen, within the decidua (Lee et al. 2011; Pollheimer et al. 2018). An important source of cytokines are cells of the maternal immune system. There is strong genetic evidence that HLA-C ligands on the surface of the interstitial cells bind with killer-like immunoglobulin receptors (KIR) on the maternal uterine natural killer (uNK) cells, causing activation of the latter (Hiby et al. 2010). Despite the terminology, there is no evidence of killing of the trophoblast, but rather the converse. Cytokines released from the uNKs, in particular GM-CSF, attract trophoblast cells (Abbas et al. 2017), and are thought to play an important role in remodeling of the spiral arteries (Moffett et al. 2015).

Interstitial trophoblast cells express a unique hyperglycosylated form of human chorionic gonadotropin (hCG-H), which functions as a cytokine rather than a hormone (Cole 2010; Evans 2016). hCG-H is a powerful stimulus for trophoblast invasion, acting in an autocrine fashion through the TGF-ß receptor rather than the LH receptor as for normal hCG. As might be expected, concentrations in maternal serum are highest during early pregnancy, and are significantly reduced in cases that go on to early pregnancy loss (Cole 2007; Guibourdenche et al. 2010). Monitoring hCG-H may therefore provide an assay of extravillous trophoblast numbers, if not invasion.

12 Remodeling of the Spiral Arteries

Connecting the placenta to the maternal arterial system poses a hemodynamic challenge due to the high pressure and velocity within the latter. The spiral arteries that ultimately supply the placenta must therefore undergo considerable remodeling to minimize the risks of damage to the delicate villous trees and mechanical disruption of the maternal–fetal interface. Remodeling, also referred to as physiological conversion or transformation, involves the loss of smooth muscle and the elastic lamina from the walls of the vessels, and its replacement with an inert fibrinoid material (Fig. 4d). These changes normally involve the endometrial portion of the artery and the portion within the inner third of the myometrium.

Remodeling starts with an endocrine priming that is associated with decidualization and affects all spiral arteries, even those in the decidua parietalis on the opposite wall of the uterus. The endothelial cells swell and there is a loosening of the smooth muscle cells (Pijnenborg et al. 2006; Whitley and Cartwright 2010; Harris 2010). More extensive changes require the presence of extravillous trophoblast, and involve the de-differentiation of the vascular smooth muscle cells and their migration away from the vessel wall. Apoptosis of the smooth muscle cells has been suggested, but it is not thought to play a major role (Bulmer et al. 2012). Initially, the endothelial cells are also lost and replaced by endovascular trophoblast, but later the vessels are re-endothelialized by maternal cells that grow along the vessels and can extend onto the inner surface of the placental basal plate (Ockleford 2010).

Remodeling has two important consequences for placental blood flow. Firstly, it is associated with dilation of the terminal parts of the vessels as they approach the basal plate (Harris and Ramsey 1966). Mathematical modeling reveals that the dilation reduces the velocity of the inflowing blood by an order of magnitude to ~10 cm/s, allowing sufficient transit time for effective maternal–fetal exchange. At the same time, it reduces the pressure in the intervillous space and preventing compression and collapse of the fetal capillary network within the villi (Burton et al. 2009). Secondly, by inactivating the hypercontractile segment of a spiral artery that lies in the junctional zone just beneath the endometrial–myometrial boundary, remodeling ensures a constant flow of blood irrespective of maternal vasomotor events. The hypercontractile segment normally contracts at the time of menstruation to prevent excessive blood loss, but contraction during pregnancy could impair the maternal–fetal supply line and lead to fluctuations in the intraplacental oxygen concentration. Experiments in vitro have demonstrated that hypoxia-reoxygenation is a powerful inducer of oxidative stress in placental tissues (Hung and Burton 2006), more so than hypoxia alone, and can recapitulate many of the changes seen in pre-eclampsia (Cindrova-Davies et al. 2007).

Failure of remodeling is associated with the most common complications of human pregnancy, including growth restriction, pre-eclampsia and pre-term delivery (Brosens et al. 2011). The effects are mediated by a combination of reduced villous development and surface area for exchange, oxidative stress, and release of pro-inflammatory factors from the placenta (Kingdom et al. 2018; Burton and Jauniaux 2018; Burton et al. 2019).

13 The Formation of the Definitive Placenta

Starting at ~eighth week of pregnancy, the villi over the superficial pole of the chorionic sac begin to appear shorter and less vascularized compared to those over the deep pole in contact with the decidua basalis (Fig. 6) (Hamilton and Boyd 1960). This is a progressive effect, so eventually only "ghosts" of villi comprising an avascular and almost acellular villous core with a thin covering of trophoblast remain in this area (Fig. 7). In the past this regression of the villi to form the chorion laeve or smooth membranes has been attributed to a lack of blood supply reaching them from the decidua capsularis. However, it is notable that the villi are often surrounded by masses of maternal erythrocytes, and an alternative hypothesis has been put forward

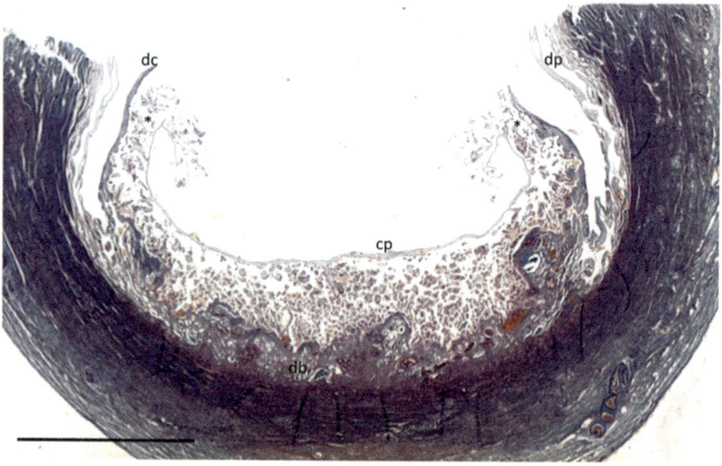

Fig. 6 Photomicrograph of a specimen at 6 weeks p-f. The chorionic sac covered by the decidua capsularis (dc) projects into the uterine lumen, and has been opened to remove the embryo. The decidua capsularis has not yet fused with the decidua parietalis (dp) lining the rest of the uterine wall. Villi (asterisk) towards the superficial pole of the chorionic sac are less extensive than those over the deep pole in contact with the decidua basalis (db). The interface with the decidua basalis is very uneven at this stage. Glands within the decidua are still active. cp, chorionic plate. Scale bar, 10 mm. Stain, Masson's trichrome

relating to the onset of the maternal arterial circulation. In normal pregnancies, onset of the circulations starts preferentially in the peripheral parts of the placenta where the endovascular aggregates in the spiral arteries are least extensive (Jauniaux et al. 2003b). Villi sampled from this region display high levels of oxidative stress and activation of the apoptotic cascade compared to their central counterparts (Jauniaux et al. 2003b; Burton et al. 2010). Thus, it has been proposed that the high local levels of oxygen induce villous regression through oxidative stress and not a lack of maternal circulation.

There are pathological parallels to this pattern of regression, for in cases of spontaneous miscarriage onset of the maternal circulation is both precocious and disorganized, occurring in both the peripheral and central regions (Jauniaux et al. 2003b). As mentioned previously, there is widespread oxidative damage to the trophoblast and other tissues, and if the placenta remains in the uterus for a period of time the villi become avascular collagenous ghosts that closely resemble those of the chorion laeve (Jauniaux et al. 2003b; Hempstock et al. 2003b). These findings support the concept that under both physiological and pathological conditions, elevated levels of oxygen cause villous regression.

Whatever the mechanism, it appears that the shape of the placenta and the centrality or otherwise of the cord insertion seen at term are determined by events taking place at the end of the first trimester (Fig. 7b) (Schwartz et al. 2011; Salafia et al. 2012).

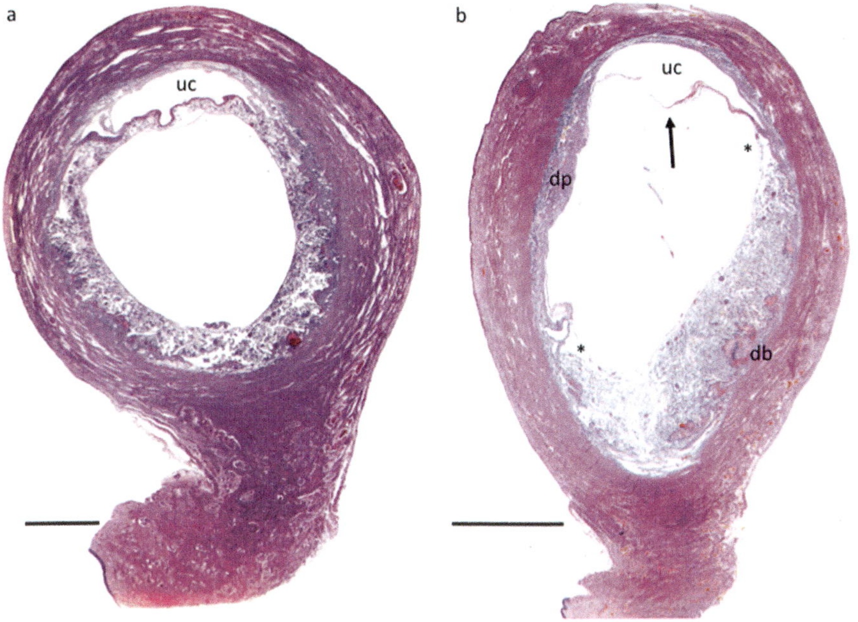

Fig. 7 Photographs of placenta-in-situ at (**a**) 7 weeks, and (**b**) 12 weeks p-f. (**a**) Villi covering the superficial pole of the chorionic sac are now less extensive and avascular compared to those over the deep pole. The decidua capsularis has not yet fused with the decidua parietalis and a small part of the uterine cavity (uc) remains. (**b**) By now the villi have fully regressed, creating the chorion laeve or smooth membranes (arrowed) that are draped over the decidua parietalis (dp) of the uterine wall opposite. The extent of the definitive placental disc is marked by the asterisks. The meshwork of placental villi within the disc is finer and denser than earlier, and the decidua basalis (db) is thinner and more even. A small portion of the uterine cavity (uc) persists. Stain, hematoxylin and eosin

14 Growth of the Definitive Placenta

Once established, the definitive placenta continues to expand over the second and third trimesters. New stem and intermediate villi are generated through the transformation of syncytial sprouts as mentioned earlier, and slowly increase in volume across gestation (Jackson et al. 1992). By contrast, elaboration of the functional units of the placenta, the terminal villi, increases exponentially from ~20 weeks until term (Fig. 8a) (Jackson et al. 1992). This elaboration ultimately provides a villous surface area of 12–14 m^2, with a diffusion distance of 7–8 μm separating the maternal and fetal circulations (Burton and Jauniaux 1995). The theoretical diffusing capacity of the placenta can be calculated from these parameters, and if expressed per kg of fetal weight the value stays remarkably constant across gestation (Mayhew et al. 1993). This finding suggests a close relationship between the development of the fetus and the structural parameters of the placenta.

The chorionic and basal plates become better defined as gestation advances. The vessels in the chorionic plate distributing umbilical blood flow to the stem villi

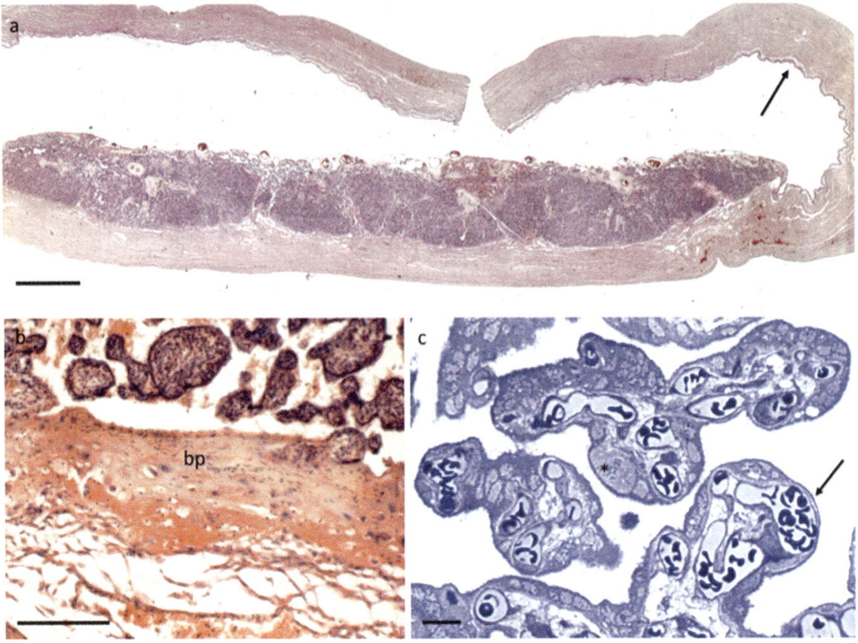

Fig. 8 Photographs of (**a**) and (**b**) placenta-in-situ at ~30.5 weeks p-f, and (**c**) placental villi at term. (**a**) The density of the villus trees continues to increase with the elaboration of intermediate and terminal villi. Infoldings of the basal plate create septae that partially separate the placenta into a series of lobes. The chorion laeve (arrowed) is apposed to the decidua parietalis. (**b**) Higher power photomicrograph of the basal plate (bp), which comprises a complex mix of extravillous trophoblast arising from the original cytotrophoblastic shell and decidual cells enmeshed in fibrinoid material. Under the basal plate is a loose meshwork of tissue that acts as a plane of cleavage at the time of delivery. (**c**) Photomicrograph of terminal villi from a term placenta embedded in resin and sectioned at 1 μm. The fetal capillaries display localized dilations that bring the endothelium close to the synctiotrophoblast to form a vasculo-syncytial membrane (arrowed). The distance between the maternal and fetal circulation is reduced to ~2 μm at these sites, which facilitate diffusional exchange. Scale bars; (**b**) 200 μm, (**c**) 20 μm. Stains; (**a**) and (**b**) hematoxylin and eosin, (**c**) methylene blue

increase in caliber (Fig. 8a). The decidua basalis continues to thin, and the remaining decidual cells become enmeshed in a fibrinoid matrix with the remnants of the cytotrophoblastic shell to form the basal plate (Fig. 8b). Infoldings of the basal plate form septa that subdivide the placenta into a series of lobes; each lobe may contain one of more lobules as below. The septa do not extend as far as the chorionic plate, but nonetheless may serve to direct and partially compartmentalize maternal blood flow (Ramsey and Donner 1980).

Despite remodeling of the spiral arteries, the maternal blood still enters the placenta in jet-like spurts that gradually diminish in velocity as gestation advances (Collins et al. 2012). The momentum of the blood is sufficient to sculpt the villous trees into a series of lobules, each with a central villus-free cavity located over the opening of an artery. Once delivered into that cavity, the maternal blood disperses through the clefts or pores between neighboring villi, before draining into the uterine veins. Each lobule therefore acts as an independent maternal–fetal exchange unit, physiologically equivalent to a cotyledon of a ruminant placenta. Measurements of the expression and activity of antioxidant enzymes indicate that an oxygen gradient exists across a lobule, with an arterial centre and more venous periphery (Hempstock et al. 2003a). This pattern of blood flow and oxygenation has recently been confirmed using magnetic resonance imaging techniques (Hutter et al. 2020).

The elaboration of the villous trees is accompanied by an increase in the volume of the trophoblast, both syncytiotrophoblast and cytotrophoblast. Although markers of stemness in trophoblast progenitors decline at the end of the first trimester, proliferation and fusion continue, possibly involving a transit amplifying population or equivalent. Consequently, the number of nuclei in both subtypes increases until term (Simpson et al. 1992). However, as villous surface area enlarges rapidly, the cytotrophoblast layer becomes discontinuous, and cells are only occasionally observed in any one section, creating a false impression of their total number. This discontinuity contributes to the reduction in villous membrane thickness and facilitates diffusional exchange.

The majority (~80%) of the nuclei within the syncytiotrophoblast are transcriptionally active, as evidenced by incorporation of fluorouracil and immunopositivity for phosphorylated upstream binding factor (pUBF) and phosphorylated cAMP response element binding protein (pCREB) that are indicative of RNA polymerase I (RNA Pol I) and RNA pol II driven transcription respectively (Ellery et al. 2009; Fogarty et al. 2011). The proportion of active nuclei remains constant across gestation.

Some of the active nuclei are also immunopositive for proliferating cell nuclear antigen (PCNA), indicating that they have only recently entered the syncytiotrophoblast; PCNA has a half-life of more than 20 h (Fogarty et al. 2011). The chromatin pattern of the syncytial nuclei varies greatly, with those that are transcriptionally inactive generally showing greater heterochromatin. Aggregates of nuclei with extremely condensed chromatin, referred to as syncytial knots (Burton and Jones 2009), are seen increasingly towards term, and in the past it has been suggested that these represent apoptotic changes (Huppertz et al. 1999; Huppertz and Kaufmann 1999). However, the majority of the nuclei are negative for TUNEL staining (Coleman et al. 2013), and it is now accepted that apoptosis does not occur in the normal syncytiotrophoblast (Longtine et al. 2012a, b). Instead, it is thought that the chromatin condensation reflects epigenetic changes, for the nuclei in syncytial knots stain strongly for 5'hydroxymethylcytosine (Fogarty et al. 2015). They are also immunopositive 8-oxo-deoxyguanosine, indicative of oxidative damage (Fogarty et al. 2013), and so the following life cycle has been proposed. Nuclei entering the syncytiotrophoblast are transcriptionally active, but a proportion gradually acquire oxidative damage and are shut-down epigenetically. Whether this is a purely temporal phenomenon or responsive to the local microenvironment within the intervillous space is not known. The effete nuclei are aggregated into syncytial knots at sites on the villous surface where they do not impinge on placental functional

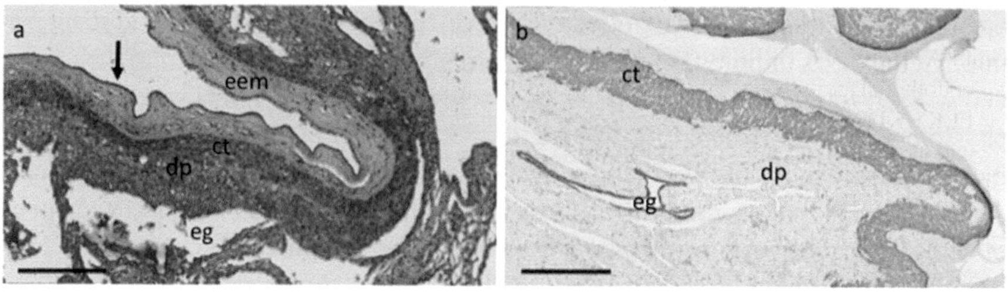

Fig. 9 Photomicrograph of placental membranes at (**a**) 20 weeks p-f, and (**b**) at term. (**a**) The membranes comprise the amniotic epithelium (arrowed), extraembryonic mesenchyme (eem), and a layer of extravillous trophoblast (ct) that abut the cells of the decidua parietalis (dp). Endometrial glands (eg) are still evident in the decidua parietalis at this stage. (**b**) Immunostaining against cytokeratin confirms the presence of the cytotrophoblast cells (ct) that abut the decidua parietalis which now only contains rudimentary endometrial glands (eg). Scale bars, 250 μm. Stain; (**a**) hematoxylin and eosin; (**b**) anti-pancytokertain

capacity, but the mechanism of aggregation is not known (Coleman et al. 2013; Calvert et al. 2016).

As the villi mature their stromal core becomes more condensed, and the channels seen during the first trimester are gradually obliterated (Fig. 8c). The fetal vasculature becomes more prominent, particularly in terminal villi where the capillaries may represent ~35% of the villous volume. The capillaries follow a tortuous course, forming a complex network with many interconnections (Jirkovska et al. 2002; Plitman Mayo et al. 2016a). The diameter of the capillaries varies along their length, with narrow sections interspersed with dilated regions, referred to as sinusoids, usually located on the point of an acute bend (Plitman Mayo et al. 2019). The dilation brings the outer wall of the capillary into close contact with the trophoblast basement membrane and the syncytiotrophoblast, which is thinned as a consequence of the distending pressure of the capillary. These areas are referred to as vasculosyncytial membranes, and the thickness of the villous membrane may be reduced to as little as 2 μm. Computational modeling reveals that they are the principal sites of diffusional exchange (Plitman Mayo et al. 2016b). Hence, the condition of terminal villus deficiency, when terminal villi fail to form, is associated with growth restriction and fetal hypoxemia (Khong et al. 2016).

With the enlargement of the chorionic sac, the decidua capsularis becomes apposed to the decidua parietalis on the opposite wall of the uterus (Fig. 7) at approximately weeks 10–12, and fuses. By now the villi have completely regressed over the chorion laeve, but the extravillous trophoblast of the original cytotrophoblastic shell in this region form a layer 5–10 cells thick (Fig. 9a). This layer stains immunopositively for HLA-G and cytokeratin and is apposed to the decidual cells, but no invasion takes place (Fig. 9b).

15 Placental Senescence

The concept of placental ageing and senescence is an old one (Martin and Spicer 1973; Rosso 1976), but has received renewed interest following the finding that cell fusion to form a syncytium induces senescence changes (Chuprin et al. 2013; Cox and Redman 2017). Senescence can also be induced by chronic oxidative stress, and as mentioned previously oxidatively damaged nuclei accumulate in the syncytiotrophoblast towards term in the form of syncytial knots (Fogarty et al. 2013). The damage is thought to arise from an increasing mismatch between maternal supply and feto-placental demand leading to fluctuations in the oxygen concentration within the intervillous space. Markers of senescence are variable in the normal term placenta, but oxidized DNA and lipids and the expression p21, p16, and cGAMP are all greatly increased in the post-mature placenta and in pathological cases (Maiti et al. 2017; Cindrova-Davies et al. 2018). Senescence and impairment of placental functions may therefore contribute to the increased risk of stillbirth and neonatal death in post-mature pregnancies.

16 Comparison with Other Primates

A comparison of placentation in the human with that across the order Primates is beyond the scope of this chapter, and detailed accounts can be found elsewhere (Mossman 1987; Wooding and Burton 2008; Carter and Pijnenborg 2011). Here we focus on two species of Old-World monkeys that have been extensively used for placental and reproductive research, the rhesus macaque and the baboon, and the great apes. Macroscopically, the structure and development of the placental villous trees are very similar across all these, but there are differences in the mode of implantation and invasion of the extravillous trophoblast, and the complexity of the immunological interactions.

Implantation in the macaque and baboon is superficial, and so no decidua capsularis forms and the blastocyst remains within the uterine lumen exposed to the endometrial secretions. Whether this influences the route for histotrophic nutri tion has not been addressed. In the baboon only one placental disc is formed at the site of implantation (Houston 1969), whereas in the macaque a bi-discoidal placenta is common, with a second placental disc developing 11–12 days later where the blastocyst impinges on the opposite uterine wall (Enders 2007; de Rijk and van Esch 2008). There is thus no equivalent of the regression of villi of the chorion frondosum to form the definitive placenta as in the great apes and human.

Nonetheless, where the trophectoderm contacts the uterine epithelium there is transformation into a primary syncytiotrophoblast that is initially invasive, and a progenitor population of cytotrophoblast cells. A thick cytotrophoblastic shell is formed in both the macaque and the baboon, and unlike in the human it has a sharp demarcation with the endometrium and persists throughout pregnancy (Enders et al.

2001). Endovascular trophoblast invasion is prolific in both, and aggregates of endovascular cells plug the spiral arteries in the macaque as in the human, restricting flow of maternal arterial blood into the placenta during early pregnancy (Ramsey and Donner 1980; Blankenship and Enders 2003). By contrast with humans, interstitial trophoblast is extremely limited, and there are no multinucleated giant cells in the placental bed (Pijnenborg et al. 1996). This calls into question the role of the interstitial trophoblast in humans, and it may be that they are more important for paracrine signaling to stimulate secretions from the endometrial glands than for arterial remodeling (Burton et al. 2020). Another significant difference is that remodeling of the spiral arteries is restricted to only the endometrial portion, and does not extend into the myometrial segment in the baboon (Pijnenborg et al. 1996). In the chimpanzee and gorilla, trophoblast invasion occurs via the endovascular and interstitial routes, and extends as far as the inner third of the myometrium, as in the human (Pijnenborg et al. 2011a, b).

These differences in extravillous trophoblast invasion are associated with marked coevolution of the KIR and HLA-C systems. Uterine NK cells are present in the placental bed of the rhesus macaque but HLA-C is absent on the trophoblast cells. An HLA-C1 epitope evolved in the orangutan from an *MHC-B* ancestor, and a C2 epitope is found in the gorilla, chimpanzee and human. The highest degree of complexity is seen in humans where the *KIR A* and *B* haplotypes have evolved (Parham and Moffett 2013; Moffett and Colucci 2015). Certain combinations of the *KIR* and *HLA-C* haplotypes predispose to complications of human pregnancy, depending on the degree of inhibition and activation of the KIR. Thus, the interactions between KIR-A/A and C2/C2 are predominantly inhibitory and associated with an increased risk of miscarriage, growth restriction, and pre-eclampsia (Hiby et al. 2010). Indeed, these immune interactions appear to regulate birth weight across the entire range, from growth restriction to macrosomia (Moffett et al. 2015), presumably acting through the degree of remodeling of the spiral arteries and the effect on utero-placental blood flow.

17 Conclusions

Placentation in the human and closely related great apes displays major morphological differences compared to other orders of mammals, in particular the degree of invasiveness of the trophoblast. Physiologically, there are greater similarities; the reliance on histotrophic nutrition during the period of organogenesis, the importance of the yolk sac during early pregnancy, and the large surface area and thin interhemal membrane to facilitate exchange. Human pregnancy is associated with a high rate of pregnancy complications, such as pre-eclampsia, and genetic analyses reveal that this predisposition may be a powerful driver of placental evolution towards a less-invasive state (Elliot and Crespi 2015).

Acknowledgements The authors gratefully acknowledge the contributions of their many colleagues to the ideas and evidence presented here. They are pleased to acknowledge the generosity of Professor Allen C Enders in making his photomicrographs of the Carnegie Collection freely available. They can be accessed at www.trophoblast.cam.ac.uk/Resources/enders.

References

Abbas Y, Oefner CM, Polacheck WJ, Gardner L, Farrell L, Sharkey A, Kamm R, Moffett A, Oyen ML (2017) A microfluidics assay to study invasion of human placental trophoblast cells. J R Soc Interface 14:20170131

Al-Lamki RS, Skepper JN, Burton GJ (1999) Are human placental bed giant cells merely aggregates of small mononuclear trophoblast cells? An ultrastructural and immunocytochemical study. Hum Reprod 14:496–504

Aplin JD, Whittaker H, Jana Lim YT, Swietlik S, Charnock J, Jones CJ (2015) Hemangioblastic foci in human first trimester placenta: distribution and gestational profile. Placenta 36:1069–1077

Arias-Stella J (2002) The Arias-Stella reaction: facts and fancies four decades after. Adv Anat Pathol 9:12–23

Benirschke K, Burton GJ, Baergen RN (2012) Pathology of the human placenta. Springer, Heidelberg

Bentin-Ley U, Horn T, Sjogren A, Sorensen S, Falck Larsen J, Hamberger L (2000) Ultrastructure of human blastocyst-endometrial interactions in vitro. J Reprod Fertil 120:337–350

Blankenship TN, Enders AC (2003) Modification of uterine vasculature during pregnancy in macaques. Microsc Res Tech 60:390–401

Boss AL, Chamley LW, James JL (2018) Placental formation in early pregnancy: how is the centre of the placenta made? Hum Reprod Update 24:750–760

Boyd JD, Hamilton WJ (1970) The human placenta. Heffer and Sons, Cambridge

Braude P, Bolton V, Moore S (1988) Human gene expression first occurs between the four- and eight-cell stages of preimplantation development. Nature 332:459–461

Brent RL, Fawcett LB (1998) Nutritional studies of the embryo during early organogenesis with normal embryos and embryos exhibiting yolk sac dysfunction. J Pediatr 132:S6–S16

Brosens I, Pijnenborg R, Vercruysse L, Romero R (2011) The "great obstetrical syndromes" are associated with disorders of deep placentation. Am J Obstet Gynecol 204:193–201

Buffe D, Rimbaut C, Gaillard JA (1993) Alpha-fetoprotein and other proteins in the huamn yolk sac. In: Nogales FF (ed) The human yolk sac and yolk sac tumours. Springer-Verlag, Heidleberg

Bulmer JN, Innes BA, Levey J, Robson SC, Lash GE (2012) The role of vascular smooth muscle cell apoptosis and migration during uterine spiral artery remodeling in normal human pregnancy. FASEB J 26:2975–2985

Burton GJ (1987) The fine structure of the human placenta as revealed by scanning electron microscopy. Scanning Microsc 1:1811–1828

Burton GJ (2018) The John Hughes memorial lecture: stimulation of early placental development through a trophoblast-endometrial dialogue. J Equine Vet 66:14–18

Burton GJ, Jauniaux E (1995) Sonographic, stereological and Doppler flow velocimetric assessments of placental maturity. Br J Obstet Gynaecol 102:818–825

Burton GJ, Jauniaux E (2017) The cytotrophoblastic shell and complications of pregnancy. Placenta 60:134–139

Burton GJ, Jauniaux E (2018) Pathophysiology of placental-derived fetal growth restriction. Am J Obstet Gynecol 218:S745–S761

Burton GJ, Jones CJ (2009) Syncytial knots, sprouts, apoptosis, and trophoblast deportation from the human placenta. Taiwan J Obstet Gynecol 48:28–37

Burton GJ, Jauniaux E, Watson AL (1999) Maternal arterial connections to the placental intervillous space during the first trimester of human pregnancy; the Boyd collection revisited. Am J Obstet Gynecol 181:718–724

Burton GJ, Watson AL, Hempstock J, Skepper JN, Jauniaux E (2002) Uterine glands provide histiotrophic nutrition for the human fetus during the first trimester of pregnancy. J Clin Endocrinol Metab 87:2954–2959

Burton GJ, Woods AW, Jauniaux E, Kingdom JC (2009) Rheological and physiological consequences of conversion of the maternal spiral arteries for uteroplacental blood flow during human pregnancy. Placenta 30:473–482

Burton GJ, Jauniaux E, Charnock-Jones DS (2010) The influence of the intrauterine environment on human placental development. Int J Dev Biol 54:303–312

Burton GJ, Jauniaux E, Murray AJ (2017) Oxygen and placental development; parallels and differences with tumour biology. Placenta 56:14–18

Burton GJ, Redman CW, Roberts JM, Moffett A (2019) Pre-eclampsia: pathophysiology and clinical implications. BMJ 366:l2381

Burton GJ, Cindrova-Davies T, Turco MY (2020) Review: histotrophic nutrition and the placental-endometrial dialogue during human early pregnancy. Placenta 102:21–26

Calvert SJ, Longtine MS, Cotter S, Jones CJ, Sibley CP, Aplin JD, Nelson DM, Heazell AE (2016) Studies of the dynamics of nuclear clustering in human syncytiotrophoblast. Reproduction 151:657–671

Caniggia I, Mostachfi H, Winter J, Gassmann M, Lye SJ, Kuliszewski M, Post M (2000) Hypoxia-inducible factor-1 mediates the biological effects of oxygen on human trophoblast differentiation through TGFbeta(3). J Clin Invest 105:577–587

Carter AM (2012) Evolution of placental function in mammals: the molecular basis of gas and nutrient transfer, hormone secretion, and immune responses. Physiol Rev 92:1543–1576

Carter AM, Mess A (2007) Evolution of the placenta in eutherian mammals. Placenta 28:259–262

Carter AM, Pijnenborg R (2011) Evolution of invasive placentation with special reference to non-human primates. Best Pract Res Clin Obstet Gynaecol 25:249–257

Castellucci M, Zaccheo D, Pescetto G (1980) A three-dimensional study of the normal human placental villous core. I. The Hofbauer cells. Cell Tissue Res 210:235–247

Cha J, Sun X, Dey SK (2012) Mechanisms of implantation: strategies for successful pregnancy. Nat Med 18:1754–1767

Chang WL, Liu YW, Dang YL, Jiang XX, Xu H, Huang X, Wang YL, Wang H, Zhu C, Xue LQ, Lin HY, Meng W, Wang H (2018) PLAC8, a new marker for human interstitial extravillous trophoblast cells, promotes their invasion and migration. Development 145:dev148932

Chuprin A, Gal H, Biron-Shental T, Biran A, Amiel A, Rozenblatt S, Krizhanovsky V (2013) Cell fusion induced by ERVWE1 or measles virus causes cellular senescence. Genes Dev 27:2356–2366

Cindrova-Davies T, Yung HW, Johns J, Spasic-Boskovic O, Korolchuk S, Jauniaux E, Burton GJ, Charnock-Jones DS (2007) Oxidative stress, gene expression, and protein changes induced in the human placenta during labor. Am J Pathol 171:1168–1179

Cindrova-Davies T, Van Patot MT, Gardner L, Jauniaux E, Burton GJ, Charnock-Jones DS (2015) Energy status and HIF signalling in chorionic villi show no evidence of hypoxic stress during human early placental development. Mol Hum Reprod 21:296–308

Cindrova-Davies T, Jauniaux E, Elliot MG, Gong S, Burton GJ, Charnock-Jones DS (2017) RNA-seq reveals conservation of function among the yolk sacs of human, mouse, and chicken. Proc Natl Acad Sci U S A 114:E4753–E4761

Cindrova-Davies T, Fogarty NME, Jones CJP, Kingdom J, Burton GJ (2018) Evidence of oxidative stress-induced senescence in mature, post-mature and pathological human placentas. Placenta 68:15–22

Clark AR, James JL, Stevenson GN, Collins SL (2018) Understanding abnormal uterine artery Doppler waveforms: a novel computational model to explore potential causes within the utero-placental vasculature. Placenta 66:74–81

Cole LA (2007) Hyperglycosylated hCG. Placenta 28:977–986
Cole LA (2010) Hyperglycosylated hCG, a review. Placenta 31:653–664
Coleman SJ, Gerza L, Jones CJ, Sibley CP, Aplin JD, Heazell AE (2013) Syncytial nuclear aggregates in normal placenta show increased nuclear condensation, but apoptosis and cytoskeletal redistribution are uncommon. Placenta 34:449–455
Collins SL, Birks JS, Stevenson GN, Papageorghiou AT, Noble JA, Impey L (2012) Measurement of spiral artery jets: general principles and differences observed in small-for-gestational-age pregnancies. Ultrasound Obstet Gynecol 40:171–178
Conrad KP, Rabaglino MB, Post Uiterweer ED (2017) Emerging role for dysregulated decidualization in the genesis of preeclampsia. Placenta 60:119–129
Cox LS, Redman C (2017) The role of cellular senescence in ageing of the placenta. Placenta 52:139–145
Dasilva-Arnold SC, Kuo CY, Davra V, Remache Y, Kim PCW, Fisher JP, Zamudio S, Al-khan A, Birge RB, Illsley NP (2019) ZEB2, a master regulator of the epithelial-mesenchymal transition, mediates trophoblast differentiation. Mol Hum Reprod 25:61–75
De Rijk EPCT, Van Esch E (2008) The macaque placenta—a mini-review. Toxicol Pathol 36:108S–111S
Deglincerti A, Croft GF, Pietila LN, Zernicka-Goetz M, Siggia ED, Brivanlou AH (2016) Self-organization of the in vitro attached human embryo. Nature 533:251–254
Demir R, Kaufmann P, Castellucci M, Erbengi T, Kotowski A (1989) Fetal vasculogenesis and angiogenesis in human placental villi. Acta Anat 136:190–203
Demir R, Kayisli UA, Celik-Ozenci C, Korgun ET, Demir-Weusten AY, Arici A (2002) Structural differentiation of human uterine luminal and glandular epithelium during early pregnancy: an ultrastructural and immunohistochemical study. Placenta 23:672–684
Dempsey EW (1972) The development of capillaries in the villi of early human placentas. Am J Anat 134:221–238
Ellery PM, Cindrova-Davies T, Jauniaux E, Ferguson-Smith AC, Burton GJ (2009) Evidence for transcriptional activity in the syncytiotrophoblast of the human placenta. Placenta 30:329–334
Elliot MG, Crespi BJ (2009) Phylogenetic evidence for early hemochorial placentation in eutheria. Placenta 30:949–967
Elliot MG, Crespi BJ (2015) Genetic recapitulation of human pre-eclampsia risk during convergent evolution of reduced placental invasiveness in eutherian mammals. Philos Trans R Soc Lond B Biol Sci 370:20140069
Enders AC (1989) Trophoblast differentiation during the transition from trophoblastic plate to lacunar stage of implantation in the rhesus monkey and human. Am J Anat 186:85–98
Enders AC (2001) Perspectives on human implantation. Infertil Reprod Med Clin N Am 12:251–269
Enders AC (2007) Implantation in the macaque: expansion of the implantation site during the first week of implantation. Placenta 28:794–802
Enders AC, King BF (1988) Formation and differentiation of extraembryonic mesoderm in the rhesus monkey. Am J Anat 181:327–340
Enders AC, Schlafke S, Hendrickx AG (1986) Differentiation of the embryonic disc, amnion, and yolk sac in the rhesus monkey. Am J Anat 177:161–185
Enders AC, Blankenship TN, Fazleabas AT, Jones CJ (2001) Structure of anchoring villi and the trophoblastic shell in the human, baboon and macaque placenta. Placenta 22:284–303
Evans J (2016) Hyperglycosylated hCG: a unique human implantation and invasion factor. Am J Reprod Immunol 75:333–340
Filant J, Spencer TE (2014) Uterine glands: biological roles in conceptus implantation, uterine receptivity and decidualization. Int J Dev Biol 58:107–116
Fogarty NME, Mayhew TM, Ferguson-Smith AC, Burton GJ (2011) A quantitative analysis of transcriptionally active syncytiotrophoblastic nuclei across human gestation. J Anat 219:601–610

Fogarty NM, Ferguson-Smith AC, Burton GJ (2013) Syncytial knots (Tenney-Parker changes) in the human placenta: evidence of loss of transcriptional activity and oxidative damage. Am J Pathol 183:144–152

Fogarty NM, Burton GJ, Ferguson-Smith AC (2015) Different epigenetic states define syncytiotrophoblast and cytotrophoblast nuclei in the trophoblast of the human placenta. Placenta 36:796–802

Freyer C, Renfree MB (2009) The mammalian yolk sac placenta. J Exp Zool B Mol Dev Evol 312:545–554

Gellersen B, Reimann K, Samalecos A, Aupers S, Bamberger AM (2010) Invasiveness of human endometrial stromal cells is promoted by decidualization and by trophoblast-derived signals. Hum Reprod 25:862–873

Genbacev O, Joslin R, Damsky CH, Polliotti BM, Fisher SJ (1996) Hypoxia alters early gestation human cytotrophoblast differentiation/invasion in vitro and models the placental defects that occur in preeclampsia. J Clin Investig 97:540–550

Goldstein J, Braverman M, Salafia C, Buckley P (1988) The phenotype of human placental macrophages and its variation with gestational age. Am J Pathol 133:648–659

Gonzalez-Crussi F, Roth LM (1976) The human yolk sac and yolk sac carcinoma. An ultrastructural study. Hum Pathol 7:675–691

Gorr TA (2017) Hypometabolism as the ultimate defence in stress response: how the comparative approach helps understanding of medically relevant questions. Acta Physiol (Oxf) 219:409–440

Guibourdenche J, Handschuh K, Tsatsaris V, Gerbaud P, Leguy MC, Muller F, Brion DE Fournier T (2010) Hyperglycosylated hCG is a marker of early human trophoblast invasion. J Clin Endocrinol Metab 95:E240–E244

Haider S, Meinhardt G, Saleh L, Fiala C, Pollheimer J, Knofler M (2016) Notch1 controls development of the extravillous trophoblast lineage in the human placenta. Proc Natl Acad Sci U S A 113:E7710–E7719

Hamilton WJ, Boyd JD (1960) Development of the human placenta in the first three months of gestation. J Anat 94:297–328

Hanna CW, Demond H, Kelsey G (2018) Epigenetic regulation in development: is the mouse a good model for the human? Hum Reprod Update 24:556–576

Harris LK (2010) Review: trophoblast-vascular cell interactions in early pregnancy: how to remodel a vessel. Placenta 31(Suppl):S93–S98

Harris JWS, Ramsey EM (1966) The morphology of human uteroplacental vasculature. Contrib Embryol 38:43–58

Hemberger M, Udayashankar R, Tesar P, Moore H, Burton GJ (2010) ELF5-enforced transcriptonal networks define an epigentically regulated trophoblast stem cell compartment in the human placenta. Mol Hum Genet 19:2456–2467

Hemberger M, Hanna CW, Dean W (2020) Mechanisms of early placental development in mouse and humans. Nat Rev Genet 21:27–43

Hempstock J, Bao Y-P, Bar-Issac M, Segaren N, Watson AL, Charnock Jones DS, Jauniaux E, Burton GJ (2003a) Intralobular differences in antioxidant enzyme expression and activity reflect oxygen gradients within the human placenta. Placenta 24:517–523

Hempstock J, Jauniaux E, Greenwold N, Burton GJ (2003b) The contribution of placental oxidative stress to early pregnancy failure. Hum Pathol 34:1265–1275

Hempstock J, Cindrova-Davies T, Jauniaux E, Burton GJ (2004) Endometrial glands as a source of nutrients, growth factors and cytokines during the first trimester of human pregnancy; a morphological and immunohistochemical study. Reprod Biol Endocrinol 2:58

Hertig AT, Rock J (1941) Two human ova of the pre-villous stage, having an ovulation age of about eleven and twelve days respectively. Contrib Embryol 29:127–156

Hertig AT, Rock J, Adams EC (1956) A description of 34 human ova within the first 17 days of development. Am J Anat 98:435–494

Hiby SE, Apps R, Sharkey AM, Farrell LE, Gardner L, Mulder A, Claas FH, Walker JJ, Redman CC, Morgan L, Tower C, Regan L, Moore GE, Carrington M, Moffett A (2010) Maternal activating KIRs protect against human reproductive failure mediated by fetal HLA-C2. J Clin Invest 120:4102–4110

Houston ML (1969) The development of the baboon (Papio sp.) placenta during the fetal period of gestation. Am J Anat 126:17–29

Hung TH, Burton GJ (2006) Hypoxia and reoxygenation: a possible mechanism for placental oxidative stress in preeclampsia. Taiwan J Obstet Gynecol 45:189–200

Huppertz B, Kaufmann P (1999) The apoptosis cascade in human villous trophoblast. Trophoblast Res 13:215–242

Huppertz B, Frank H-G, Reister F, Kingdom J, Korr H, Kaufmann P (1999) Apoptosis cascade progresses during turnover of human trophoblast: analysis of villous cytotrophoblast and syncytial fragments in vitro. Lab Invest 79:1687–1702

Hustin J, Schaaps JP, Lambotte R (1988) Anatomical studies of the utero-placental vascularisation in the first trimester of pregnancy. Trophoblast Res 3:49–60

Hustin J, Jauniaux E, Schaaps JP (1990) Histological study of the materno-embryonic interface in spontaneous abortion. Placenta 11:477–486

Hutter J, Harteveld AA, Jackson LH, Franklin S, Bos C, Van Osch MJP, O'Muircheartaigh J, Ho A, Chappell L, Hajnal JV, Rutherford M, De Vita E (2020) Perfusion and apparent oxygenation in the human placenta (PERFOX). Magn Reson Med 83:549–560

Jackson MR, Mayhew TM, Boyd PA (1992) Quantitative description of the elaboration and maturation of villi from 10 weeks of gestation to term. Placenta 13:357–370

Jauniaux E, Gulbis B (2000) Fluid compartments of the embryonic environment. Hum Reprod Update 6:268–278

Jauniaux E, Watson AL, Hempstock J, Bao Y-P, Skepper JN, Burton GJ (2000) Onset of maternal arterial bloodflow and placental oxidative stress; a possible factor in human early pregnancy failure. Am J Pathol 157:2111–2122

Jauniaux E, Watson AL, Burton GJ (2001) Evaluation of respiratory gases and acid-base gradients in fetal fluids and uteroplacental tissue between 7 and 16 weeks. Am J Obstet Gynecol 184:998–1003

Jauniaux E, Gulbis B, Burton GJ (2003a) The human first trimester gestational sac limits rather than facilitates oxygen transfer to the fetus-a review. Placenta 24(Suppl. A):S86–S93

Jauniaux E, Hempstock J, Greenwold N, Burton GJ (2003b) Trophoblastic oxidative stress in relation to temporal and regional differences in maternal placental blood flow in normal and abnormal early pregnancies. Am J Pathol 162:115–125

Jauniaux E, Cindrova-Davies T, Johns J, Dunster C, Hempstock J, Kelly FJ, Burton GJ (2004) Distribution and transfer pathways of antioxidant molecules inside the first trimester human gestational sac. J Clin Endocrinol Metab 89:1452–1459

Jauniaux E, Hempstock J, Teng C, Battaglia F, Burton GJ (2005) Polyol concentrations in the fluid compartments of the human conceptus during the first trimester of pregnancy; maintenance of redox potential in a low oxygen environment. J Clin Endocrinol Metab 90:1171–1175

Jirkovska M, Kubinova L, Janacek J, Moravcova M, Krejci V, Karen P (2002) Topological properties and spatial organization of villous capillaries in normal and diabetic placentas. J Vasc Res 39:268–278

Johns J, Jauniaux E, Burton GJ (2006) Factors affecting the early embryonic environment. Rev Gynaecol Perinat Pract 6:199–210

Jones CJP, Jauniaux E (1995) Ultrastructure of the materno-embryonic interface in the first trimester of pregnancy. Micron 26:145–173

Jones CJ, Choudhury RH, Aplin JD (2015) Tracking nutrient transfer at the human maternofetal interface from 4 weeks to term. Placenta 36:372–380

Khong TY, Mooney EE, Ariel I, Balmus NC, Boyd TK, Brundler MA, Derricott H, Evans MJ, Faye-Petersen OM, Gillan JE, Heazell AE, Heller DS, Jacques SM, Keating S, Kelehan P, Maes A, Mckay EM, Morgan TK, Nikkels PG, Parks WT, Redline RW, Scheimberg I, Schoots MH, Sebire NJ, Timmer A, Turowski G, Van Der Voorn JP, Van Lijnschoten I, Gordijn SJ

(2016) Sampling and definitions of placental lesions: Amsterdam placental workshop group consensus statement. Arch Pathol Lab Med 140:698–713

Kingdom JC, Audette MC, Hobson SR, Windrim RC, Morgen E (2018) A placenta clinic approach to the diagnosis and management of fetal growth restriction. Am J Obstet Gynecol 218:S803–S817

Knoth M, Larsen JF (1972) Ultrastructure of a human implantation site. Acta Obstet Gynecol Scand 51:385–393

Lee CL, Lam KK, Koistinen H, Seppala M, Kurpisz M, Fernandez N, Pang RT, Yeung WS, Chiu PC (2011) Glycodelin-A as a paracrine regulator in early pregnancy. J Reprod Immunol 90:29–34

Lee CQE, Gardner L, Turco M, Zhao N, Murray MJ, Coleman N, Rossant J, Hemberger M, Moffett A (2016) What is trophoblast? A combination of criteria define human first-trimester trophoblast. Stem Cell Rep 6:257–272

Lee CQE, Turco MY, Gardner L, Simons BD, Hemberger M, Moffett A (2018) Integrin alpha2 marks a niche of trophoblast progenitor cells in first trimester human placenta. Development 145:dev162305

Lees JG, Gardner DK, Harvey AJ (2017) Pluripotent stem cell metabolism and mitochondria: beyond ATP. Stem Cells Int 2017:2874283

Lindenberg S, Hyttel P, Sjøgren A, Greve T (1989) A comparative study of attachment of human, bovine and mouse blastocysts to uterine epithelial monolayer. Hum Reprod 4:446–456

Longtine MS, Barton A, Chen B, Nelson DM (2012a) Live-cell imaging shows apoptosis initiates locally and propagates as a wave throughout syncytiotrophoblasts in primary cultures of human placental villous trophoblasts. Placenta 33:971–976

Longtine MS, Chen B, Odibo AO, Zhong Y, Nelson DM (2012b) Caspase-mediated apoptosis of trophoblasts in term human placental villi is restricted to cytotrophoblasts and absent from the multinucleated syncytiotrophoblast. Reproduction 143:107–121

Luckett WP (1978) Origin and differentiation of the yolk sac and extraembryonic mesoderm in presomite human and rhesus monkey embryos. Am J Anat 152:59–97

Maiti K, Sultana Z, Aitken RJ, Morris J, Park F, Andrew B, Riley SC, Smith R (2017) Evidence that fetal death is associated with placental aging. Am J Obstet Gynecol 217:441 e1–441 e14

Martin RD (2008) Evolution of placentation: implications of mammalian phylogeny. Evol Biol 35:125–145

Martin BJ, Spicer SS (1973) Ultrastructural features of cellular maturation and aging in human trophoblast. J Ultrastruct Res 43:133–149

Martinoli C, Castellucci M, Zaccheo D, Kaufmann P (1984) Scanning electron microscopy of stromal cells of human placental villi throughout pregnancy. Cell Tissue Res 235:647–655

Maruo T, Matsuo H, Murata K, Mochizuki M (1992) Gestational age-dependent dual action of epidermal growth factor on human placenta early in gestation. J Clin Endocrinol Metab 75:1362–1367

Mayhew TM, Jackson MR, Boyd PA (1993) Changes in oxygen diffusive conductances of human placental during gestation (10-41 weeks) are commensurate with the gain in fetal weight. Placenta 14:51–61

Moffett A, Colucci F (2015) Co-evolution of NK receptors and HLA ligands in humans is driven by reproduction. Immunol Rev 267:283–297

Moffett A, Hiby SE, Sharkey AM (2015) The role of the maternal immune system in the regulation of human birthweight. Philos Trans R Soc Lond B Biol Sci 370:20140071

Moser G, Weiss G, Gauster M, Sundl M, Huppertz B (2015) Evidence from the very beginning: endoglandular trophoblasts penetrate and replace uterine glands in situ and in vitro. Hum Reprod 30:2747–2757

Mossman HW (1987) Vertebrate fetal membranes: comparative ontogeny and morphology; evolution; phylogenetic significance; basic functions; research opportunities. Macmillan, London

Nakamura T, Okamoto I, Sasaki K, Yabuta Y, Iwatani C, Tsuchiya H, Seita Y, Nakamura S, Yamamoto T, Saitou M (2016) A developmental coordinate of pluripotency among mice, monkeys and humans. Nature 537:57–62

Niakan KK, Eggan K (2013) Analysis of human embryos from zygote to blastocyst reveals distinct gene expression patterns relative to the mouse. Dev Biol 375:54–64

Nogales FF, Beltran E, Gonzalez F (1993) Morphological changes of the secondary human yolk sac in early pregnancy wastage. In: NOGALES FF (ed) The human yolk sac and yolk sac tumours. Springer-Verlag, Berlin

Ockleford CD (2010) The Allo-epi-endothelial lining of the intervillous space. Placenta 31:1035–1042

Parham P, Moffett A (2013) Variable NK cell receptors and their MHC class I ligands in immunity, reproduction and human evolution. Nat Rev Immunol 13:133–144

Pereda J, Niimi G (2008) Embryonic erythropoiesis in human yolk sac: two different compartments for two different processes. Microsc Res Tech 71:856–862

Petropoulos S, Edsgard D, Reinius B, Deng Q, Panula SP, Codeluppi S, Reyes AP, Linnarsson S, Sandberg R, Lanner F (2016) Single-cell RNA-Seq reveals lineage and X chromosome dynamics in human preimplantation embryos. Cell 167:285

Pijnenborg R, D'Hooghe T, Vercruysse L, Bambra C (1996) Evaluation of trophoblast invasion in placental bed biopsies of the baboon, with immunohistochemical localisation of cytokeratin, fibronectin, and laminin. J Med Primatol 25:272–281

Pijnenborg R, Vercruysse L, Hanssens M (2006) The uterine spiral arteries in human pregnancy: facts and controversies. Placenta 27:939–958

Pijnenborg R, Vercruysse L, Carter AM (2011a) Deep trophoblast invasion and spiral artery remodelling in the placental bed of the chimpanzee. Placenta 32:400–408

Pijnenborg R, Vercruysse L, Carter AM (2011b) Deep trophoblast invasion and spiral artery remodelling in the placental bed of the lowland gorilla. Placenta 32:586–591

Plitman Mayo R, Charnock-Jones DS, Burton GJ, Oyen ML (2016a) Three-dimensional modeling of human placental terminal villi. Placenta 43:54–60

Plitman Mayo R, Olsthoorn J, Charnock-Jones DS, Burton GJ, Oyen ML (2016b) Computational modeling of the structure-function relationship in human placental terminal villi. J Biomech 49:3780–3787

Plitman Mayo R, Abbas Y, Charnock-Jones DS, Burton GJ, Marom G (2019) Three-dimensional morphological analysis of placental terminal villi. Interface Focus 9:20190037

Pollheimer J, Vondra S, Baltayeva J, Beristain AG, Knofler M (2018) Regulation of placental Extravillous trophoblasts by the maternal uterine environment. Front Immunol 9:2597

Ramsey EM, Donner MW (1980) Placental vasculature and circulation. In: Anatomy, physiology, radiology, clinical aspects, atlas and textbook. Georg Thieme, Stuttgart

Reyes L, Wolfe B, Golos T (2017) Hofbauer cells: placental macrophages of fetal origin. Results Probl Cell Differ 62:45–60

Roberts VHJ, Morgan TK, Bednarek P, Morita M, Burton GJ, Lo JO, Frias AE (2017) Early first trimester uteroplacental flow and the progressive disintegration of spiral artery plugs: new insights from contrast-enhanced ultrasound and tissue histopathology. Hum Reprod 32:2382–2393

Robin C, Bollerot K, Mendes S, Haak E, Crisan M, Cerisoli F, Lauw I, Kaimakis P, Jorna R, Vermeulen M, Kayser M, Van Der Linden R, Imanirad P, Verstegen M, Nawaz-Yousaf H, Papazian N, Steegers E, Cupedo T, Dzierzak E (2009) Human placenta is a potent hematopoietic niche containing hematopoietic stem and progenitor cells throughout development. Cell Stem Cell 5:385–395

Rosso P (1976) Placenta as an aging organ. Curr Concepts Nutr 4:23–41

Saghian R, Bogle G, James JL, Clark AR (2019) Establishment of maternal blood supply to the placenta: insights into plugging, unplugging and trophoblast behaviour from an agent-based model. Interface Focus 9:20190019

Salafia CM, Yampolsky M, Shlakhter A, Mandel DH, Schwartz N (2012) Variety in placental shape: when does it originate? Placenta 33:164–170

Schiessl B, Innes BA, Bulmer JN, Otun HA, Chadwick TJ, Robson SC, Lash GE (2009) Localization of angiogenic growth factors and their receptors in the human placental bed throughout normal human pregnancy. Placenta 30:79–87

Schwartz N, Mandel D, Shlakhter O, Coletta J, Pessel C, Timor-Tritsch IE, Salafia CM (2011) Placental morphologic features and chorionic surface vasculature at term are highly correlated with 3-dimensional sonographic measurements at 11 to 14 weeks. J Ultrasound Med 30:1171–1178

Seppälä M, Jukunen M, Riitinen L, Koistinen R (1992) Endometrial proteins: a reappraisal. Hum Reprod 7(Suppl. 1):31–38

Shahbazi MN, Jedrusik A, Vuoristo S, Recher G, Hupalowska A, Bolton V, Fogarty NM, Campbell A, Devito LG, Ilic D, Khalaf Y, Niakan KK, Fishel S, Zernicka-Goetz M (2016) Self-organization of the human embryo in the absence of maternal tissues. Nat Cell Biol 18:700–708

Simpson RA, Mayhew TM, Barnes PR (1992) From 13 weeks to term, the trophoblast of human placenta grows by the continuous recruitment of new proliferative units: a study of nuclear number using the disector. Placenta 13:501–512

Soncin F, Khater M, To C, Pizzo D, Farah O, Wakeland A, Arul Nambi Rajan K, Nelson KK, Chang CW, Moretto-Zita M, Natale DR, Laurent LC, Parast MM (2018) Comparative analysis of mouse and human placentae across gestation reveals species-specific regulators of placental development. Development 145:dev156273

Teh WT, McBain J, Rogers P (2016) What is the contribution of embryo-endometrial asynchrony to implantation failure? J Assist Reprod Genet 33:1419–1430

Turco MY, Gardner L, Hughes J, Cindrova-Davies T, Gomez MJ, Farrell L, Hollinshead M, Marsh SGE, Brosens JJ, Critchley HO, Simons BD, Hemberger M, Koo BK, Moffett A, Burton GJ (2017) Long-term, hormone-responsive organoid cultures of human endometrium in a chemically defined medium. Nat Cell Biol 19:568–577

Velicky P, Meinhardt G, Plessl K, Vondra S, Weiss T, Haslinger P, Lendl T, Aumayr K, Mairhofer M, Zhu X, Schutz B, Hannibal RL, Lindau R, Weil B, Ernerudh J, Neesen J, Egger G, Mikula M, Rohrl C, Urban AE, Baker J, Knofler M, Pollheimer J (2018) Genome amplification and cellular senescence are hallmarks of human placenta development. PLoS Genet 14:e1007698

Whitley GS, Cartwright JE (2010) Cellular and molecular regulation of spiral artery remodelling: lessons from the cardiovascular field. Placenta 31:465–474

Wildman DE, Chen C, Erez O, Grossman LI, Goodman M, Romero R (2006) Evolution of the mammalian placenta revealed by phylogenetic analysis. Proc Natl Acad Sci U S A 103:3203–3208

Wong FTM, Lin C, Cox BJ (2019) Cellular systems biology identifies dynamic trophoblast populations in early human placentas. Placenta 76:10–18

Wooding FP, Burton GJ (2008) Comparative placentation. Structures, functions and evolution. Springer, Berlin

Woollett LA (2008) Where does fetal and embryonic cholesterol originate and what does it do? Annu Rev Nutr 28:97–114

Zybina TG, Kaufmann P, Frank HG, Freed J, Kadyrov M, Biesterfeld S (2002) Genome multiplication of extravillous trophoblast cells in human placenta in the course of differentiation and invasion into endometrium and myometrium. I. Dynamics of polyploidization. Tsitologiia 44:1058–1067

Zybina TG, Frank HG, Biesterfeld S, Kaufmann P (2004) Genome multiplication of extravillous trophoblast cells in human placenta in the course of differentiation and invasion into endometrium and myometrium. II. Mechanisms of polyploidization. Tsitologiia 46:640–648

Correction to: Canine Endotheliochorial Placenta: Morpho-Functional Aspects

Mariusz P. Kowalewski, Ali Kazemian, Karl Klisch, Tina Gysin, Miguel Tavares Pereira, and Aykut Gram

Correction to:
Chapter 8 in: R. D. Geisert, T. Spencer (eds.), *Placentation in Mammals*, Advances in Anatomy, Embryology and Cell Biology 234, https://doi.org/10.1007/978-3-030-77360-1_8

Chapter 8 was inadvertently published with errors and the following corrections were updated:

On page 159, the word "see" in italics was changed to roman type in the line "(for details see in...".

On page 160, the phrase "The immunosuppressive Treg cells appear" was changed to "The immunosuppressive regulatory T and natural killer (NK) cells appear".

On page 160, hyperlink was added for the citation (Tavares Pereira et al. 2021).

On page 161, the phrase "The lumen of the uterine glands develops..." was changed to "The lumina of the uterine glands develop...".

On page 161, the phrase "expands into the extraembryonic coelom..." was changed to "expands into the chorionic cavity...".

On page 162, Figure 1 was replaced with a high-resolution figure.

On page 163, the part labels (**a**, **b**) in Figure 2 did not match with the figure and caption, and were updated uniformly.

On page 165, Figure 3 was replaced with a high-resolution figure.

On page 165, the caption of Figure 3, the text was "Histological appearance of the canine placenta" changed to "Histological appearance of the canine fully developed placenta".

The updated online version of this chapter can be found at
https://doi.org/10.1007/978-3-030-77360-1_8

R. D. Geisert, T. Spencer (eds.), *Placentation in Mammals*, Advances in Anatomy, Embryology and Cell Biology 234, https://doi.org/10.1007/978-3-030-77360-1_12

On page 166, Figure 4 was replaced with the corrected figure.

On page 170, the following sentence was added: "Furthermore, the uterine immune milieu appears to be associated with an increased presence of immune cells during parturition, involving macrophages involved in tissue remodeling, as well as Th and Treg (own unpublished data).

On page 179, doi links were removed from the below two references.

"Nowak M, Rehrauer H, Ay SS, Findik M, Boos A, Kautz E et al (2019) Gene expression profiling of the canine placenta during normal and antigestagen-induced luteolysis. Gen Comp Endocrinol 282:113194.

and

"Tavares Pereira M, Nowaczyk R, Payan-Carreira R, Miranda S, Aslan S, Kaya D, Kowalewski MP (2021) Selected uterine immune events associated with the establishment of pregnancy in the dog. Front Vet Sci 7:625921".

GPSR Compliance

The European Union's (EU) General Product Safety Regulation (GPSR) is a set of rules that requires consumer products to be safe and our obligations to ensure this.

If you have any concerns about our products, you can contact us on ProductSafety@springernature.com

In case Publisher is established outside the EU, the EU authorized representative is:

Springer Nature Customer Service Center GmbH
Europaplatz 3
69115 Heidelberg, Germany

Batch number: 10732628

Printed by Printforce, the Netherlands

Advances in Anatomy, Embryology and Cell Biology publishes critical reviews and state-of-the-art research in the areas of anatomy, developmental and cellular biology. Founded in 1891, this book series has a long standing tradition of publishing focused and condensed information on a given topic with a special emphasis on biomedical and translational aspects.

The series is open to both contributed volumes (each collecting 7 to 15 focused reviews written by leading experts) and single-authored or multi-authored monographs (providing a comprehensive overview of their topic of research).

Advances in Anatomy, Embryology and Cell Biology is indexed in BIOSIS, Journal Citation Reports/Science Edition, Medline, SCImago, SCOPUS, Science Citation Index Expanded (SciSearch), Zoological Record.

234
Advances in Anatomy, Embryology and Cell Biology

More information about this series at
http://www.springer.com/series/102